Wireless Networks and Industrial IoT

Nurul Huda Mahmood • Nikolaj Marchenko
Mikael Gidlund • Petar Popovski
Editors

Wireless Networks and Industrial IoT

Applications, Challenges and Enablers

Springer

Editors
Nurul Huda Mahmood
6G Flagship, Center for Wireless Communications (CWC)
University of Oulu
Oulu, Finland

Nikolaj Marchenko
Corporate Research and Advanced Engineering
Robert Bosch GmbH (Germany)
Stuttgart, Germany

Mikael Gidlund
Department of Information Systems and Tech.
Mid Sweden University
Västernorrlands Län
Sundsvall, Sweden

Petar Popovski
Connectivity Section, Department of Electronic Systems
Aalborg University
Aalborg Øst, Denmark

ISBN 978-3-030-51472-3 ISBN 978-3-030-51473-0 (eBook)
https://doi.org/10.1007/978-3-030-51473-0

This Springer imprint is published by the registered company Springer Nature Switzerland AG.
The registered company address is: Gewerbestrasse 11, 6330 Cham, Switzerland

Preface

The recent decade has been marked by an unprecedented development of wireless networks, smartphones, and cloud computing technologies. Today, more than five billion people are connected with each other and use a vast number of Internet services. In addition to that, the volume of Internet-of-Things (IoT) devices started to proliferate during recent years, resulting in a vast variety of "things" connected wirelessly to the Internet, such as home appliances, lights, cars, shipping containers, environmental sensors, etc. The availability of wireless connectivity brings a new quality to these physical objects, allowing them to contribute to digital optimization of various processes and systems.

IoT represents now an exponentially growing technology sector, with the number of connected devices expected to rise from around 11 billion in 2019 to around 25 billion by the end of 2025.[1] This growth is mainly driven by the push towards greater efficiency in various vertical domains, such as industrial automation, transportation, agriculture, smart city management, smart homes, and building automation.

Wireless Communication and Industrial IoT

Industry 4.0, or Fourth Industrial Revolution, is the term coined for the paradigm of inter-connecting different machines, devices, objects, and processes to easily collect and process relevant data and, thus, further automate and optimize manufacturing and delivery of goods. This ongoing trend promises to address the need for high production efficiency, growing product customization, shortening of the production cycle, and dynamic global supply chain.

Future smart factories are envisioned to be highly automated and flexible to react quickly to changes in supply chain and market demand. Enhanced mobile robots can transport goods and spare parts form modular production islands and

[1] https://www.ericsson.com/en/mobility-report/reports/november-2019/iot-connections-outlook

flexibly reconfigure the production. Production steps are digitally represented in real-time in a digital twin. For that, tight integration with enterprise IT-systems becomes absolutely necessary, and massive data collection across devices, sensors, and actuators is needed.

In this vision, a large part of the machinery control is done at the edge cloud, where the computing resources are easier to scale and orchestrate in real-time compared to the classic distributed controllers. Human workers are not excluded from the manufacturing, but will perform more sophisticated and diverse tasks, operate multiple complex machines, and require additional human-machine interfaces, e.g., in a form of augmented reality.

Industrial IoT is a specific segment of IoT that aims to enable and accelerate the vision of Industry 4.0. It is characterized by specific industrial requirements on communication and operation of the IoT-devices. Some aspects that set Industrial IoT apart include a high level of resilience, communication availability, security, precision, automation, and compatibility. Moreover, Industrial IoT use cases are expected to provide a measurable return on investment and value for the original equipment manufacturer (OEM) and their customer, which is not always the case for consumer IoT applications.

Wireless connectivity is considered to be one of the main enabling technologies in Industrial IoT vision, as it can provide the required flexibility, efficiency, scale, and mobility support for the manufacturing world. Although the existing consumer-oriented wireless technologies, such as WiFi, Bluetooth, and 4G, are already used for certain industrial applications, they can only address a very limited set of applications on a shopfloor. Such performance metrics as network coverage, capacity, power consumption, and data downlink/uplink throughput, which played defining roles in consumer WiFi and cellular networks, although still relevant, are not sufficient to cover the main industrial applications. In particular, these technologies cannot enable critical control applications since they are not designed to satisfy the dedicated challenging requirements (e.g., latency, reliability, low jitter, etc.) in this domain.

Reliable Low-Latency Communication

The ultimate frontier for wireless connectivity is to enable closed-loop machine-to-machine control systems over the air. This removes the need for physical connections among the robots and modules, while keeping them logically interconnected and capable to cooperate and coordinate. Control applications compound the core of industrial automation and require guaranteed message delivery time and very high communication availability. For example, in motion control applications, the required end-to-end message delivery time can reach under 1 ms, while at the same time, a failure to deliver several consecutive messages leads the system to 'emergency stop'.

Factory automation with wireless connectivity has been dominated so far by proprietary industrial solutions such as ABB WISA (based on proprietary modifications of Bluetooth standard IEEE 802.15.1) and Siemens Industrial WLAN (based on proprietary modifications of IEEE 802.11 MAC protocol). Although these solutions are an important development for wireless connectivity in factories, due to their proprietary nature, they only allow isolated single vendor networks.

One example of the standardization effort in this area is IO-Link Wireless. Similar to WISA, it is based on the IEEE 802.15.1, uses unlicensed frequency band in 2.4 GHz, and aims for short-distance low-power communication with latency time down to 5 ms. IO-Link Wireless is specified as an extensions of IO-Link, which is a popular factory automation fieldbus-independent communication standard dedicated to connecting sensors and actuators as described in IEC 61131-1 standard.

The need for an open wireless communication standard addressing low-latency and high-reliability requirements has also been recognized early by the 3GPP standardization community. Significant standardization efforts have been made to define ultra-reliable low-latency communication (URLLC) service class as one of the main aspects of the fifth generation (5G) cellular communication system (see chapter "Overview of 3GPP New Radio Industrial IoT Solutions" for more details).

3GPP New Radio (NR) Release 16, finalized in the second half of 2020, addresses new verticals and deployment scenarios for intelligent transport systems (ITS), vehicle-to-everything (V2X) communication, and Industrial IoT.[2] Taken together, the proposed improvements in Release 16 significantly enhance NR for URLLC and also add capabilities to replace wired Ethernet and tightly integrate wireless 5G with Time Sensitive Networking (TSN) on the shopfloor (cf. chapter "Time-Sensitive Networking for Industrial Control Networks"). Further enhancements for Industrial IoT are already in planning for Release 18 of 3GPP.

The operation of wireless cellular networks for Industrial IoT also implies deployments and operation models different to those typically used by large mobile telecom operators for wide area networks. Many factory operators across the world have shown interest in dedicated frequency usage for wireless networks in Industrial IoT for critical applications. The benefits of such exclusive local spectrum licensing for factory owners are twofold: (a) high control of spectrum usage leads to higher communication reliability compared to the use of unlicensed or shared bands, and (b) spectrum ownership prevents certain operator lock-in and enables fully private 5G enterprise networks.

From this perspective, the potential for local private licensed spectrum has also been recognized by many frequency regulation authorities across the world. For example, in Germany, USA, UK, and Japan licensing of local dedicated spectrum for factory owners became possible under certain country-specific regulations. Many other countries have also made or are considering to make this decision in the near

[2] A. Ghosh, A. Maeder, M. Baker, and D. Chandramouli, "5G Evolution: A View on 5G Cellular Technology Beyond 3GPP Release 15," IEEE Access, 2019.

future. The advances of 3GPP and regulators encourage emergence of new players and new operation models for wireless networks in Industrial IoT domain.

Although there is a significant progress in the standardization of URLLC networks, it is still open to see how such networks will perform in real-world factories, how efficiently they can be integrated and operated with existing industrial systems, and what levels of communication reliability can be provided, and with which measures. Furthermore, the use of new 5G frequency bands in 28 GHz seems to be attractive for factory automation due to additional capacity, beam-forming capabilities, and better protection against jamming, but also requires additional evaluation and optimization.

Part I of this book focuses on various aspects of reliable low-latency communication for Industrial IoT. URLLC, however, is not limited to Industry 4.0. Intelligent Transportation Systems and Unmanned Aerial Vehicles (UAVs) can be seen as domains adjacent to Industrial IoT, e.g., with applications in warehouse and factory logistics, and can benefit from URLLC services. Overview chapters of Intelligent Transportation and UAV Systems are collected in Part IV of this book.

Low-Power Wide Area Networks

Although Industrial IoT is commonly associated with challenging URLLC use cases, another important pillar of Industrial IoT is the use of massive sensor/actuator networks to collect data and/or perform control at a time scale that is much larger than the one considered in URLLC. The data can be environmental (temperature, pressure, humidity, smoke/gas detectors, vibration, etc.) and/or related to metering information and status (current state, location, error logs, etc.), while control applications include process automation, building automation, lights, valves, e-paper tags, etc.

Typically, the transmitted data are not large, ranging from few bytes to few kilobytes per measurement or command. And in some use cases, few transmissions per hour or even per day might be sufficient (e.g., e-paper tags, metering). The term Massive IoT is used by some sources in the R&D community and accommodates other domains for such applications besides Industry 4.0 with similar performance requirements, e.g., in logistics, agriculture, or smart cities.

Dedicated wireless systems for a large number of low-power sensors across very large areas, referred as Low-Power Wide Area Networks (LPWAN), have been developed and found successful application in recent years. One of the main principles for low-power consumption is the use of narrow frequency bands, typically 125–500 kHz, used at frequencies below 1 GHz. In addition, these systems often use transmissions based on low coding-modulation rates in order to increase the coverage area. This results in very low data rates (<250 Kbps), which is acceptable, as the data rate is not the key metric in focus for such applications. Prominent examples of such systems are LoRA and SigFox and are covered in chapters "Wireless Communications for Industrial Internet of Things: The LPWAN

Solutions" and "Pervasive Listening: A Disruptive Network Design for Massive Low-Power IoT Connectivity," respectively.

The 3GPP standardization community refers to communication for such applications by the term massive Machine Type Communications (mMTC). To address applications with maximal required peak data rates <250 Kbps, 3GPP specified the standalone Narrow-Band-IoT (NB-IoT) system. In addition, 3GPP standardizes Cat-M radio technology as the extension of LTE cellular technology for LPWAN applications, which demand data rates up to 7 Mbps, but are also less critical in terms of power, complexity, and cost constraints than NB-IoT applications. These applications are referred to as enhanced MTC (eMTC) and include, for example, wearables, trackers for fleet management, and terminals with voice support. Both, commercial NB-IoT and LTE Cat-M systems have recently been deployed in many countries across the world.

Process automation (e.g., for chemical processes in a refinery) represents an application area where more frequent transmissions of wireless low-power sensors/actuators are typically required—in the range from around 50 ms to several seconds. This area has been initially addressed by multi-hop wireless sensor network technologies such as WirelessHART, ISA100.11a, and IEEE 802.15.4e, all working in licence-free sub-GHz or 2.4 GHz bands and relying on IEEE 802.15.4 PHY layer specification.

Although there is a broad range of existing radio technologies for LPWAN applications, not all challenges have been addressed so far. Further optimizations are required taking into account the expected massive growth of such devices in the future. In addition, some novel low-power applications will require higher data rates (e.g., camera or radar sensors). Other applications, while keeping low-power requirement, also need low-latency response. LPWAN community in the academia and the industry brings out innovations to address these challenges, and 3GPP plans already further enhancements for mMTC and eMTC in its future releases. Part II of this book gives a detailed overview of LPWAN aspects in Industrial IoT and other areas.

Beyond Communication

As explained above, wireless access technologies play a critical enabling role in Industrial IoT. However, recent technological shifts, such as Edge Computing, Machine Learning, and Cyber Security, certainly play critical roles in Industrial IoT as well. It is not difficult to imagine significant cross-synergy effects for these technologies, which will also benefit wireless connectivity for factory automation.

Edge Computing can enable industrial wireless control applications hosted in virtual instances in edge cloud, which is much easier to scale, orchestrate, and maintain compared to the classic approaches used today. Machine Learning is already in use today for predictive maintenance and anomaly detection in manufacturing. And with the use of massive sensors with mMTC-type wireless connectivity, better results will be possible with lower costs. In addition, Machine Learning can help to optimize wireless networking and its operation in Industrial IoT.

Security has always been a sensitive topic for manufacturing companies, which often leads to the use of very conservative and rigid solutions. When wireless networks become a part of critical infrastructure the factory operation relies on, security and protection against attacks across such networks becomes absolutely critical as well.

Chapters addressing the aforementioned topics are covered in Part III of this book.

Book Motivation and Structure

The potential and challenges of wireless communication in Industrial IoT have been recognized across academic and industrial research communities. A remarkable amount of work came out in recent years, addressing different aspects of this important area. Given the wide scope of the topic and its cross-domain importance, the editorial team of this book was motivated to provide was to provide a comprehensible overview of the most relevant research and standardization results in the area of wireless networking and Industrial IoT.

Most chapters in this book are intended to serve as short tutorials of particular topics, highlighting the main developments and solutions as well as giving an outlook of the upcoming research challenges. Due to such format, detailed methodology explanations and elaborated results analysis, common in journal publications, come short in these chapters but nevertheless can be found in the referenced works. In contrast, this book also provides a systematic analysis and deep-dives into selected connectivity topics of Industrial IoT.

We hope that, on one side, this book can bring important insights to the readers who are interested but are not yet very familiar with particular topics. On the other side, by collecting various research aspects from highly experienced authors, we hope the book can provide an inspiring and multifaceted perspective for researchers and engineers already actively working on some of the topics presented in this book.

The contributed chapters are grouped into four parts, as explained below.

Part I: Reliable Low-Latency Communication in Industrial IoT

The first part consists of four chapters and focuses on challenges, enablers, and standardization efforts for reliable low-latency communication in Industrial IoT networks.

Chapter "Overview of 3GPP New Radio Industrial IoT Solutions," provides the details on how Industrial IoT needs and requirements are addressed in 3GPP 5G NR

URLLC standardization. The specific focus is on the Radio Access Network and seamless integration with the legacy Industrial Ethernet technologies that are already in use for factory automation. The authors also give an outlook on challenges and enhancements for further 3GPP standardization efforts in this area.

In chapter "Selected Aspects and Approaches on Improving Dependability in Industrial Radio Networks," the authors focus on the reliability aspects of wireless communication. They discuss the main dependability metrics relevant for wireless networking and control applications. The chapter dives deeper into multi-connectivity as one measure to improve communication availability. In addition, joint control-communication design of control applications over wireless links is elaborated.

The authors of chapter "Time-Sensitive Networking for Industrial Control Networks," provide an overview of real-time control applications. Furthermore, they describe the main principles of Time Sensitive Networking (TSN), an open standard that is still under development, for wired communication based on Ethernet extensions. Approaches for future integration of TSN and 5G, as well as challenges for practical deployments, are also discussed in detail in this chapter.

The final chapter of this part, chapter "Random Access Protocols for Industrial Internet of Things: Enablers, Challenges, and Research Directions," is dedicated to random access protocols for low-latency and low-overhead Industrial IoT communications. The authors break random access protocols into typical building blocks and introduce the design challenges that need to be addressed to make these building blocks Industrial IoT-ready.

Part II: Low-Power Wide Area Networks for Massive IoT

The second part of this book focuses on Massive IoT, which requires highly cost- and energy-efficient technology components to connect a massive number of low-cost IoT devices. This part consists of the following five chapters:

The authors of chapter "Wireless Communications for Industrial Internet of Things: The LPWAN Solutions," provide an overview of the main LPWAN technologies, including NB-IoT, SigFox, and LoRa/LoRaWAN, and their potential for Industrial IoT applications. LoRa is then presented in more details, together with various simulation tools useful for further evaluation and optimization of LPWAN.

In chapter "Power Measurement Framework for LPWAN IoT," the authors focus on the most critical metric of all LPWAN technologies, which is the power consumption of end-devices. They propose a practical approach for modeling the power consumption based on real-world measurements on NB-IoT devices and various networking scenarios, and show how a 10-year battery lifetime on a certain traffic profile can be achieved.

Chapter "Dynamic Resource Management in Real-Time Wireless Networks," addresses an important problem of efficient resource management for real-time wireless applications. Since disturbances in a wireless medium are impossible to avoid, the challenge is to find out which time and frequency resources are needed to overcome a potential packet loss and provide required delivery guarantees. An assumption of worst-case scenario, although more reliable, also means very high over-provisioning of the valuable RF and energy resources. The authors of this chapter argue for a dynamic resource management. They present several practical methods and explain the trade-offs on an example of the IEEE 802.15.4e system, suitable for process automation control, with time delivery requirements of 10–100 ms.

Next, chapter "Pervasive Listening: A Disruptive Network Design for Massive Low-Power IoT Connectivity," focuses on the problem of massive device coordination, where the required signaling and computational complexity end up draining a significant amount of critical energy resources in mMTC-type of applications. The authors focus on SigFox LPWAN and introduce a new pervasive listening and cooperative reception approach, where coordination for uplink transmissions by IoT devices is not necessary since a single frequency band is used for uplink and multiple coordinated base stations are able to receive the data. The downlink coordination is also done at the network core. The approach creates gains in terms of efficient implementation with the use of cognitive Software-Defined-Radios at the base stations.

Finally, chapter "Information-Centric Networking for the Industrial Internet of Things," discusses the inefficiencies of end-to-end communication sessions, use of transport TCP/UDP for IoT application with small data transmissions, and high energy constraints. The authors elaborate on two disruptive information-centric approaches, Content-Centric Networking and Named Data Networking (NDN), based on decoupling of the content objects from its origins. Important aspects of ICN: information caching, quality-of-service, and security are also discussed in the chapter. Finally, the authors provide an overview of a practical experimental ICN solution RIOT and discuss further challenges in this domain.

Part III: Enabling Technologies for Industrial IoT

This part covers three enabling technologies beyond communication for Industrial IoT that are important to both—critical and massive IoT—namely, *Security*, *Machine Learning/Artificial Intelligence*, and *Edge Computing*.

Important security aspects of Industrial IoT are discussed in detail in chapter "Security Challenges for Industrial IoT." Here, the authors provide an overview of relevant security standards and requirements. Such security principles as physical security, trusted execution, isolation, attestation, and cryptography are elaborated for the application in IIoT and complemented by relevant examples.

Chapter "Machine Learning/AI as IoT Enablers," gives an overview of the application of machine learning/AI for intelligent connectivity. The role of AI and big data in IoT networks are explored, followed by related use cases and architecture. The chapter also provides a discussion on the future technologies of this type in the context of IoT.

The last chapter of this part, chapter "Edge Computing for Industrial IoT: Challenges and Solutions," addresses multi-access edge computing topics as an important part of Industrial IoT and elaborates upon how edge computing complements wireless communication. The chapter also gives a basic overview of critical aspects of edge computing, such as security, resource management, and optimization.

Part IV: Wireless IoT-Networks for Transportation and UAV Systems

Connected industries is a far-reaching concept that also includes peripheral verticals, such as connected transportation and logistics. This last part of the book covers aspects of Industrial IoT that are important in, for example, warehouse and port logistics, product delivery, and transportation among industries.

Details of wireless networks in vehicular-to-everything (V2X) communication are presented in chapter "Intelligent Transport System as an Example of a Wireless IoT System." The authors present the main technologies of V2X, such as IEEE 802.1p and 3GPP-based ones, elaborate use case examples and performance indicators, and discuss future development challenges in this domain, relevant also for Industrial IoT, for example, in warehouse or port logistics.

Last but not least, chapter "UAV-Enabled IoT Networks: Architecture, Opportunities, and Challenges," discusses the IoT applications enabled by UAV in smart cities, crowd surveillance, emergency disaster assistance, agricultural application, airborne sensing, etc. The architecture of UAV-enabled IoT networks, their communication schemes, challenges, and opportunities are detailed in this chapter, providing important guidelines on the design of UAV-assisted IoT infrastructure.

Acknowledgments

Like other processes and projects in 2020, the preparation of this book was challenged and delayed by the COVID-19 pandemic. We wish that, at the time of publishing, the pandemic will be defeated and left behind.

The editorial team would like to thank all the authors for their efforts and contributions, which made this book possible. It was an honor to cooperate with all of you in materializing this book.

We also would like to thank the reviewers for their time and effort in reviewing individual chapters and providing valuable comments and suggestions. Also, a big thank you to the editorial and support team at Springer. We also thank our colleagues, who supported the idea of bringing a book on Industrial IoT.

Finally, yet most importantly, thanks to our close family members and friends, who always provide the strongest support of all. Nurul Huda Mahmood would especially like to thank his wife for her never-ending support.

Oulu, Finland	Nurul Huda Mahmood
Stuttgart, Germany	Nikolaj Marchenko
Sundsvall, Sweden	Mikael Gidlund
Aalborg Øst, Denmark	Petar Popovski

Contents

Part I
Reliable Low-Latency Communication in Industrial IoT

Overview of 3GPP New Radio Industrial IoT Solutions

Klaus Pedersen and Troels Kolding

Abbreviations

3GPP	3rd Generation Partnership Project
5G-ACIA	5G Alliance for Connected Industries and Automation
5GS	5G System
5QI	5G QoS Identifier
BAT	Burst Arrival Time
CB	Code Block
CG	Configured Grant
CN	Core Network
CNC	Centralized Network Controller
CQI	Channel Quality Indicator
CRAN	Centralized RAN
CU	Centralized Unit
CUC	Centralized User Configuration
DCI	Downlink Control Information
DFT-s-OFDMA	Discrete Fourier Transform spread OFDMA
DL	Downlink
DU	Distributed Unit
E2E	End-2-End
eMBB	Enhanced MBB
eURLLC	Enhanced URLLC
FTP	File Transfer Protocol
GBR	Guaranteed Bit Rate

K. Pedersen (✉) · T. Kolding
Nokia Bell Labs in NOVI-8, Alfred Nobels Vej 27, DK-9220, Aalborg East, Denmark
e-mail: klaus.pedersen@nokia-bell-labs.com; troels.kolding@nokia-bell-labs.com

N. H. Mahmood et al. (eds.), *Wireless Networks and Industrial IoT*,
https://doi.org/10.1007/978-3-030-51473-0_1

gNB	NR base station
HARQ	Hybrid Automation Repeat request
IIoT	Industrial IoT
IMT2020	International Mobile Telecommunications 2020
IoT	Internet of Things
LDPC	Low-Density Parity Check
LTE	Long Term Evolution
MAC	Medium Access Control
MBB	Mobile BroadBand
MCS	Modulation and Coding Scheme
MPQUIC	Multi-Path QUIC
NR	New Radio
OFDMA	Orthogonal Frequency Division Multiple Access
PDCCH	Physical Downlink Control CHannel
PDCP	Packet Data Convergence Protocol
PDSCH	Physical Downlink Shared CHannel
PDU	Packet Data Unit
PHY	PHYsical layer
PRB	Physical Resource Block
PUCCH	Physical Uplink Control Channel
PUSCH	Physical Uplink Shared Channel
QoS	Quality of Service
QUIC	Quick UDP Internet Connections
RAN	Radio Access Network
RLC	Radio Link Control
RRC	Radio Resource Control
SDAP	Service Data Application Protocol
SIB	System Information Block
SINR	Signal to Interference Noise Ratio
SPS	Semi-Persistent Scheduling
TB	Transport Block
TRP	Transmission Reception Point
TSC	Time-Sensitive Communication
TSCAI	TSC Assistance Information
TSN	Time-Sensitive Network
TTI	Transmission Time Interval
UE	User Equipment
UL	Uplink
URLLC	Ultra-Reliable Low Latency Communication
VIAPA	Video, Imaging and Audio for Professional Applications

1 Introduction

In this chapter we present a compact overview of the 3GPP designed New Radio (NR) system with emphasis on the Radio Access Network (RAN) part, and related innovations that enable mission critical communication for Industrial Internet of Things (IIoT) including the use of 5G seamlessly in an Industrial Ethernet scenario. We focus on Ultra-Reliable Low-Latency Communication (URLLC), including its evolution towards enhanced URLLC (eURLLC), and Time-Sensitive Communications (TSC). The chapter is organized by first shortly introducing the addressed IIoT use cases and requirements addressed by the NR (e)URLLC and TSC enablers. To understand how 5G enables wireline-like performance while retaining the flexibility of wireless deployment, we then provide a brief overview of the NR RAN protocol stack and architecture options followed by a summary of main enablers for (e)URLLC in 3GPP NR Release-15 and 16. Then, the seamless integration of 5G into an IIoT factory setup with TSC is explained and main enablers introduced in 3GPP NR Release-16 are introduced. Finally, the chapter is closed with an outlook of further enhancements being considered for Release-17 standardization. Throughout the chapter, we provide pointers to the most relevant 3GPP Technical Reports (TRs) and Technical Specifications (TS), as well as selected publications where more details can be found.

2 New Use Cases and Requirements

The key requirement for URLLC is a user plane 1 ms one-way latency for the RAN part with five-nines (99.999%) reliability for payloads ranging from 32 to 200 bytes. The term reliability is defined as follows by 3GPP: "percentage value of the amount of sent network layer packets successfully delivered to a given system entity within the time constraint required by the targeted service, divided by the total number of sent network layer packets". Furthermore, there is a 20 ms control plane latency requirement for the time needed to establish an active RRC Connected mode link. 3GPP have concluded that NR Release-15 fulfil those requirements, as well as all the other IMT2020 requirements as defined in [1, 2]. Moreover, 5G NR Release-16 standardization includes enhancements to further improve the performance for e.g. factory automation use cases by meeting even stricter requirements like 0.5 ms latency with 99.9999% reliability with eURLLC and TSC. Further enhancements for specific requirements of time-sensitive applications, such as survival time indicating the number of consecutive errors the system can manage before a critical failure, is considered in Release-17 [3].

Apart from meeting those strict latency, jitter, and reliability requirements, 3GPP has defined several related use cases, see [4] and [5]. Those use cases are the result of collection of input from different industries and organizations such as e.g. 5G-ACIA (5G Alliance for Connected Industries and Automation). Table 1 shows a

Table 1 Overview of selected use cases

Use case	Reliability (%)	Latency	Data packet size and traffic model
Power distribution	99.9999	5 ms (end to end latency) Note: 2–3 ms air interface latency	DL & UL: 100 bytes ftp model 3 with arrival interval 100 ms
	99.999	15 ms (end to end latency) Note: 6–7 ms air interface latency	DL & UL: 250 bytes Periodic and deterministic with arrival interval 0.833 ms Random offset between UEs
Factory automation including TSC	99.9999	2 ms (end to end latency) Note: 1 ms air interface latency	DL & UL: 32 bytes Periodic deterministic traffic model with data arrival interval 2 ms
	99.9999	0.5 ms air interface latency for Industrial Ethernet use-cases Additional: UE must be time synchronized to 1 microsecond accuracy	DL & UL: 20 bytes Periodic deterministic traffic model with data arrival interval 0.5 ms
Release-15 enabled use case (e.g. AR/VR)	99.999	1 ms (air interface delay) for 32 bytes 1 ms and 4 ms (air interface delay) for 200 B	DL & UL: 32 and 200 bytes FTP model 3 or periodic with different arrival rates
	99.9	7 ms (air interface delay)	DL & UL: 4096 and 10 kbytes FTP model 3 or periodic with different arrival rates
Transport Industry	99.999	5 ms (end to end latency) Note: 3 ms air interface latency	UL: 2.5 Mpbs; Packet size 5220 bytes DL: 1Mbps; Packet size 2083 bytes Note: Data arrival rate 60 packets per second for periodic traffic model
	99.999	10 ms (end to end latency) Note: 7 ms air interface latency	UL&DL: 1.1 Mbps; Packet size 1370 bytes Note: Data arrival rate 100 packets per second for periodic traffic model

sub-set of the 3GPP adopted use cases that require different latency and reliability targets, as well their traffic characteristics in terms of payload size packet arrival process. As part of the 3GPP Release-16 Study of physical layer enhancements for NR URLLC, several system-level performance results were produced for those scenarios as can be found in [6]. Moreover, it is worth mentioning that 3GPP also recently developed a new radio propagation channel model for indoor industrial scenarios, being representative for different factory scenarios for conducting both realistic link- and system-level simulations (see details in [7]).

It is seen that E2E latency is the part important to the time-sensitive application, whereas NR specification work addresses only the delay over the air interface including the processing in the 5G base station (gNB) and the end-device (UE). Notable delays in traditional wireless networks include the transport and core network delays. 5G supports effective deployment of core network (CN) user plane functions at the edge, e.g. close to the application including factory premises, near or integrated to the gNB, etc. With optimized CN processing functions for IIoT, the delay contribution can be minimized to typically negligible values (e.g. <100 microseconds) which means that majority of the E2E delay contribution comes from NR.

3 NR Basics

The NR specifications define the basic QoS architecture, interfaces, and RAN protocol design [8, 9]. The NR RAN user plane protocol stack as pictured in Fig. 1 include several enhancements as compared to LTE. In particular, the Layer-2 user plane protocols, which include SDAP (Service Data Application Protocol), PDCP (Packet Data Convergence Protocol), RLC (radio link control), and MAC (medium access control), have been carefully designed to be processing-friendly and support

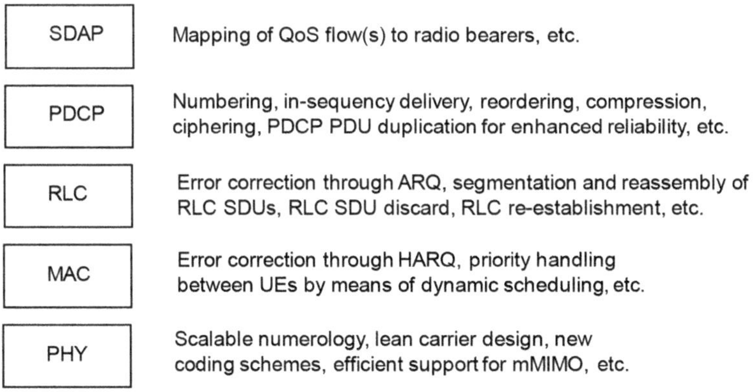

Fig. 1 Overview of the NR user plane protocol stack and its primary functionalities

very low latencies. In this context, the following Layer-2 protocol design principles for the NR are worth highlighting: (*i*) decoupling of higher layer functions from real-time constraints, (*ii*) avoiding duplicated functions, (*iii*) supporting front-haul interface splits, and (*iv*) flexible QoS, not necessarily constrained by the CN. The SDAP offers improved QoS handling [10]. As a new functionality, the PDCP supports duplication of PDUs (Packet Data Units) over two different RLC legs for enhanced reliability [11]. The latter comes in addition to options for PDU split options as known also from LTE Dual Connectivity to enhance the data rate of MBB services [12]. Removing concatenation and reordering from the RLC-layer allows RLC PDU pre-creation at the Transmitter (Tx) side as the knowledge of available grants is not required, while removing reordering allows immediate submission to PDCP entity when full RLC PDU is received. This is contributing to significantly improved pre-processing as compared to LTE. The PHY and MAC build on a flexible radio frame structure with scalable physical layer numerology (subcarrier spacing), short transmission time intervals (TTIs), and reduced gNB and UE processing times. The reduced PHY layer processing times are obtained with an optimized design tailored to support so-called pipeline processing by appropriate arrangement of control, reference, and data symbols. NR uses OFDMA (Orthogonal Frequency Division Multiple Access) as its primary waveform and multiple access method for both link directions, but also supports DFT-s-OFDMA (Discrete Fourier Transform spread OFDMA) for the uplink; see more details in [13]. Furthermore, NR adopts a user-centric lean carrier design, where advanced massive MIMO is an integrated part of the design. Also new PHY encoding schemes as compared to LTE are adopted, namely LDPC (Low-Density Parity Check) for data channels and Polar codes for control channels. The NR RAN protocol design, and the associated QoS architecture enable efficient E2E service deliverable and a rich set of options for multiplexing (aka scheduling) of diverse services with different QoS requirements [14]. 3GPP is also in the process of extending NR so it supports deployments for unlicensed spectrum bands [15], which e.g. is relevant for private IIoT deployments. For more information, a study of latency-reliability performance of NR unlicensed is available in [16], including effects of the clear channel access procedures that apply for unlicensed spectrum operation.

In addition, NR also comes with a new design of the control plane, where the RRC (Radio Resource Control) protocol is enhanced as compared to LTE, including introducing a new intermediate RRC state called RRC Inactive, in addition to having RRC Idle and Connected states [17]. Among others, the NR RRC design offers attractive trade-offs between UE power consumption, signalling overhead, and experienced latencies; see e.g. the system-level performance study in [18]. It also fulfils the aforementioned IMT2020 requirement of a 20 ms control plane latency to establish an active RRC Connected Mode connection where user plane data transmissions can take place.

The NR design comes with a number of functional split options, and implementation alternatives, for the RAN part [19]. Those include two options for realization of CRAN (centralized RAN) implementations. The so-called higher-layer split option with a CU (Centralized Unit) that hosts the RRC, SDAP, and PDCP and may be deployed in a distributed cloud environment. The CU is connected to a potentially large number of DUs (Distributed Units) that host the RLC, MAC, and PHY). Secondly, a lower layer split option is possible where the CU includes nearly all the RAN protocol stack, except the lower-layer PHY functions that is located in the remote radio heads. The latter option is often referred to as the multi-TRP (Transmission Reception Point) case. Among others, those CRAN options for NR offers attractive possibilities to improve the overall URLLC system performance; e.g. by means of efficient centralized multi-cell queuing- and radio channel aware scheduling as studied in [20].

4 URLLC Enablers

One of the enablers for URLLC is the flexible frame structure that comes with the NR, which allows fast scheduling to obtain low latency. The NR operates with a 10 ms radio frame composed of ten 1 ms subframes. Furthermore, the notation of 14-symbol slots, and mini-slots of 1–13 symbols are introduced, as illustrated in Fig. 2. NR supports dynamic scheduling with variable transmission time intervals (TTIs), including slot and mini-slot resolution. The symbol duration depends on the selected subcarrier spacing configuration.

Table 2 show examples of some of the possible TTI sizes for different subcarrier spacing when using either slot- or mini-slot TTI resolution. Here it should be noticed that the option with aggregated slot transmissions is achieved with bundled slot transmissions, i.e. is obtained as automatic non-adaptive retransmissions. NR also comes with reduced processing times for UEs to decode transmissions (downlink) and preparation of new transmissions (uplink) [21]. As an example, this means that a first transmission, and one HARQ (Hybrid Automatic Repeat reQuest) retransmission is supported within the 1-ms latency limit for URLLC if using 30 kHz subcarrier spacing and 2-symbol mini-slot transmission for the downlink.

As illustrated in Fig. 3, for each dynamic scheduled transmission, the gNB sent the corresponding scheduling grant to the UE on the PDCCH (aka Downlink Control Information – DCI). The scheduling grant contains information such as resource allocation, e.g. on which PRB (Physical Resource Blocks) as well as the used MCS

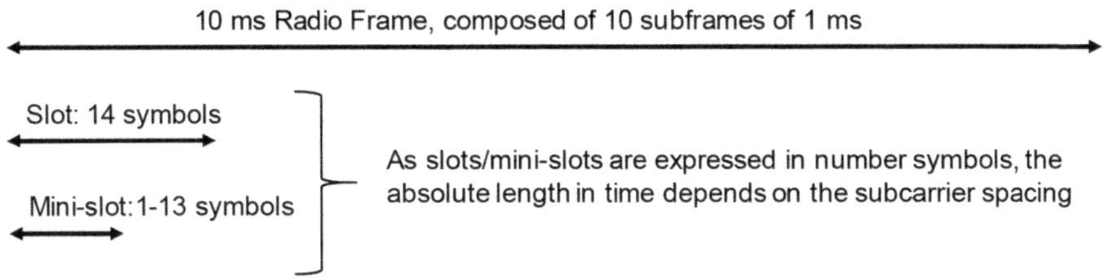

Fig. 2 Simple illustration of NR radio frame, subframes, slots, and mini-slots

Table 2 Overview of options for TTI sizes

TTI size / subcarrier spacing	15 kHz	30 kHz	60 kHz
2-symbol mini-slot	0.14 ms	0.07 ms	0.035 ms
4-symbol mini-slot	0.28 ms	0.14 ms	0.07 ms
7-symbol mini-slot	0.50 ms	0.25 ms	0.125 ms
14 symbol slot	1.0 ms	0.5 ms	0.25 ms
2 slot aggregation	2.0 ms	1 ms	0.5 ms

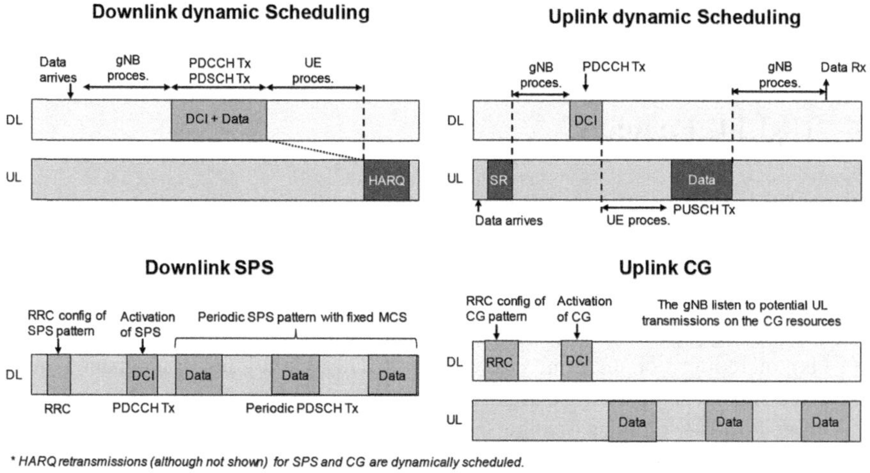

Fig. 3 High-level illustrations of downlink and uplink dynamic scheduling principles, as well as downlink SPS and uplink CG

(Modulation and Coding Scheme). Enhanced dynamic link adaption (i.e. selection of MCS) is supported to fulfil the ultra-reliability requirements by introducing a new CQI (Channel Quality Indicator) feedback options that corresponds to 1e-5 BLER for URLLC. As an example, this opens for service specific joint link adaptation and scheduling implementations [22]. Designing efficient gNB packet scheduling policies to efficiently serve URLLC traffic (potentially in coexistence with eMBB traffic) is a challenging problem, although being addressable with algorithms of modest complexity as shown in [23]. The scheduling policy should take the following into account; head of line delay metrics for the URLLC payloads, HARQ effects, overhead from scheduling grants (i.e. PDCCH DCI), as well as avoiding segmenting smaller URLLC payloads over multiple TTIs.

As in LTE, also SPS (Semi-Persistent Scheduling) is supported for NR where periodic transmission resources are configured for a UE. SPS is particularly attractive for deterministic traffic flows (including TSC), and also overcomes potential errors related to decoding of dynamic scheduling grants. For meeting the 1-ms URLLC target the uplink, there is often not enough time for the gNB to await reception of a scheduling request from a UE, followed by transmission of a dynamic scheduling, and finally the UE's uplink transmission (see Fig. 3). The option of CG

(Configured Grant) is therefore supported the uplink, where the gNB may configure periodic transmission resources for a UE. The UE thereafter immediate transmit to the gNB on those pre-configured resources whenever it has data available, thereby achieving lower latencies. The former is also known as grant-free transmission as is analyzed in [24] for URLLC use cases. The configuration of SPS and CG resources happens via RRC signalling, followed by activation by means of DCI on PDCCH as illustrated in Fig. 3 and summarized in Table 3.

For efficient co-existence of eMBB and (e)URLLC/TSC services in the downlink, preemptive scheduling has been introduced. In short, preemptive scheduling offers improved downlink multiplexing of eMBB and latency critical traffic, where the gNB may partially overwrite (i.e. pre-emption) an ongoing eMBB transmission with a shorter urgent URLLC transmission. The gNB may announce such pre-emptions to the affected eMBB UEs by sending them an interrupted transmission indication, such that the UE(s) know that part of their transmission has been overwritten. In order to minimize the impact on the eMBB UEs that are subject to pre-emption, code block group based HARQ retransmissions are introduced as a smart recovery mechanism, where only the effected code block groups (i.e. a subset of the full transmission block) is retransmitted [25]. The latter is one of the HARQ enhancements that are introduced for NR [26]. For the uplink, and equivalent solution is standardized for Release-16, where the gNB can inform eMBB UEs to cancel an ongoing transmission to quickly release radio resources for urgent URLLC transmissions. The gNB cancellation indication is send using a group common DCI signalling (known as format 2–4 [3GPP TS 38.213]). For gNBs with eight or more antenna ports, so-called opportunistic spatial pre-emptive scheduling for efficient co-existence of URLLC and eMBB is recommended as studied in [27] by utilizing advanced multi-user MIMO techniques. More details on the NR scheduling schemes can be found in [14] as well as in Table 3.

For achieving ultra-reliability, NR includes PDCP PDU duplication over two RLC legs where the same data packet is sent over two different channels to the UE. Thereby introducing additional redundancy, and thus achieving higher probability of correct reception assuming there is low correlation of errors on the two legs. PDCP PDU duplication may e.g. be applied to a UE that is in dual connectivity mode with two different gNBs such as a macro and small-cell operating at different frequencies as studied in [28]. However, it should be noticed that in a multi-user, multi-cell system, use of PDCP PDU duplication comes with the risk of creating more interference and additional queuing at the gNBs as payloads are now transmitted twice, thereby causing increased load. As studied in greater details [29], careful operation of dual node connectivity with PDCP PDU duplication is therefore important to leverage the tradeoffs between reliability gains of such techniques versus the cost that it causes in terms of additional load and interference.

At the MAC/PHY layer, the multi-TRP scenario also offers advanced options for transmission to the same UE from different nodes. One such example is non-coherent joint transmission, where the same MAC PDU is transmitted from two different nodes to a UE, followed by combining of the transmissions in the UE. Such multi-TRP transmissions required a CRAN architecture with MAC and higher-layer

Table 3 Overview of (e)URLLC-related NR features

Feature	Description
Reduced processing times	Processing-friendly RAN protocol design for low latency, including decoupling of higher layer functions from real-time constraints at lower layer, and pipeline processing. Significantly reduced PHY processing times as compared to LTE (details in [21]).
Short TTI	Variable TTI size, including mini-slot resolution TTI size of 1–13 symbols. Symbol duration scaled with the subcarrier spacing options of 15 kHz, 30 kHz, 60 kHz, 120 kHz, and 240 kHz.
HARQ	Fast asynchronous HARQ with support for CBG-based retransmissions. E.g. allowing one HARQ retransmission within the 1-ms latency budget for URLLC when using 30 kHz subcarrier spacing and short TTIs of 2-symbols.
Dynamic scheduling	Dynamic scheduling grant with flexible indication of allocated time-frequency resources for the user. For the uplink, only contiguous frequency domain resource allocation is supported for PUSCH with DFT-s-OFDM waveform.
Semi-persistent scheduling (SPS)	The time-domain periodicity is configured by RRC signalling. The corresponding frequency-domain allocation, and starting time, is given with the DCI activation. Assuming fixed MCS for the allocations until new DCI is given.
Configured Grant (CG)	UE is configured to immediately transmit on configured resources whenever it has data (aka grant-free transmissions). Type-1: Resource allocation and MCS is configured by RRC (Time-frequency allocation grid, MCS, etc.) Type-2: Similar as SPS.
Link adaptation	Dynamic link adaptation for PDCCH (aggregation level adjustment) and PDSCH/PUSCH (MCS selection). Options for configuring UEs with CQI feedback corresponding to 1e-2 and 1e-5 target BLER for the PDSCH are supported.
Preemptive scheduling (downlink)	Efficient downlink multiplexing of eMBB and URLLC UEs, where the gNB may partially overwrite (i.e. pre-emption) an ongoing eMBB transmission with a shorter urgent URLLC transmission. If a UE receives the interrupted transmission indication from the gNB, the UE may assume that its transmission has been pre-empted, such that no transmission to the UE is present on the indicated PRBs and symbols.
Uplink cancellation	Option where the gNB can cancel an ongoing uplink (eMBB) transmission to quickly unleash uplink resources for urgent URLLC transmissions. Uplink cancellation indication signalled from the gNB with group common DCI (known as format 2–4).

(continued)

Table 3 (continued)

Feature	Description
PDCP duplication	Packet duplication at the PDCP protocol layer. Transmitted over two different RLC legs for enhanced reliability.
Multi-TRP	Physical layer multi-node transmission and reception scheme for enhanced reliability (e.g. DL non-coherent joint transmission and UL multi-site reception and combining).
MIMO	Enhanced diversity mechanisms (i.e. reduced probability channel fades) and massive MIMO with grid-of-beams for improved SINR.

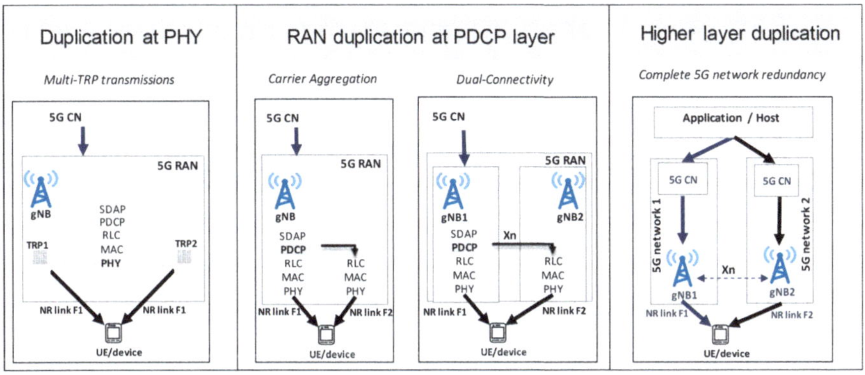

Fig. 4 Overview of options for multi-link connectivity with packet duplication for enhanced reliability

PHY in the same place, while higher-layer PDCP PDU duplication is supported also from two separate gNBs. The different options for redundancy transmission modes are pictured in Fig. 4. Showing the PHY layer multi-TRP option to the left, the two options for PDCP duplication (using either carrier aggregation or dual connectivity) in the middle, and higher layer duplication transmitted over two separate 5G NR network paths with complete 5G network redundancy, including multi-UE device capabilities for better availability performance. The latter option may e.g. be realized by using advanced transport layer network protocols, for instance Multipath QUIC, which has shown to improve wireless communication reliability and availability for autonomous vehicles in e.g. [30]. Such protocols are very efficient even when deployed by the same operator leveraging two or more different UE, but even better gains are possible when conducted across network segments that do not leverage site, RAN, or network sharing.

Finally, MIMO diversity schemes are of high importance for achieving ultra-reliable transmission as they help reduce the probability of experiencing deep fades, and thus low SINRs. The closed- and open-loop MIMO schemes are therefore instrumental in achieving the URLLC targets. Moreover, massive MIMO options with grid-of-beams also offer benefits for URLLC as those help further improve the experienced SINRs. Table 3 provides a summary of the primary (e)URLLC enablers for NR; see also [9] and [31].

In line with the communication theory, serving URLLC traffic with strict latency and reliability constraints naturally comes with a cost of reduced spectral efficiency as compared to eMBB [32]. The lower spectral efficiency of serving URLLC is mainly contributed by: (i) using more conservative MCS for each transmission to achieve low BLER, (ii) using short TTIs that represent larger relative control overhead from scheduling grant (PDCCH) and headers, (iii) higher interference sensitivity that may put restrictions on how much traffic load the system can tolerate while still fulfilling the strict latency-reliability constraints. As a few examples, the studies in [22, 23, 25] show how the eMBB capacity of NR is affected when URLLC traffic is carried.

5 Seamless Integration of 5G Into IIoT Environment with TSC

A major use case in 3GPP Release-16 is to bring the flexibility of 5G NR seamlessly into Industrial Ethernet environments that use IEEE time-sensitive networks (TSN) mechanisms. TSN is an openly standardized layer-2 solution that adds full determinism to Industrial Ethernet, with key features that include network-wide scheduling with strict traffic reservation, shaping, and pre-emption mechanisms as well as redundancy methods, see ex. [33–36]. To ensure that all networking components and devices operate synchronously, a common time synchronization is achieved using IEEE802.1AS mechanisms [37].

By requiring direct support of TSN with 5G NR Release-16, it is ensured that the manufacturing and automation industry can easily augment their wired environments with 5G, e.g. for new use-cases or migrating existing ones. A key contribution comes from the German BMBF research project TACNET 4.0 [38], where more than 20 manufacturing and automation use-cases were investigated to derive requirements for the 5G wireless communication service. In NR Release-16, mainly the closed loop automation use cases in factories are in focus while video, imaging and audio for professional applications (VIAPA) and other applications will be a focus point in NR Release-17 onwards. The enabling 5G component is denoted as time-Sensitive communications (TSC) and it provides means for both deterministic transport as well as absolute time synchronization of devices which is an essential enabler for e.g. TSN.

For a 5G TSC deployment to be able to integrate into an existing managed and wired Industrial Ethernet and TSN environment, a centralized configuration model is assumed which at the time of specification was the dominant configuration method found in TSN networks. This means that all network components and end-devices are configured by a single centralized network controller (CNC) and single/multiple centralized user configuration (CUC) instances, respectively. A first

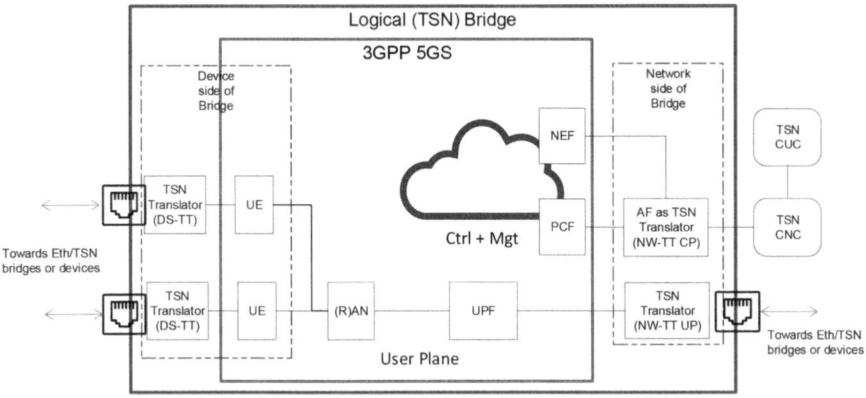

Fig. 5 Simple illustration on how 5GS integrates seamlessly into IEEE TSN environment by mimicking an Ethernet bridge

enabler in Release-16 was to allow for the 5G System (5GS) to mimic an Ethernet bridge to the factory, where each UE represents a port on the bridge and where additional port(s) are available on the CN side. This equivalent model is shown in Fig. 5. A second step is to let the 5GS be managed by the CNC of the TSN network which basically reads the requirements from all services (e.g. via the CUC), derives end-2-end schedules, and then breaks down this schedule into the individual bridges and their ports, e.g. including the 5GS virtual bridge. In order to map between IEEE domain parameters and 3GPP internal mechanisms for flow control and QoS, there are so-called TSN translators on both the device (each UE) and network side (CN port) of the 5GS. On the CN side, the translator functionality is divided into C-plane components and U-Plane components respectively interfacing with the network elements of the 3GPP system. Details of this integration and mapping between the IEEE TSN domain and the 5G System model and QoS Framework are described in [39], while basics of the IEEE management mechanisms are described in [34].

Each bridge, including the 5GS virtual one, may be programmed in terms of gate control mechanisms as described in [35]. Those mechanisms apply at the ports of the bridge. TSN offers means to specify specific time-windows where a certain traffic class can be scheduled. This ensures that resources are reserved for the most critical flows in a tic-toc fashion across the complete network encompassing multiple bridges. This allows for both deterministic delivery and jitter control to be better than one microsecond. There are many additional features, for example a pre-emption mechanism where lower priority traffic transmission is stopped if higher priority traffic comes in. A description of what IEEE mechanisms are supported in Release-16 of 5GS can be found in [40]. Most of such features are implemented in the translator functions, leaving a more traditional approach to data transmission viable inside the 5GS as will be described later in this chapter. E.g. the 5GS system may be considered as the internal wiring between the ports of a bridge and IEEE level gate and scheduling controls are conducted at the translator end-points (UE and CN).

6 TSC Enablers

In the following, the 5G NR enablers for TSC are discussed. As denoted earlier, de-jittering and time alignment to CNC controlled gate timings is done on the "outside" of the 5GS, e.g. as part of the Device Side and CN TSN Translators respectively (DS-TT, NW-TT) as shown in 6. As such, a packet may experience some jitter over the air interface which is much larger than a microsecond but at either of the translators there will be a "hold-and-forward" buffer that will hold and release the packet according to the configured gate schedule, e.g. it puts each packet back on the microsecond schedule, effectively providing de-jittering. Also, features such as IEEE pre-emption methods are not applied at the air interface but at the translator end-points. The 5G TSC is built heavily on the URLLC and eURLLC solution as the enabler to ensure short and guaranteed packet delays between the ports of the bridge (e.g. UE to CN, or UE to UE). However, due to the special nature of TSN traffic flows, some additional features are introduced.

The first feature is the ability to configure the RAN with detailed a priori information of the deterministic traffic flows. With (e)URLLC, 5GS has tackled the most complicated use-case where it cannot be assumed when a certain traffic flow has data to transmit, e.g. when packets arrive to the 5GS. While this case is still supported also for deterministic flows, many TSN traffic flows are typically strictly periodic and have a known time of arrival. This is a significant benefit to the 5GS since the system can pre-reserve resources, not only for TSN traffic but also for freeing up resources for other non-TSN traffic. This is a key feature in all deterministic networking in order to increase the possible offered load for critical flows with strict requirements. However, to leverage this information in the scheduling, the gNB must be made aware of such detailed characteristics of the TSN service flow. Like (e)URLLC, TSN service flows are mapped to the Delay Critical GBR QoS category, which informs the RAN about expected packet burst size, etc. However, to have more information regarding deterministic and periodic flows, 3GPP has introduced TSC assistance information (TSCAI) [39] that the CN can use to configure the RAN on top of normal QoS flow parameters. As shown in Fig. 6a., the TSCAI contains key traits of the service flows, including flow-direction, periodicity, and absolute time offset that the scheduler can effectively prepare resources for the flow in advance and without delay. For uplink flows, the burst arrival time (BAT) is defined at the egress interface of the UE. For downlink flows, the BAT is defined at the ingress of the RAN (gNB). It should be noted that the periodicity of vertical services may not fit well with the 5GS numerology, thus periodicity or BAT values are not necessarily an integer of symbols or slots in the 3GPP domain for instance. The 5GS monitors the difference between clock domain in order to adjust internal resourcing correspondingly. Within each defined period (defined by Periodicity parameter), TSC QoS Flows are required to transmit only one burst of a defined maximum size (maximum data burst value or MDBV) as set in the 5GS QoS Indicator (5QI) profile of the Delay Critical GBR resource type [39]. Knowing the exact timing of the incoming bursts, the gNB scheduler can prepare

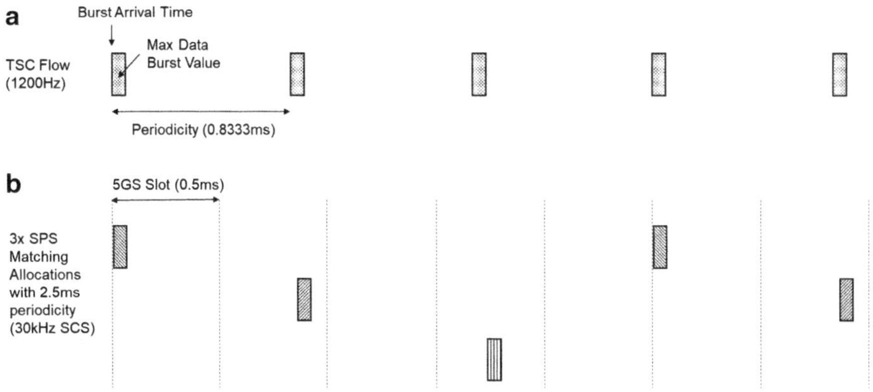

Fig. 6 Illustration of (**a**) TSC flow parameters and (**b**) how multiple SPS patterns can be combined to fit 5G nomenclature to "odd" timing from vertical industries

its resource to reduce the latency in both uplink and downlink, and it can notably produce proactive grants in uplink to reduce the time of the scheduling process and thus the experienced delays.

While dynamic scheduling is improved with TSCAI, further improvements are introduced in Release-16 to simplify handling and improve performance of TSN and TSC type services. Being strictly periodic with very low periodicity down to 0.5 ms and delay requirements for the air interface lower than 0.5 ms, semi-persistent traffic allocation methods are very efficient compared to dynamic scheduling leveraging TSCAI. Semi-persistent allocations (i.e. SPS) also bring further benefits including less reliance on high control channel reliability for carrying the information regarding dynamic allocations. A drawback is spectral efficiency as a more robust modulation and coding scheme on fixed frequency resources needs to be selected compared to dynamic scheduling. However, at very low latency requirements, dynamic scheduling must operate without hybrid automatic repeat request (H-ARQ) and here semi-persistent methods support a higher capacity in terms of number of TSC traffic flows supported. In order to improve the feasibility of semi-persistent allocation schemes, the supported periodicity in downlink has been reduced to one slot time (from 10 ms in Release-15). Further, for uplink and downlink, multiple CG or SPS configurations are supported per UE in order to support multiple TSN flows effectively. Besides being able to support more simultaneous TSC flows by a single UE, a larger number of CG/SPS allocations allows the network also to reduce jitter for traffic flows that have an odd periodicity compared to the 5GS nomenclature while having lower overhead (e.g. number of unutilized CG or SPS configurations). This is illustrated in Fig. 6b where three SPS flows with 5G compliant periodicity of 2.5 ms are combined to match the TSC flow periodicity of 1200 Hz without wasting resource allocations and with very low resulting jitter (only waiting for next available TTI opportunity). Further, additional enhancements are made to handle collisions including prioritization in NR Release-16 and it is also considered to have simpler configurations for odd-periodicity flows. With these features, even strict delay and bandwidth requirements from motion control and other tight Industry 4.0 use-cases can be met with the 5GS system as concluded in [41].

The third feature relates to absolute time synchronization. A UE was already in Release-15 able to synchronize relatively to the 5GS, e.g. identifying System Frame Number boundaries. However, in Release-16 the network can inform the UE about what the absolute time was during such boundary, allowing the UE to set its time with high accuracy compared to the clock of the 5GS, e.g. typically UTC. Having end-devices accurately time synchronized is a key feature for them to collaborate with a high accuracy within Industry 4.0 applications as well as for the network to synchronize end-points to comply with IEEE TSN mechanisms. To provide the reference time, NR Release-16 includes two different methods. One is a new System Information Block (SIB9) message that can be broadcast to all UE [17]. Additionally, the network may provide this information via an RRC message as well as in a unicast fashion. The achievable accuracy of the UE time synchronization depends on several factors including the channel environment, cell size, mobility, and implementation factors. However, as shown in [41] 5GS Release-15 can provide a time synchronization much better than 1 microsecond for the identified use-cases for wireless TSC. In order to enable a vertical to synchronize devices to an own preferred clock grand master different from that of the 5GS, TSC also supports synchronization over the 5GS according to IEEE802.1AS mechanisms. Here, gPTP messages are sent across the 5GS in a transparent manner, but at the end points, the messages are compensated for their residence time or delay experienced through the 5GS. In order to estimate the residence time, both UE and CN rely on having the same understanding of time, and here the internal 5GS timing method is used as discussed previously.

7 Outlook

With (e)URLLC and TSC, 5GS starts its journey into IIoT. During next years, availability and reliability of wireless communications will be under scrutiny from vertical players, and future releases will add features based on achieved experience from deployments. Also, as verticals start seeing benefits of wireless freedom and more agile connectivity opportunities, it is expected that applications will adapt driving new requirements and use-cases for the upcoming 5GS releases.

Among others, 3GPP is set to work on further link adaptation enhancements (e.g. through enhanced UE CSI feedback), introducing enhancements NR unlicensed to improve latency-reliability, additional features for intra-UE multiplexing of URLLC/eMBB, options for new QoS-related parameters. Although a significant step for integration into Industrial Ethernet with TSN has been accomplished in Release-16, there are further features to be considered. One is the optimization of UE-UE traffic via the network, both in terms of reducing latency and also ensuring high synchronization accuracy when the time synchronization source sits on the UE side of the network. Such issues are considered in 3GPP Release-17. In further releases, other targets may emerge following along enhancements developed for TSN in IEEE. Such enhancements may include support for distributed configuration model as supplement to the centralized model used in Release-16. Finally, as 3GPP in Release-17 and onwards focus on use-cases beyond factory applications, e.g. audio-video production and smart grid applications, new requirements emerge for TSC support also for IP applications and wide-area networks where solutions beyond seamless IEEE integration are required. Finally, time-sensitive applications use additional measures of reliability compared the average reliability of packet errors mainly addressed in wireless networks. One example is that many such applications operate according to a Survival Time where consecutive packet errors are the most critical to the system and stand-alone errors are less critical. The ability to configure the network as well as optimizing multi-connectivity radio procedures for Survival Time is considered in Release-17.

References

1. 3GPP TR 38.913 (2016, March) Study on scenarios and requirements for next generation access technologies, Version 14.1.0
2. IMT Vision (2015, Feb) Framework and overall objectives of the future development of IMT for 2020 and beyond, International Telecommunication Union (ITU), Document, Radiocommunication Study Groups
3. 3GPP TR 22.832 (2019, Dec) Study on enhancements for cyber-physical control applications in vertical domains
4. 3GPP TR 22.804 (2018, Sept) Study on communication for automation in vertical domains (Release 16)", Version 16.1.0
5. 3GPP TS 22.261 (2018, Dec) Service requirements on the 5G system; Stage 1 (Release 16)", version 16.6.0
6. 3GPP TR 38.824 (2019, Feb) Study on physical layer enhancements for NR ultra-reliable and low latency case (URLLC) (Release 16)
7. 3GPP TR 38.901 (2019, Sept) Study on channel model for frequencies from 0.5 to 100 GHz (Release 15)
8. 3GPP TS 38.300 (2017, Dec) NR and NG-RAN overall description; stage-2," Version 2.0.0
9. Chandramouli D, Liebhart R, Pirskanen J (eds) (2019, Apr) 5G for the connected world, Wiley, ISBN: 978-1-119-24708-1
10. 3GPP TS 37.324 (2019, Sept) E-UTRA and NR; Service Data Adaptation Protocol (SDAP) specification (Release 15)
11. 3GPP TS 37.340 (2017, Dec) E-UTRA and NR; Multi-connectivity; Stage-2, v:15.0.0
12. Rosa C et al (2016) Dual connectivity for LTE small cell evolution: functionality and performance aspects. IEEE Commun Mag 54(6):137–143
13. Berardinelli G et al (2016) Generalized DFT-spread-OFDM as 5G waveform. IEEE Commun Mag 54(11):99–105
14. Pedersen KI et al (2018) Agile 5G scheduler for improved E2E performance and flexibility for different network implementations. IEEE Commun Mag 56(3):210–217
15. 3GPP Technical Report 38.889 (2018, Dec) Study on NR-based access to unlicensed spectrum (Release 16)" Version 16.0.0
16. Maldonado R et al (2020, April) Analysis of high-reliable and low-latency communication enablers for new radio unlicensed. In: IEEE Proc. wireless communications network conference (WCNC)

17. 3GPP TS 38.331 (2019, Sept) NR; Radio Resource Control (RRC) protocol specification (Release 15)
18. Khlass A et al (2019, Sept) On the flexible and performance-enhanced radio resource control for 5G NR networks. In: IEEE Proc. VTC2019-fall, Honolulu, USA
19. 3GPP 38.401 (2018, March) Technical specification group radio access network; NG-RAN; Architecture description, Version 15.1.0, Release 15
20. Karimi A et al (2019) 5G centralized multi-cell scheduling for URLLC: algorithms and system-level performance. IEEE Access 6:72253
21. 3GPP TR 38.214 (2018, Dec) Physical layer procedures for data (Release 15)
22. Pocovi G et al (2018) Joint link adaptation and scheduling for 5G ultra-reliable low-latency communications. IEEE Access 6:28912
23. Karimi A et al (2019, May) Efficient low complexity packet scheduling algorithm for mixed URLLC and eMBB traffic in 5G. In: IEEE proc. vehicular technology conference (VTC2019-spring)
24. Berardinelli G et al (2018) Reliability analysis of uplink grant-free transmission over shared resources. IEEE Access 6:23602–23611
25. Pedersen KI et al (2018, June) Preemptive scheduling of latency critical traffic and its impact on mobile broadband performance. In: IEEE Proc. VTC-2017 spring
26. Pedersen KI et al (2017, June) Rethink hybrid automatic repeat reQuest design for 5G: five configurable enhancements. IEEE Wirel Commun Mag 24:154
27. Esswie A et al (2018) Opportunistic spatial preemptive scheduling for URLLC and eMBB coexistence in multi-user 5G networks. IEEE Access 6:38451
28. Mahmood N et al (2019, April) On the resource utilization of multi connectivity transmission for URLLC services in 5G new radio. In: IEEE Proc. IEEE WCNC
29. Centenaro M et al (2020) System-level study of data duplication enhancements for 5G downlink URLLC. IEEE Access 8:565
30. Mogensen R et al (2019, May) Selective redundant MP-QUIC for 5G mission critical wireless applications. In: IEEE proc. vehicular technology conference (VTC2019-spring)
31. Pocovi G et al (2018, March-April) Achieving ultra-reliable low-latency communications: challenges and envisioned system enhancements. IEEE Netw Mag 32(2):8–15
32. Soret B et al (December 2014) Fundamental tradeoffs among reliability, latency and throughput in cellular networks. In: IEEE Proc. Globecom
33. Farkas J (2018, March) Introduction to IEEE802.1 – Focus on the Time-Sensitive Networking Task Group
34. IEEE 802.1Qcc-2018 Standard for local and metropolitan area networks - bridges and bridged networks, "Amendment 31: Stream Reservation Protocol (SRP) enhancements and performance improvements," 2018, Oct
35. IEEE 802.1Qbv-2015 Standard for local and metropolitan area networks - bridges and bridged networks, "Amendment 25: enhancements for scheduled traffic,"2015, Dec
36. IEEE 802.1CB Standard for local and metropolitan area networks - Frame Replication and Elimination for Reliability, 2017, Oct
37. IEEE802.1AS-Rev Standard for local and metropolitan area networks - timing and synchronization for time-sensitive applications, 2019, Oct
38. Gundall M et al (2018) 5G as enabler for Industrie 4.0 use cases: challenges and concepts. IEEE international conference on emerging technologies and factory automation (ETFA), Torino, Italy, 2018, Sept
39. 3GPP TS 23.501 (2019, Sept) System architecture for the 5G System (Release 16), see e.g. clauses 4.4.8, 5.27, 5.28
40. 3GPP S2-1908630 (2019, June) LS on 3GPP 5G system support for integration with IEEE TSN networks (Release 16)
41. 3GPP (2018, Nov) TR 38.825 study on NR industrial Internet of Things (IoT) (Release 16)

Selected Aspects and Approaches on Improving Dependability in Industrial Radio Networks

Norman Franchi, Tom Hößler, Lucas Scheuvens, Nick Schwarzenberg, Waqar Anwar, Andreas Traßl, and Gerhard P. Fettweis

1 Dependability Metrics for Wireless Communications Systems

Dependability theory is a powerful framework, involving the main attributes availability, reliability, maintainability, safety, integrity, and security [4]. However, only availability and reliability are quantifiable as probabilities for correct service and its continuity, respectively. These metrics were proposed in the 1960s to analyze the life cycles and failures of technical systems [5]. Although the ITU has transferred the definitions to communications (cf. [15]), they are often used colloquially and incorrectly in this sector. Understanding and leveraging the following fundamental dependability metrics and their differences refine the discussion on ultra-reliable low latency communications (URLLC), which will help mastering key challenges of future wireless communications systems, e.g., in wireless factory automation and real-time remote control. In this section, basic dependability quantities are introduced before they are applied to multi-connectivity for wireless industrial communications.

Dependability theory distinguishes between an operational state, which is indicated as "up" and a failed state or in repair, if repairs are possible, indicated as "down." In the following, the term item represents a structural unit, e.g., a component of a system, a system itself, or a service.

N. Franchi (✉) · T. Hößler · L. Scheuvens · N. Schwarzenberg · W. Anwar · A. Traßl
G. P. Fettweis
Technische Universität Dresden, Dresden, Germany
e-mail: norman.franchi@tu-dresden.de; tom.hoessler@tu-dresden.de; lucas.scheuvens@tu-dresden.de; nick.schwarzenberg@tu-dresden.de; waqar.anwar@tu-dresden.de; andreas.trassl@tu-dresden.de; gerhard.fettweis@tu-dresden.de

N. H. Mahmood et al. (eds.), *Wireless Networks and Industrial IoT*,
https://doi.org/10.1007/978-3-030-51473-0_2

Definition 1 *Availability* A is the probability that an item is able to perform as required at a given point in time [11].

Availability can be interpreted as an average success probability, equivalent to the complement of the packet loss rate (PLR) or outage probability P^{out} in communications, $A = 1 - P^{\text{out}}$. Correspondingly, availability expresses the mean proportion of time an item is operational,

$$A = \frac{\text{MUT}}{\text{MUT} + \text{MDT}} \tag{1}$$

with MUT and MDT denoting the mean uptime and mean downtime, respectively. These key performance indicators (KPIs) are defined as follows.

Definition 2 *Mean uptime* MUT is defined as the average duration from a transition to an up state until the first transition back to a down state.

Definition 3 *Mean downtime* MDT is the counterpart of MUT and, thus, defined as the average duration from a transition to a down state until the first transition back to an up state.

Definition 4 *Mean time between failures* MTBF is the average time duration between consecutive transitions from an up state to a down state, equivalent to the sum

$$\text{MTBF} = \text{MUT} + \text{MDT}. \tag{2}$$

For Rayleigh fading channels, which will be evaluated in Sect. 2.5, the influence of the carrier frequency and mobility is cancelled out in this quotient [12]. This is one reason why the KPIs availability and PLR alone are of limited benefit for specifying URLLC. A further shortcoming of these metrics is the lack of reference to individual time intervals. Characterizing the probability of continuing a failure-free operation throughout a time interval is of major interest for critical machine-type communications, e.g., during maneuvers of wirelessly controlled robots or automated guided vehicles (AGVs), addressed by the following dependability quantity.

Definition 5 *Mission reliability* $R(\Delta t)$ is the probability that an item is able to perform as required throughout a mission time interval Δt [13].

In contrast to the traditional reliability definition, the term "mission" is added here, emphasizing the failure-free operation throughout this time interval. In general, it is not possible to convert between reliability and availability because mission reliability depends on the mission duration as opposed to availability. Since a failure-free operation is practically impossible for long missions due to random processes causing failures, the limiting value of the mission reliability $R(\Delta t)$ as Δt approaches infinity is zero, $\lim_{\Delta t \to \infty} R(\Delta t) = 0$. It is important to understand that referring to a certain reliability value without specifying the corresponding mission duration Δt is not a valid statement. In wireless systems, however, reliability is usually defined as the amount (in %) of sent packets successfully delivered to a given node divided by the total number of sent packets [1]. This interpretation is related to PLR or outage probability without reference to the time dimension [11]. Hence, it corresponds to the concept of availability (cf. Definition 1). If short downtimes can be accepted, which is the case for today's communications systems, the concept of mission reliability can be extended to the following dependability KPI.

Definition 6 *Mission availability* $M\left(\Delta t, t_{\mathrm{d}}^{\max}\right)$ is the probability that all downtimes are not longer than the threshold $t_{\mathrm{d}}^{\max}$ during a mission of duration Δt [14].

This metric specifies the success probability of a mission, in the case that interruptions can be tolerated. Obviously, mission reliability is a special case of mission availability given by $R(\Delta t) = M\left(\Delta t, t_{\mathrm{d}}^{\max} = 0\right)$.

2 Physical Layer Multi-connectivity

Diversity is widely accepted to be key in order to improve the dependability of a system, e.g., by introducing backup components corresponding to the redundant transmission of messages over different wireless channels. Multi-connectivity using multi-link diversity is a promising approach to enable URLLC and highly dependable IRSs, respectively. This section outlines why multi-connectivity on the physical layer should be implemented, discusses selected implementation aspects, and explains how to establish an abstraction layer for generalized link quality assessment and higher layer evaluations such as system-level simulations.

2.1 Fading in Multipath Channels

While signal processing and coding on the physical layer (PHY) of modern wireless communications systems have remarkably evolved over the past decades, their performance still depends completely on local radio conditions. Multipath propagation due to reflections from surrounding objects causes rapid random power level fluctuations (fading) at the receiver. The received signal-to-noise ratio (SNR) determines how much data may be transmitted error-free with a certain probability. In reverse, to achieve dependability on a wireless link, it needs to be ensured that the SNR stays above the level required by the desired target data rate. In practice, said SNR requirement is a function of the modulation and coding scheme (MCS). Taking receiver noise into account, the determined SNR translates to a minimum received power which is referred to as the sensitivity level.

Figure 1 shows a trace of received power over time which has been captured in a factory hall with plenty of reflective walls and machinery [7]. The power levels of two frequencies in the 5.8 GHz ISM band both exhibit typical major

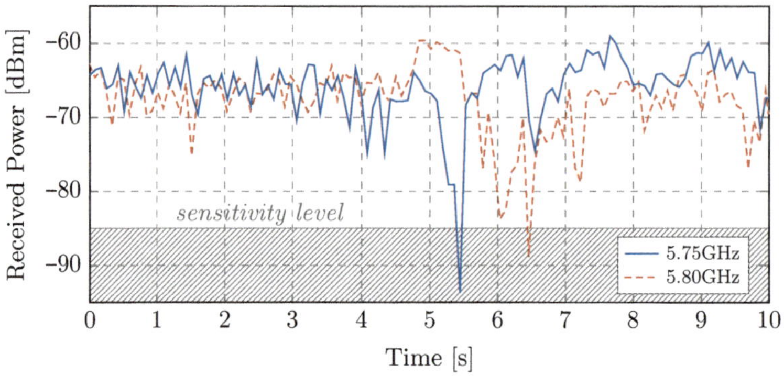

Fig. 1 Received power over time on different frequencies while a pedestrian walks by. Plotted from obstructed line of sight measurements in a factory hall with a resolution bandwidth of 1 MHz [7]. The sensitivity level has been set to this particular value for illustration purposes only

drops (deep fades). Assuming an arbitrary sensitivity level of -85 dBm, it can be observed that the received power on each frequency falls below this threshold once within the depicted time frame. A state-of-the-art single connectivity wireless system using only one of the two frequencies would experience an outage in terms of one or more packet errors. While an increased transmit power can mitigate such occasional outages, it comes at the cost of reduced energy efficiency, causing additional interference to other wireless devices. In fact, this can be mitigated by using multiple frequencies, as outlined below.

2.2 *Multi-connectivity as Source of Diversity*

Physical layer multi-connectivity solutions aim to reduce fading by utilizing multiple weakly correlated links in parallel. The idea is that it is less likely to have deep fades on all links compared to the probability of a deep fade on a single link. Physical layer multi-connectivity solutions influence the dependability of the communications system by reducing the overall PLRs. By physical layer multi-connectivity, we refer to any system where users are connected via multiple links. Both connections to a single access point (AP) and connections to multiple APs are viable. Multi-connectivity design always needs to be aligned with the wireless channel, since the channel determines the correlation between the links and therefore the multi-connectivity performance.

When connecting to a single AP over multiple links, users can be served on different frequencies, in different time instances or over spatially separated antennas. When designing a multi-connectivity system which is serving on different frequencies, the behavior of the channel over the communications bandwidth is of crucial importance for the performance of the system. The coherence bandwidth is the frequency range over which the wireless channel is highly correlated. The links of a multi-connectivity system utilizing different frequencies therefore have to be separated at least by the coherence bandwidth to ensure a weak correlation. Thus, when designing a frequency separated multi-connectivity system, usually better performance is achieved when the serving frequencies are not located next to each other. Instead, several other users are allocated in between the serving frequencies. In general, factories benefit from larger space and the use of highly reflective materials. This leads to larger propagation delays and therefore a more frequency-selective channel with a lower coherence bandwidth. However, factory halls are different from each other and frequency separated multi-connectivity performance depends on the respective factory environment [21]. Time-separated links need to be delayed at least by the coherence time of the channel which induces an additional latency source, especially in static environments. These high latency values are usually unacceptable for industrial use cases and often prevent the use of time separated multi-connectivity links in practice. Multiple antennas have to be spatially separated such that individual channel variations arise.

In the case of multiple APs serving a single mobile station (MS), the additional spatial separation is beneficial compared to the single AP case as it further decorrelates the small-scale fading of the links. Furthermore, spatially separated APs also help in disadvantageous large-scale fading and high single link path loss scenarios.

When considering diversity, it seems appropriate to compare physical layer multi-connectivity to traditional spatial diversity using Multiple-Input Multiple-Output (MIMO). Multi-connectivity can be interpreted as a multi-antenna system with a diagonal channel matrix, i.e., a system where each input connects to exactly one output. Taking a multi-connectivity setup with multiple links on separate frequencies, for example, this would require additional resources in terms of spectrum for every added link, while MIMO is able to separate links by space-time block coding. However, such multi-connectivity setup poses less constraints on the synchronicity in case of distributed APs: while time offsets between individual transmissions cause inter-symbol interference in MIMO systems, multi-connectivity is able to compensate for such offsets if each transmission is received on a separate frequency [20].

2.3 A System Model for Multi-connectivity

In multi-connectivity systems, the MS is simultaneously connected over L links to a total of N access points as depicted in Fig. 2. Redundant data is transmitted to enhance the average PHY dependability. The l-th link is characterized by its instantaneous SNR γ_l. In the following, a downlink topology between one or multiple APs and a single MS is assumed for simplicity.

At the MS the redundant data from each link is again combined to a single bit stream. The combining can be implemented at several points of the transmission chain. In Fig. 3 reception of multiple links and combining after the demodulator is

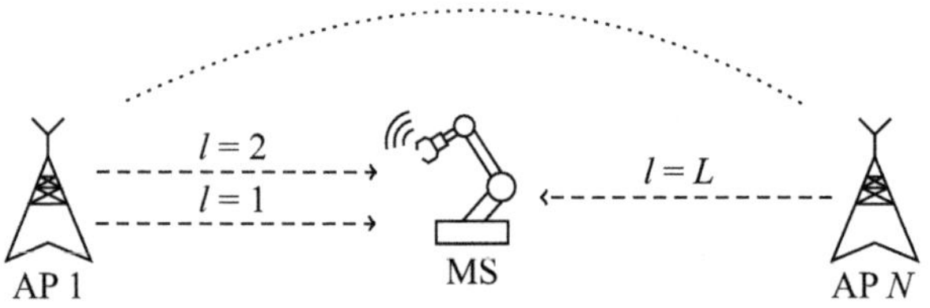

Fig. 2 Example downlink topology for multi-connectivity with distributed APs

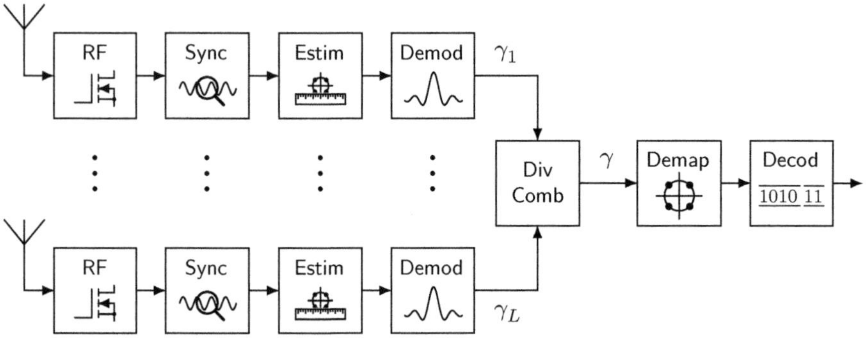

Fig. 3 Flowchart for multi-connectivity with diversity combining of L receiver chains

depicted. Regardless of whether the links were sent from only one AP or multiple APs, several combining options are available:

- *Selection combining*: Only the link with the best instantaneous SNR is selected for data recovery. Selection combining is the least complex of all combining schemes.
- *Equal gain combining*: Links are phase corrected and averaged regardless of their SNR. Therefore, every link has the same weight within the combined signal, independently of their link quality.
- *Maximum ratio combining*: Links are weighted according to their SNR before they are averaged. Links with a good average SNR are higher weighted and have a higher influence on the combined signal. In this case the SNR after the diversity combiner will be the sum of the link SNRs.

Complementing the simple and well-known schemes from above, an alternative method with deferred combining is recently being discussed. Joint decoding with distributed turbo codes, originally conceived for relaying applications [23], can be employed as coding and combining scheme [19]. Each link is being decoded in parallel, while decoders exchange soft bit information between links iteratively. In contrast to the receiver structure depicted in Fig. 3, joint decoding requires individual demappers and decoders per link, and the point of combining moves further downstream to the decoders. Joint decoding comes at the cost of higher complexity but enables different interleaving patterns per link which can be used to construct a stronger chained code [23]. It has been shown theoretically that joint decoding outperforms maximum ratio combining and selection combining in terms of outage probabilities in Rayleigh fading channels [24]. Simulation results confirm these findings as well for frequency-selective fading and indicate growing benefits as fading gets more selective and imperfections such as unavailable or noisy SNR estimates are considered [19]. However, making use of different coding parameters per link implies different transmit symbols for the same user data and is therefore not compatible with the symbol-level combining schemes listed above. For ease of understanding, the following discourse on abstraction shall be limited to the former combining schemes.

2.4 Physical Layer Abstraction for Multi-connectivity

To simulate a large network in a computationally efficient way, physical layer abstraction (PLA), or link-to-system mapping, is required, as modeling physical layer processing of each node is computationally expensive. Furthermore, to evaluate high reliability as required for industrial communications, a large number of packets are required to be transmitted, which may take months to simulate [3]. Therefore, PLA for multi-connectivity is essential to investigate various trade-offs and performance goals in industrial environments. PLA predicts the performance by mapping the received SNR to a PLR or throughput. In the case of coded modulation, no closed-form expression exists for PLR or capacity in terms of SNR. Therefore, simulation-based look-up tables are generated per MCS in the presence of additive white Gaussian noise (AWGN) fading. To evaluate performance in fading channels, an effective SNR is computed, which is equivalent to the AWGN SNR. For example, in wide-band orthogonal frequency division multiplexing (OFDM) systems, each subcarrier could have different SNRs due to frequency-selective fading. Therefore, effective SNR mapping is required to predict the performance. Moreover, for multi-connectivity network where multiple links could have variable SNRs on their subcarriers, a combined effective SNR of all assigned links is required. The effective SNR can be obtained using the SNR mapping algorithms such as receive bit information rate (RBIR) and exponential effective SNR mapping (EESM) or enhance EESM (eEESM). The authors in [2] showed that eEESM outperforms other algorithms in terms of accuracy; therefore it is considered here for PLA.

To obtain a combined link quality metric (LQM) for multi-connectivity, the post-combined subcarrier SNRs $\underline{\gamma} = [\gamma_1, \gamma_2, \ldots, \gamma_G]$, where G is the total number of subcarriers, are mapped to an effective SNR, given as

$$\gamma_{\text{eff}} = \frac{\beta}{2}\mathcal{W}\left(\frac{2}{\beta}\left(\frac{1}{G}\sum_{g=1}^{G}\frac{1}{\sqrt{\gamma_g}}e^{-\gamma_g/\beta}\right)^{-2}\right), \tag{3}$$

where $\mathcal{W}(\cdot)$ is Lambert-W function [9], g is the subcarrier index, and β is a modulation and channel-dependent parameter which can be optimized as

$$\beta = \arg\min_{\beta} |\gamma_{\mathrm{AWGN}} - \gamma_{\mathrm{eff}}(\beta)|^2 . \quad (4)$$

The default values of β are 1, 2, 10, 42, and 170 for BPSK, QPSK, 16-QAM, 64-QAM, and 256-QAM, respectively, obtained from upper bounds on symbol error probability [6].

After having an accurate LQM, system-level simulation or link adaptation can be performed simply by mapping SNR to expected PLR or throughput using look-up tables. These tables are generated per MCS under AWGN fading conditions.

The concept of system-level simulations using PLA is explained in Fig. 4. In the system-level simulator, a random frequency response of each link is generated, and per subcarrier SNRs are obtained. The multi-user dependencies can be added such as scheduling, resource allocation, access schemes, and interference. Then, these SNRs are mapped to an effective SNR, and the performance is determined using look-up tables. The resultant LQM can be used for multiple purposes such as performance evaluation, link adaptation, or automatic repeat request (ARQ).

2.5 Applying Dependability Metrics on Multi-connectivity

In this section, a multi-connectivity scenario is evaluated with regard to the dependability quantities introduced in Sect. 1. The focus is on small-scale fading due to multi-path propagation as a major cause of failure for wireless communications systems. Channels are modeled as repairable components. A channel is denoted as "up" (operational), if it can successfully transmit and receive messages; otherwise the channel is called "down." This interpretation of the wireless channel as a repairable item complies with the Gilbert-Elliot model [10]. The fading margin is

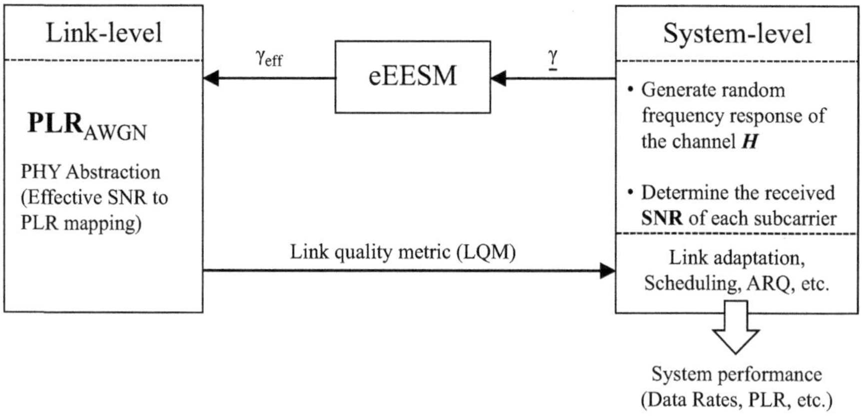

Fig. 4 Link-to-system-level mapping using physical layer abstraction

defined as $F = p_{\mathrm{avg}}/p_{\mathrm{min}}$ with the average receive power p_{avg} and the required minimum receive power. The maximum Doppler frequency $f_{\mathrm{D}} = vf/c$ reflects the influence of the carrier frequency f and the relative velocity v between transmitter, receiver, and scatterers with c denoting the speed of light. Level crossing analysis obtains the average (non-)fade duration of a Rayleigh faded signal [12]. The reciprocals characterize the transition rates between the up and down state, which are referred to as failure rate and repair rate. These rates are assumed to be constant implying that the random fading process is stationary.

The multi-connectivity scheme selection combining is considered, where the user's communication is successful if at least 1 out of n wireless channels is operational. This wireless communications system can be modeled as an irreducible and homogeneous continuous-time Markov chain (CTMC) and evaluated with respect to the introduced dependability metrics. Detailed evaluations can be found in [12–14]. In the discussion below, the index n relates to the number of independent Rayleigh fading channels a single user is simultaneously connected to, performing selection combining. For readability reasons, the complementary availability and mission reliability are depicted on a reversed axis. This presentation takes advantage of the logarithmic scale in the relevant range, emphasizing that values on top are superior, which is common in dependability theory.

Evaluations of the availability A_n are shown for different values of F and n in Fig. 5. Higher degrees of redundancy, which are equivalent to higher numbers of simultaneous connections n, increase the availability. It is important to note that the availability solely depends on the fading margin F for any selected multi-connectivity order n in the considered scenario. Hence, this metric cannot reflect the influence of mobility aspects or the carrier frequency on the communication performance.

Subsequently, the KPIs MDT, MUT, and availability are studied jointly, confining the concentration on the exemplary fading margin $F = 20\,\mathrm{dB}$. Table 1 provides

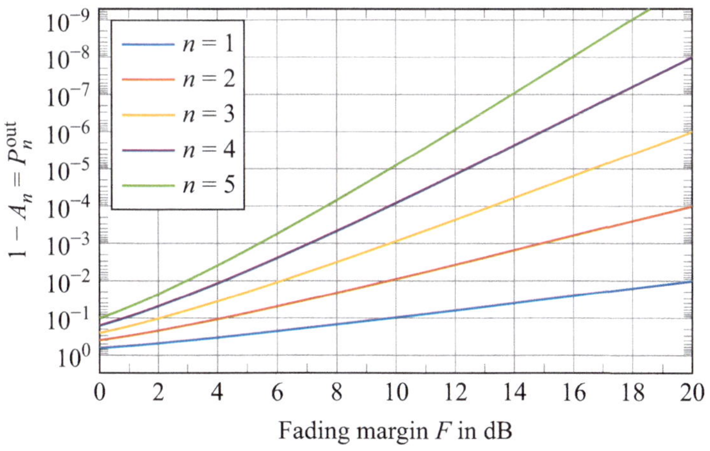

Fig. 5 Complementary availability, which is independent of the Doppler frequency f_{D}

Table 1 Exemplary comparison of MUT, MDT, and availability of n selection combined links for $F = 20\,\text{dB}$

n	$1 - A_n$	v [m/s]	f [GHz]	MDT_n	MUT_n
3	10^{-6}	5	3.75	2.1 ms	3.6 min
3	10^{-6}	5	5.7	1.4 ms	2.4 min
3	10^{-6}	80	3.75	133.5 µs	13.6 s
3	10^{-6}	80	5.7	87.9 µs	8.9 s
5	10^{-10}	5	3.75	1.3 ms	15.2 d
5	10^{-10}	5	5.7	843.5 µs	10.0 d
5	10^{-10}	80	3.75	80.1 µs	22.8 h
5	10^{-10}	80	5.7	52.7 µs	15.0 h

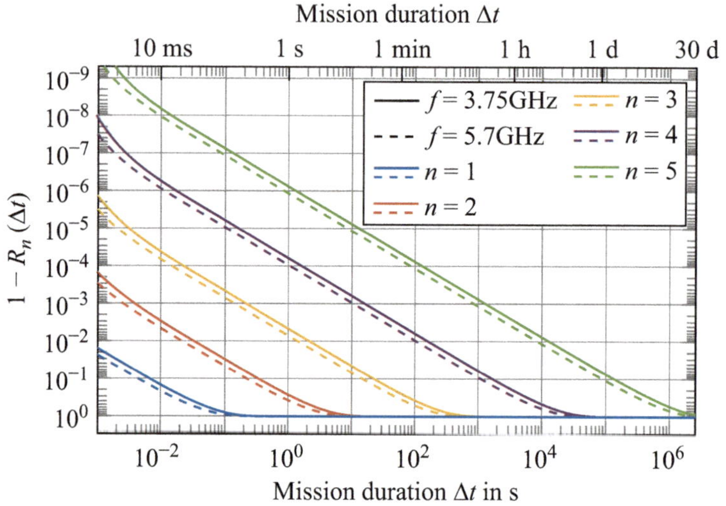

Fig. 6 Complementary mission reliability for $v = 5\,\text{m/s}$, $F = 20\,\text{dB}$

evaluations combining low and high velocity v with different carrier frequencies f for $n = 3$ and $n = 5$ redundant links. Multiple system designs with the same availability can exhibit significantly varying MUT and MDT. For $n = 3$ links, an availability $A_3 = 1 - 10^{-6}$ is obtained, which appears promising for many URLLC use cases, but the MUT_3 varies in the range of a few minutes and less. The corresponding MDT_3 values are comparable to latency requirements for URLLC applications, e.g., wireless industrial automation. Current systems can tolerate short downtimes. However, the strict requirements of many industrial applications cannot be permanently satisfied if the MDT is in the range of their latency constraints. As expected, two additional redundant links improve the availability A_5 by a factor of 10,000 due to the selected fading margin $F = 20\,\text{dB}$. The MUT_5 also increases significantly the expected uptime of more than 2 weeks for the low mobility and low frequency scenario may be comparable to the maintenance cycle in factory automation. However, the MDT_5 is not even reduced by half.

Figure 6 reveals the trade-off between mission duration, mission reliability, and number of links, which could be applied for industrial radio networks. In accordance to basic dependability theory, all n components (channels) are usually assumed to be operational at the beginning of the observation. In contrast to the KPIs availability or PLR, performance requirements can be defined by a target mission duration Δt and the corresponding mission reliability $R(\Delta t)$. For instance, if a wirelessly controlled robot in a factory is required to continue a failure-free operation during a mission duration $\Delta t = 10$ s with a probability of more than 99.999%, at least $n = 5$ selection combined links will be necessary. The offsets between the mission reliability curves of the two considered carrier frequencies are caused by their impact on the Doppler frequency. It turns out that the frequency $f = 3.75$ GHz, which is under discussion for industrial applications, slightly outperforms the unlicensed frequency $f = 5.7$ GHz.

3 Joint Design of Wireless Control and Communications

Past and current wireless communications systems strongly focus on enabling broadband services such as video, image, audio, and text data. For all these services, a human is commonly the addressee, consequently increasing the demand for high data rates (mainly caused through video streams [8]) and posing less stringent requirements regarding latency and reliability of the wireless connection. These requirements on a wireless connection were fundamentally redefined with the advent of industrial services. In contrast to human-centered services, the industrial setting features particularly machine-to-machine communications, i.e., with no human involved. Many of these services were defined in a multitude of national and international projects that promise great benefits from incorporating wireless communications instead of/or in addition to state-of-the-art cable solutions. One example is the management of automated guided vehicle (AGV) fleets. Thereby, each AGV is steered and controlled over the air without provisioning path-related infrastructure such as wiring or tape on the ground, thus providing the highest degree of freedom of movement. This also greatly reduces capital expenditure (CapEx) and operational expenditure (OpEx) as restructuring the whole fleet can be done solely in software.

3.1 Networked Control Systems and URLLC

Implementing closed-loop control over wireless communications systems is a challenging task as wireless systems are faulty, slow, and inaccurate compared to wired alternatives such as Ethercat, Profinet I/O, SERCOS III, etc. There are currently two major streams of research toward enabling closed-loop applications over wireless communications systems. The first exists since the late 1990s and is termed networked control system (NCS). Researchers around NCS view the challenge of indeterministic, erroneous, delayed, and quantized data from the control domain, developing sophisticated control algorithms that can tolerate such non-idealities. A recent survey of results can be found in [25]. The second major stream of research is termed URLLC and views the challenge solely from a communications perspective. Broadly, the goal in URLLC research is to "replace the cable" by providing latency, determinism, reliability, and data rates comparable to cable solutions.

With the goal in mind to design the best possible wireless communications system that ensures acceptable control application behavior, both approaches do not provide enough input on the interdependencies of the control and communications domain. The approach of URLLC targets latency times in the order of 1 ms at packet reliability values of 99.9999%. In [16], fundamental availability analysis has shown that in order to realize outage durations larger than 10 ms at probabilities $<10^{-4}$ for a fading margin $F = 10\,\text{dB}$, at least four independent fading links need to be deployed simultaneously (with selection combining under Rayleigh-fading conditions). For two more orders of magnitude improvement, even more links need to be deployed in parallel. It shows that this approach does not scale since the required resources are tremendous. Although the research on NCS gives a good indication of which communications non-idealities have a negative impact on control performance, there exists no control-communications co-design approach that can perform a cost-benefit assessment to provide design recommendations.

NCS and URLLC both make a great effort in solving the problem in their own respective domains. However, it is believed that a CoCoCo approach will alleviate the required efforts in both domains for a constant quality of control (QoC). Figure 7 offers an abstract explanation. Thereby, on the spectrum of only optimizing the control domain (left side) and only the communications domain (right side), the middle ground, which tackles the challenge jointly from both sides, achieves a co-design gain as the sum effort is reduced.

Driven by the need for scalable yet reliable communications systems for industrial applications, the ultimate goal is to develop a deep understanding of the requirements closed-loop control applications pose on a communications system. Furthermore, design recommendations for wireless communications systems are to be made that ensure correct application behavior while minimizing system

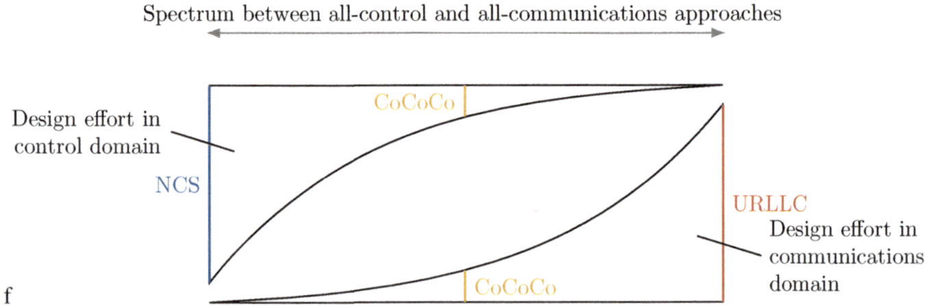

Fig. 7 The CoCoCo approach promises to reduce the required complexity in both domains, also leading to an overall reduced sum complexity

resources. Thereby, CoCoCo is a broad term that covers everything in which both domains are jointly adapted to one another. For instance, in [18], it was shown how a cross-domain manager (CDM) is able to translate in real-time the state information (e.g., latency in the network) from the communications domain in order to optimize the performance of the control domain (e.g., through adaptive controller redesign). In [22], it was shown how very accurate times-tamps can be generated over standard WiFi systems, enabling a global clock synchronization of both domains. This in turn enables sophisticated control algorithms as timestamped messages and an accurate clock in the receiver enable to calculate precise delay-adjusted estimation values. These examples show that there are many interconnections between the two domains that can be jointly optimized. In the following, the co-design of a dynamic resource allocation scheme is presented that evaluates and exploits the necessity of successfully transmitting a packet with control information.

3.2 Joint Design Requires Co-design Performance Metrics

Co-designing in two engineering areas simultaneously requires performance metrics that are able to describe how design choices in one domain affect the other domain.

In [17], it was shown on a fundamental level that packet losses do not always have fatal consequences for the application. Doubling the sampling rate for a (equidistant sampling) control application allows for every other packet to be lost without degradation of control performance (but requires double the amount of wireless resources). This simple thought experiment also demonstrates the necessity to include the time instant of packet loss to the evaluation. While especially communications engineers still use the PLR as the main KPi for reliability assessment, this metric holds no information about the temporal occurrences of packet loss and therefore is not sufficient when transmitting time-sensitive data (as is the case for control applications). Hence, a metric termed *control communications availability (CCA)* is introduced in order to emphasize the (un-)availability of the control application based on communications availability. The time of a lost packet is critical to determine whether the control application is working or not. Therefore, CCA is a completely new concept.

To illustrate this with an extreme example, consider a control application that can tolerate single packet losses. This means that as long as every lost packet is followed by a successfully transmitted packet, the application is still operational. The performance of the control application might be degraded but is still deemed good enough by the control application engineer. Therefore, in the worst case, the PLR can reach values of 50% (every other packet) and still fulfill the control applications' requirements. On the other hand, even extremely low PLR values are meaningless as soon as two consecutive packets are lost and the application consequently stops.

Having a tolerable number of consecutive packets that can be lost effectively adds time diversity to the system design *on an application level*. Time diversity on

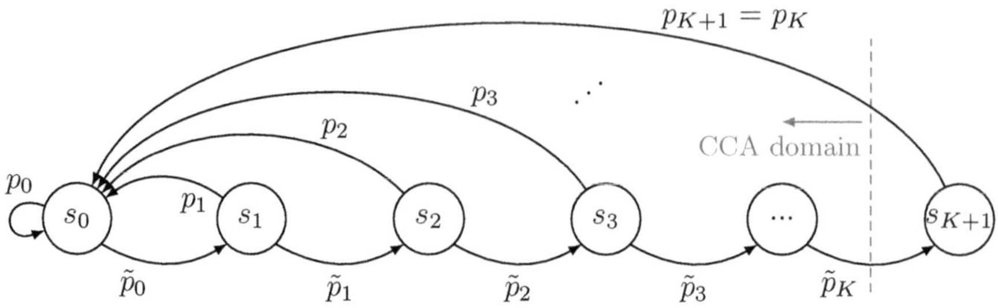

Fig. 8 Markov model of the new CCA approach

a communications level is well-known to date and is featured in re-transmission schemes in all major wireless technologies. However, it must be emphasized that the proposed method comprises a try-once-discard (TOD) approach on the communications level since the system knows that on a control level the lost packet can be tolerated.

3.3 *Fault Tolerance Modeling*

Modeling the interdependencies between control and communications domain is essential for a co-design. For the goal of developing a dynamic resource allocation scheme, the statistics of packet losses need to be modeled.

For the communications system at hand, the frequency spacing of the system is assumed larger than the coherence bandwidth of the channel, and the sampling period is assumed larger than the coherence time of the channel. Both assumptions are reasonable for industrial environments as the sampling period and the multipath delay spread are large. For simplicity, it is also assumed that the system is capable of multi-connectivity with diversity in frequency, combined with selection combining. Hence, links can be assigned in parallel and they as well as consecutive transmissions on the same link fade independently. Since this leads a memoryless system, the Markov chain in Fig. 8 can be considered for modeling. K describes the number of consecutive packet losses a control application can tolerate. The application jumps from a state s_k to s_{k+1} when a packet is lost, with k describing the current number of consecutive packet losses. Whenever a packet is successfully received, the application jumps back to s_0. Only when $K + 1$ consecutive packets have been lost, the application is considered "down" (rightmost state). All values $k \leq K$ span the "CCA domain" in which the application is "up."

Deriving a meaningful value for K is left to the application engineer. An exemplary derivation was performed in [17] for the AGV use case, yielding $K = 3$ for a sampling rate $T_s = 30\,\text{ms}$.

The transition probabilities in Fig. 8 denote the probabilities that (in a certain state) the transmission succeeds/fails. These probability values can be adjusted through many different approaches, e.g., adaptive modulation and coding or multi-connectivity.

3.4 Deriving an Appropriate Radio Resource Allocation Scheme

Co-design implies continuous scrutinization of how design choices will affect the other domain. In this paragraph, a radio resource allocation scheme that maximizes the CCA will be designed. This can be achieved through negatively correlating packet loss because it was found that in the context of control applications, negative temporal packet loss correlation increases QoC tremendously while burst errors degrade it [17]. With this knowledge, a resource allocation can be developed that features such negative correlation without requiring a large amount of resources.

The time diversity and frequency diversity can be exploited to a great extent by adjusting the transition probability values p_k for every k. Instead of deploying a static resource allocation scheme which leads to the same values p_k for all k, it is proposed to save resources in "early" states, whereas in "late" states (k close to K), many concurrent resources are spent in order to avoid the imminent application outage. This approach is termed *state-aware resource allocation (SARA)*. In order to efficiently describe different resource allocation schemes, the general notation S_l^j is introduced, with l indicating the base number of links, i.e., the number of links allocated after a successful transmission, and j indicating the number of links added for each lost packet. Hence, schemes S_l^0 denote static resource allocation schemes with l links in parallel and are also presented here for comparison.

3.5 Impact on the More Precise Understanding of KPIs

Newly developed KPIs on the interface of control and communications target a high informative value on the interdependencies of the two domains. At the same time, these new metrics also allow a sharp performance comparison between design choices made in one domain and how these choices affected the other domain.

With the help of the Markov model for the steady state, the results in Table 2 are obtained for a per-link packet loss probability of $p_{\text{loss}} = 10\%$, which constitutes a comparable per-link packet loss probability to many state-of-the-art wireless communications systems.

The table demonstrates that even with high link failure probabilities of 10%, extremely high CCA values can be achieved, yielding MTBF values that are in the order of years instead of minutes without spending significantly more resources; compare S_1^0 and S_1^1 that use 1.0 and 1.09 links on average but diverge by 6 orders of magnitude in terms of CCA. It also shows that spending resources when they

Table 2 Resulting KPIs for $K = 3$ tolerable packet losses at $T_s = 30\,\text{ms}$

Scheme	Average packet loss rate PLR	Control-comm. unavailability $1 - \text{CCA}$	Mean time betw. failures MTBF	Average links $\bar{c}$
S_1^0	100.0×10^{-3}	1.0×10^{-4}	5 min	1.00
S_2^0	10.0×10^{-3}	1.0×10^{-8}	35 days	2.00
S_3^0	1.0×10^{-3}	1.0×10^{-12}	10^3 years	3.00
S_1^1	91.7×10^{-3}	9.1×10^{-11}	10 years	1.09
S_2^1	9.9×10^{-3}	9.9×10^{-15}	10^5 years	2.01
S_1^2	91.0×10^{-3}	9.1×10^{-17}	10^7 years	1.18
S_2^2	9.9×10^{-3}	9.9×10^{-21}	10^{11} years	2.02

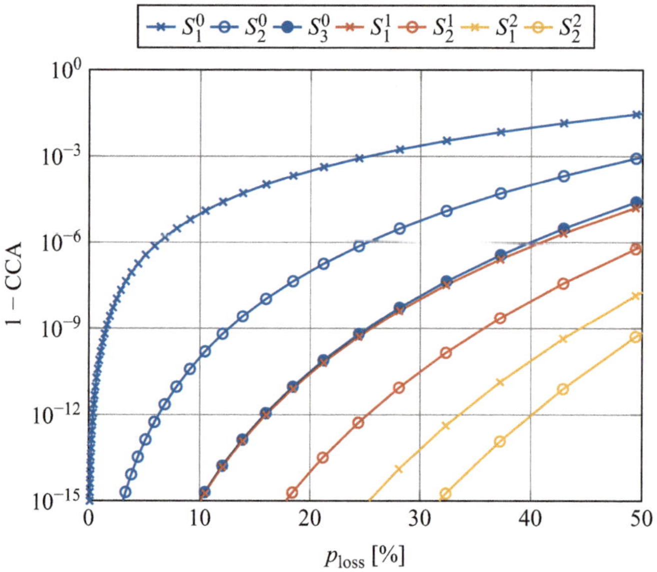

Fig. 9 Relationship between p_{loss} and CCA for all example schemes at $K = 4$

are actually needed, i.e., shortly before an application fails, instead of all the time, greatly reduces the average number of links for each scheme while still providing exceptional CCA values; compare S_3^0 and S_2^1 with 3 and 2.01 used links on average but a 100-fold increase in CCA.

This clearly demonstrates the benefit of incorporating SARA into industrial wireless communications systems. Figure 9 demonstrates that a high tolerance against consecutive packet loss ($K = 4$) enables exceptionally high CCA values, even for high per-link packet loss probabilities p_{loss} of 20–30%. It furthermore enables to tune p_{loss} to just the right value in order to achieve a targeted CCA. This allows for high spectral efficiency on the physical layer, reducing the required resources even further.

References

1. 3GPP: TR 22.891 V14.2.0: feasibility study on new services and markets technology enablers (2016)
2. Anwar W, Kulkarni K, Franchi N, Fettweis G (2018) Physical layer abstraction for ultra-reliable communications in 5G multi-connectivity networks. In: IEEE International Symposium on Personal, Indoor and Mobile Radio Communications
3. Anwar W, Kumar A, Franchi N, Fettweis G (2019) Performance analysis using physical layer abstraction modeling for 5G and beyond waveforms. In: IEEE Global Communications Conference
4. Avizienis A, Laprie J, Randell B, Landwehr C (2004) Basic concepts and taxonomy of dependable and secure computing. IEEE Trans Dependable Secure Comput 1:11–33
5. Barlow R, Proschan F (1965) Mathematical theory of reliability. Wiley, New York
6. Barry J, Messerschmitt D, Lee E (2003) Digital communication, 3rd edn. Kluwer Academic Publishers, Dordrecht
7. Block D, Fliedner N, Meier U CRAWDAD dataset init/factory (v. 2016-06-13). https://crawdad.org/init/factory/20160613/factory1-channel-gain
8. Cisco: Cisco visual networking index: global mobile data traffic forecast update, 2017–2022 (2019). https://www.cisco.com/c/en/us/solutions/collateral/service-provider/visual-networking-index-vni/white-paper-c11-741490.pdf
9. Corless R, Gonnet G, Hare D, Jeffrey D, Knuth D (1996) On the Lambert W function. Adv Comput Math 5:329–360
10. Gilbert E (1960) Capacity of a burst-noise channel. Bell Syst Tech J 39:1253–1265
11. Hößler T, Scheuvens L, Franchi N, Simsek M, Fettweis GP (2017) Applying reliability theory for future wireless communication networks. In: IEEE International Symposium on Personal, Indoor and Mobile Radio Communications
12. Hößler T, Simsek M, Fettweis G (2018) Joint analysis of channel availability and time-based reliability metrics for wireless URLLC. In: IEEE Global Communications Conference
13. Hößler T, Simsek M, Fettweis GP (2018) Mission reliability for URLLC in wireless networks. IEEE Commun Lett 22:2350–2353
14. Hößler T, Schulz P, Simsek M, Fettweis G (2019) Mission availability for wireless URLLC. In: IEEE Global Communications Conference
15. ITU: E.800: definitions of terms related to quality of service (2008)
16. Öhmann D, Fettweis G (2015) Minimum duration outage of wireless Rayleigh-fading links using selection combining. In: IEEE Wireless Communications and Networking Conference
17. Scheuvens L, Hößler T, Noll-Barreto A, Fettweis G (2019) Wireless control communications co-design via application-adaptive resource management. In: IEEE 5G World Forum
18. Scheuvens L, Simsek M, Noll-Barreto A, Franchi N, Fettweis G (2019) Framework for adaptive controller design over wireless delay-prone communication channels. IEEE Access 7:49726–49737
19. Schwarzenberg N, Wolf A, Franchi N, Fettweis G (2018) Quantifying the gain of multi-connectivity in wireless LAN. In: European Conference on Networks and Communications
20. Schwarzenberg N, Burmeister F, Wolf A, Franchi N, Fettweis G (2019) Joint synchronization in macro-diversity multi-connectivity networks. In: IEEE 90th Vehicular Technology Conference (VTC Fall), Honolulu

21. Traßl A, Hößler T, Scheuvens L, Franchi N, Fettweis G (2019) Deriving an empirical channel model for wireless industrial indoor communications. In: IEEE International Symposium on Personal, Indoor and Mobile Radio Communications
22. Trsek H (2016) Isochronous wireless network for real-time communication in industrial automation, 1st edn. Springer Vieweg, Berlin
23. Valenti M, Zhao B (2003) Distributed turbo codes: towards the capacity of the relay channel. In: IEEE Vehicular Technology Conference (VTC Fall)
24. Wolf A, Schulz P, Öhmann D, Dörpinghaus M, Fettweis G (2017) On the gain of joint decoding for multi-connectivity. In: IEEE Global Communications Conference
25. Zhang X, Han Q, Yu X (2016) Survey on recent advances in networked control systems. IEEE Trans Ind Inf 12:1740–1752

Time-Sensitive Networking for Industrial Control Networks

David Ginthör, René Guillaume, Naresh Nayak, and Johannes von Hoyningen-Huene

1 Introduction

Manufacturing domain in today's era is striving for efficiently manufacturing highly individualized products. One of the requirements to achieve the "lot size one" goal (i.e. customization of manufactured products on an individual basis) is a flexible and re-configurable manufacturing shop floor. Existing communication architectures separating the information technology (IT) infrastructure from the operation technology (OT) infrastructure hinder flexible networks. A converged networking infrastructure shared between the IT and OT applications is rather desirable. Deploying such a networking however is far from trivial. The diverse applications executed in the manufacturing infrastructure have varying demands from the underlying communication network. On the one side are applications implementing industrial process control on the field level, which require hard real-time communication guarantees like upper bounds on communication latency and latency variance, reliable packet delivery, etc. In contrast are the Manufacturing Execution System (MES) and Enterprise Resource Planning (ERP) applications, which are rather soft real-time in nature and are focused on bandwidth and throughput. Thus, one of the prerequisites of the networking technology that can be deployed in smart factories is the capability to provide different levels of quality of service (QoS) for the applications. While many networking technologies meet this requirement, most of them are not interoperable with each other. Moreover, a few of them are proprietary and result in vendor lock-in. Hence, several standards organizations like the IEEE, 3GPP , etc. are working on standardizing networking technologies, wired

D. Ginthör (✉) · R. Guillaume · N. Nayak · J. von Hoyningen-Huene
Robert Bosch GmbH, Stuttgart, Germany
e-mail: david.ginthoer@de.bosch.com; rene.guillaume@de.bosch.com; nayak.naresh@de.bosch.com; johannes.hoyningen-huene@de.bosch.com

N. H. Mahmood et al. (eds.), *Wireless Networks and Industrial IoT*,
https://doi.org/10.1007/978-3-030-51473-0_3

as well as wireless, for a vendor-independent converged network in the context of manufacturing shop floors.

The IEEE Time-Sensitive Networking (TSN) Task Group (TG) is working on extending the IEEE 802.1Q standards to equip Ethernet (IEEE 802.3) with features (cf. Sect. 3.1), enabling it to handle traffic differentiation in their QoS requirements. The widespread penetration of Ethernet makes it a natural choice as a basis technology for developing a standardized converged network. On the wireless side, the 3GPP is in the process of specifying the standards for the fifth generation (5G) of mobile communication technology. Various mechanisms (cf. Sect. 3.2) are being incorporated in the 5G standards for making it compatible with TSN and, thus, provide end-to-end (across wired and wireless sub-domains) seamless QoS guarantees for applications.

1.1 Contribution

The aforementioned networking technologies – TSN and TSN over 5G – are seen as enabling technologies for converged networks and deterministic vertical integration. However, the interplay between these technologies (parts of which are either recently standardized or are being standardized) throws a lot of open questions. In this chapter, we provide a brief overview of the different mechanisms being standardized as a part of TSN and TSN over 5G. We also discuss the different challenges that need to be addressed before these technologies can be deployed to achieve a converged network.

This chapter is structured as follows. In Sect. 2, we introduce several challenging use cases for the "Factory of the Future" and derive requirements with respect to communication networks. We briefly introduce the networking technologies being conceived to meet these requirements in Sect. 3. Finally, in Sect. 4, we discuss open challenges we need to address before deploying these technologies on a manufacturing shop floor.

2 Use Case Analysis

The main drivers for the introduction of new network technologies in the industrial environment are the changing requirements from new applications and deployment scenarios. To address new manufacturing paradigms and processes developed as part of Industry 4.0, innovative network technologies such as 5G and TSN are finding their way into the shop floor. To better understand the requirements from an application point of view, we summarize relevant use cases and infer their implications on the network infrastructure.

2.1 Overview of Relevant Application Scenarios

Several representative scenarios and use cases are described from different perspectives by the TSN and 5G standardization TG, respectively. The IEEE 802 TG discusses TSN applications and deployment scenarios in the TSN industrial profile IEC/IEEE 60802 [1]. The 3GPP TG deals with applications utilizing the mobile 5G technology and particularly considers extended networks combining 5G and TSN, e.g., in TR 22.804 [2]. In the following, we group relevant use cases to clusters that shall provide a simplified yet distinctive overview of representative application scenarios, as depicted in Fig. 1.

Vertical Integration

Vertical integration supports business processes through remote control and analysis, enabling holistic management of the production for optimized performance. One

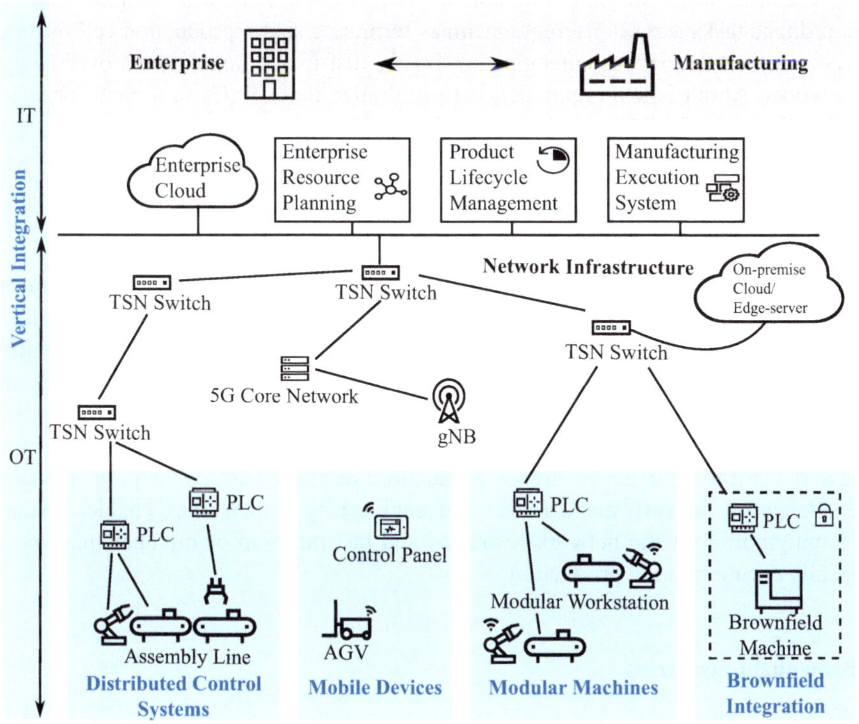

Fig. 1 Converged network infrastructure of a Factory of the Future supporting the main use case clusters vertical integration, distributed control systems, mobile devices, modular machines, and brownfield integration

primary use case for seamless vertical integration is the exchange of data between an enterprise service (e.g., running on a cloud server) down to the desired machine component (e.g. an actuator or sensor) for status monitoring, failure prediction, or optimization purposes.

(Distributed) Control Systems

Process automation deals with the handling of goods and control of manufacturing processes. Programmable logic controls (PLCs) are responsible for operating field devices, e.g., by using sensor measurements to steer actuators in a closed- or open-loop fashion. Between the PLC and field devices, isochronous communication for motion control and measuring tasks is often required. Isochronous use cases are characterized by very short communication cycles and usually require network synchronization to enable precise operation and coordination of actuators and sensors with very low latency and jitter. For control-to-control (C2C) use cases, these requirements are more relaxed. Nonetheless, end-to-end latency must be guaranteed with high reliability to ensure fail-proof operation. A possible scenario for C2C is, for example, when multiple programmable logic controls (PLCs) communicate to coordinate tasks across different machines forming a single production cell or line. This can be especially challenging for highly distributed applications over larger networks. Another novel approach is to centralize these PLCs in a cloud or edge server for more flexibility and maintainability.

Modular Machine Systems

Modular manufacturing processes are becoming more and more essential to meet the requirements of an increasingly volatile market, demanding higher individualization and flexibility to maintain resource and cost efficiency. Nowadays, for highly customizable goods, the production and assembly lines are already changing on daily basis, requiring high modularity of the production assets in order to operate efficiently. A quick configuration of the network between the modular units means less down-time and hence higher production volume. This could pose a major challenge on network management and engineering, which must enable flexible reconfiguration of the network resources and fast ramp-up of modular machines, ideally during running production.

Brownfield Scenarios

Due to long life cycles of industrial machines and production lines, different generations of systems likely need to coexist in a common network infrastructure. This may even refer to machine-internal modules or components within a production cell. Hence, different types of network traffic need to be isolated in such a way that they cannot affect each other. Also, to allow a soft migration path toward innovative technologies such as TSN, there should be ways to integrate legacy technologies into a common network infrastructure.

Mobile Devices

Mobile devices in factory and process automation have played only a negligible role in the past due to limited need in a static manufacturing environment. Nowadays, mobile and versatile production assets are playing an increasingly vital role in providing the needed flexibility on the shop floor. Mobile robots are deployed to perform tasks including transportation of goods as automated guided vehicles (AGV) or assisting in manufacturing processes. Another class of applications is human-machine interfaces (HMIs) that enable interaction between people and the production environment, including control panels, IT devices, or augmented/virtual reality (AR/VR) applications. HMIs may address different tasks such as maintenance, control, monitoring, or safety functionalities on multiple machines in the factory. Depending on the application, operation of mobile devices can impose stringent requirements across the wired and wireless network in terms of reliability and availability in order to guarantee uninterrupted service or fail-safe operation for critical use cases.

2.2 Requirements on the Network Infrastructure

To enable the previously discussed use cases, certain network properties are required to ensure their performance. This section gives an overview of the main aspects.

Quality of Service Supporting a holistic manufacturing process in a factory environment, as described in the vertical integration use case, requires a unified communication system traversing IT and OT. This holistic integration depends on a converged network meeting strongly diverse requirements from different network participants in terms of latency, jitter, and bandwidth. The network must hence support different QoS classes and provide sufficient end-to-end service guarantees over the entire network.

Time Synchronization A common understanding of time among all devices within the network may have two motivations. On the one hand, different devices of a distributed, time-critical applications need to be synchronized *via* the network to perform tasks at the same point of time or to track and control the sequential operation of tasks. On the other hand, some QoS mechanisms can only ensure the lowest latency, when the devices are precisely synchronized *with* the network infrastructure devices.

Reliability Depending on the use case, the demands for reliability may differ as well. Non-time-critical applications may apply reliability concepts like ARQ based on retransmission to ensure sufficient end-to-end reliability at the cost of latency. However, a highly time-critical application may have higher demands that require seamless redundancy over independent network paths. Possible threats to reliable communication range from random communication errors that lead to packet drops, malfunctioning devices that flood the network unintentionally, up to explicit attacks from devices on the ongoing communication, e.g., to cause congestion. Resilience measures such as path diversity and isolation between critical traffic and non-reserved traffic can mitigate those effects.

Network Configuration and Deployment To feasibly operate a holistic and vertical network as envisioned for Industry 4.0, flexibility and expandability play an important role. A converged industrial network integrating OT and IT enables higher flexibility but comes at the price of increased configuration complexity. Highly individualized and demand-driven manufacturing results in frequent reorganizations of production lines on the shop floor. The provision of common interfaces for dynamic configuration of network devices, endpoints, and services is indispensable for an efficient implementation. From an administrator perspective, availability of engineering tools allowing fast network reconfigurations with minimal down-time is necessary for feasible operation. This includes generic mechanisms to calculate and deploy routing and resource configurations to each device.

3 Enabling Converged Networks

In Sect. 2, various requirements and challenges to implement a converged network for the shop floor have been derived from a use case perspective. In this section, we discuss how each of these requirements/challenges is addressed by the networking technologies, viz., TSN and TSN over 5G, to implement a converged network.

3.1 Time-Sensitive Networking (TSN)

The Ethernet (IEEE 802.3) networking technology was conceived primarily as a best-effort network, i.e., the network attempts to transport frames to their destination without any service guarantees [3]. The TSN TG (earlier known as the Audio/Video Bridging (AVB) Task Group) has been working on incorporating the notion of real time in Ethernet networks. By means of extensions to the IEEE 802.1Q standard, TSN TG addresses the different demands of a converged network [4]. In the following, we provide an overview of the main TSN features.

Quality of Service The basis of ensuring QoS is the Strict Priority (SP). Here, a virtual LAN (VLAN) tag including the Priority Code Point (PCP) is inserted into

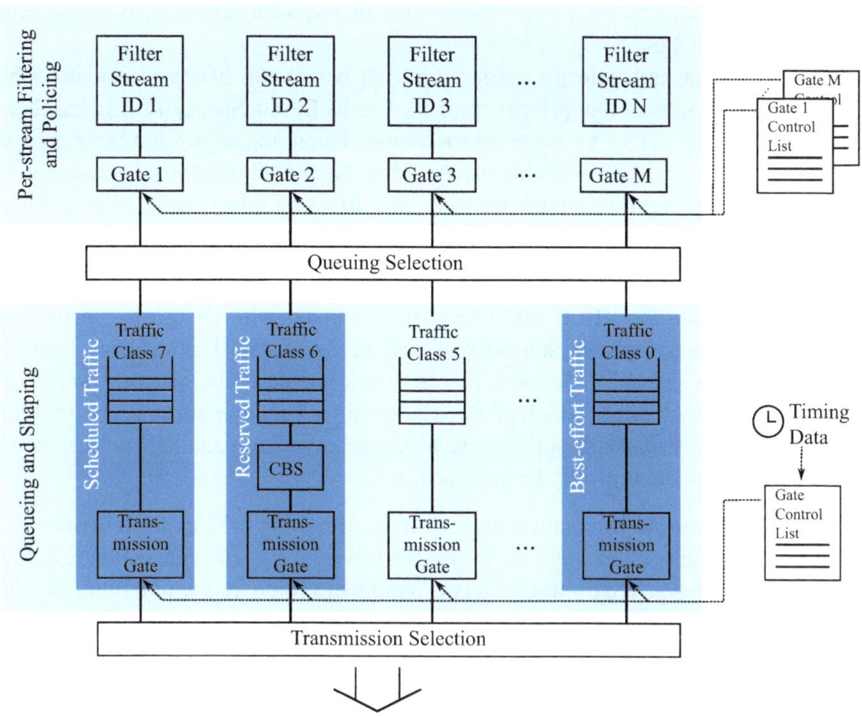

Fig. 2 Exemplary configuration of the egress port of a TSN switch supporting Per-stream Filtering and Policing and traffic shaping. It employs a Time-Aware Shaper driven by a Gate Control List according to a synchronized network time. Here, three possible configurations of a high-priority reserved traffic queue for scheduled traffic, a Credit-Based Shaper for reserved traffic, and best-effort traffic are shown

time-critical frames and used to differentiate up to eight different traffic classes, realized with individual egress queues. With SP, frames with higher PCP are transmitted prior to low-priority frames. However, since high-priority frames may still be affected by queuing delay while waiting on ongoing transmission to be finished and burst of high-priority traffic may lead to resource starvation of lower priority classes, more advanced TSN mechanisms are needed. A non-exhaustive list of such features is as follows:

Time-Aware Shaper (TAS) introduces a programmable gating mechanism, which regulates the transmission selection. Ethernet frames in a queue can be transmitted only if the corresponding gate is open. The cyclic schedule can be programmed by means of port-specific Gate Control Lists (GCL), as depicted in Fig. 2. The requirements of isochronous traffic can be addressed with exclusive gating, where only the gate of this traffic class is opened during specific time slots, hence preventing any interference.

With frame preemption, the transmission of a low-priority frame can be interrupted in favor of a high-priority frame. For a line rate of 1 Gbps, frame preemption can reduce the worst-case queuing delay without TAS of a high-priority frame from 12.3 μs to about 1 μs.

To avoid that high-priority traffic with high bandwidth like audio/video (A/V) applications overruns low-priority frames, Credit-Based Shaper (CBS) has been introduced. CBS strives to space out the frames belonging to A/V traffic classes in the network based on the available bandwidth to prevent bursts of the corresponding streams propagating through the network and affecting other traffic classes. Thus, the use of CBS may slightly increase the latency of the corresponding traffic classes (compared to SP only) but improves the performance of low-priority traffic classes.

Time Synchronization IEEE 802.1AS was conceived with the goal of achieving clock synchronization with sub-microsecond accuracy [5]. The standard includes mechanisms to determine the most precise clock source in the network (known as the grandmaster clock) to which all devices within a TSN domain synchronize their respective clocks to. With appropriate hardware, synchronization with a residual error of <100 ns can typically be achieved.

Reliability There are two main sources of frame dropping in Ethernet networks, i.e., link or switch failures leading to loss of connectivity and traffic congestion resulting in buffer overflows. TSN introduces two features to improve the reliability of Ethernet networks, namely, Frame Replication and Elimination for Reliability (FRER) and Per-Stream Filtering and Policing (PSFP) [6]. The former addresses reliable frame delivery even in the presence of link and switch failures by transmitting replicated frames through redundant paths within the network. PSFP offers different mechanisms to identify potential congestion in the network and provide strategies to protect the intended traffic.

Network Configuration and Deployment With the addition of aforementioned features in Ethernet as a part of TSN, the complexity of managing and configuring networks has increased manifold. For deployment and configuration of these features, TSN defines the Stream Reservation Protocol (SRP), which can be used to configure network parameters based on the stream requirements either in a centralized or distributed manner. While mechanisms, like SP and CBS, may be configured with the distributed model, more advanced mechanism like TAS, PSFP, and FRER require a global view of the network, which is given in the centralized model (cf. Fig. 3). With new TSN mechanism in development, more enhancements to SRP are likely to appear in the future.

3.2 TSN Over 5G

As the industry transforms toward flexible and highly connected environments with novel use cases, the need for wireless communication on the shop floor becomes more and more apparent. Several efforts have been made to bring wireless connectivity to the harsh industrial environment, most notably in the 3GPP, where a

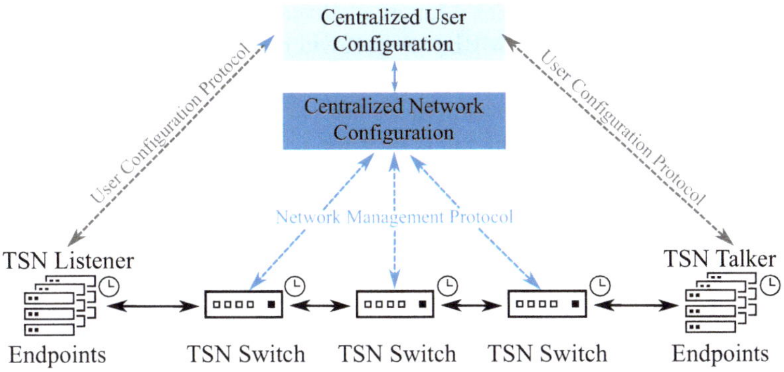

Fig. 3 Centralized configuration model of a TSN network. While the centralized user configuration (CUC) configures endpoints based on their application requirements, the Centralized Network Configuration (CNC) communicates with all TSN switches and configures end-to-end streams between talker and listener

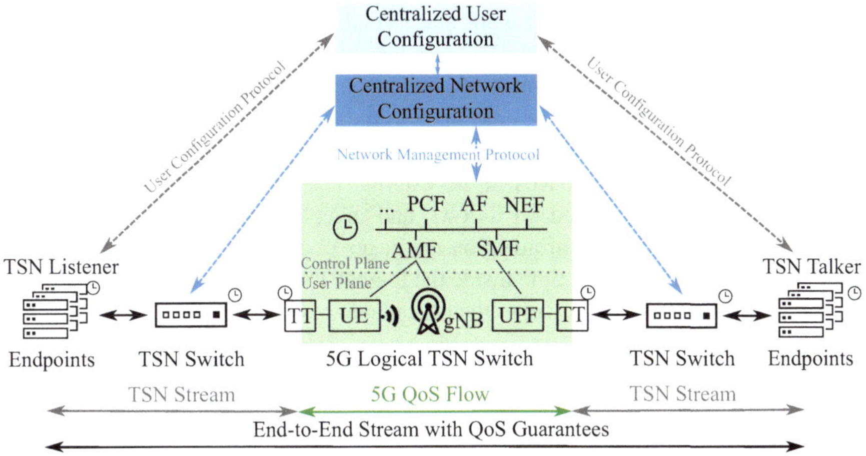

Fig. 4 Integration of a 5G mobile network as a logical TSN switch in a TSN network. The 5GS transmits TSN streams with suitable 5G QoS flows that are configured according to the centralized configuration model over a CNC

seamless integration of the cellular 5G system (5GS) and TSN is aimed for. While different approaches have been evaluated, the integration of the 5G network as a logical switch inside a TSN network is considered as the most feasible option [7]. To the outer network, 5G provides the necessary interfaces to transport TSN frames but uses its 5G-specific framework to guarantee services inside the logical switch. The concept is depicted in Fig. 4. In the following, we evaluate how the different requirements on QoS, synchronization, reliability, and configuration are addressed.

Quality of Service To support the transport of TSN frames over 5G, the mobile network relies on the 5G-specific QoS framework. Its end-to-end architecture defined in the 3GPP Release 15 enables support of diverse traffic classes ranging from enhanced mobile broadband (eMBB) for bandwidth-intensive applications to ultra-reliable and low-latency communications (URLLC) enabling real-time applications with short and periodic data transmissions. The key to supporting consistent QoS guarantees with different requirements over the wireless network is a mapping of TSN streams to suitable 5G QoS flows. The basis to support TSN streams serving time-critical applications is the 5G URLLC enhancements. The new flexible MAC layer enables shorter latencies by having an adjustable resource granularity. While eMBB users are usually scheduled on a slot basis, URLLC users can be scheduled within a fraction of it, with so-called mini-slots. Together with further enhancements such as preemption capabilities and improved decoding performance, a one-way latency of <1 ms is achievable. However, consistently guaranteeing such a low latency with minimal jitter is challenging. Especially reservation-based scheduling and radio access network (RAN) slicing concepts play a major role in supporting deterministic communication with TSN over 5G.

Time Synchronization To support the synchronization of endpoints to coordinate applications or to enable time-aware end-to-end scheduling as used for TAS in IEEE 802.1Q, each network device must be able to convey timing information according to a defined time source. To enable synchronization with external TSN switches, the 5G system must be compliant to the IEEE 802.1AS standard. For this purpose, the TSN translator devices (TT) are deployed at the network and device side to synchronize with devices in the external network. The 5G system acts as a time-aware system – based on its own and independent 5G time – to support the forwarding of time synchronization messages over the air. This is achieved by accounting for the residence time inside the logical TSN switch at the TT devices in the synchronization procedure over the user plane.

Reliability Compared to wired communication, wireless solutions suffer from high unreliability of the link, mainly due to unpredictable variations in the channel. Furthermore, state-of-the-art cellular systems operate with relatively high error ratios and achieve reliability by applying hybrid ARQ schemes where undecodable data is restored with retransmissions. For most of the TSN applications, this approach is not suitable as it leads to high latency and jitter. To guarantee the successful delivery of data with high reliability within a strict deadline, several enhancements for URLLC-type communication are defined by the 3GPP that can be employed to achieve the necessary performance. For example, the 5G system allows using more robust modulation and coding schemes that improve the error ratio at the cost of lowered spectral efficiency. Furthermore, diversity schemes are highly anticipated to bring the necessary reliability [8]. Especially diversity schemes such as massive MIMO, Coordinated Multipoint (CoMP), and multi-connectivity methods are currently being discussed as a suitable mechanism to achieve reliability targets as required by TSN.

Network Configuration and Deployment The logical TSN switch based on the 5G system is suitable to be operated in the fully centralized model, as described in 3.1.

For this purpose, the 5G system employs a TSN-specific Application Function (AF) in the 3GPP core network that provides the necessary control plane interfaces to interact with the external TNS CNC. The tasks of this TSN AF in the 5G system are twofold. First, it must act as a manageable bridge to the outer network and provide bridge capabilities and topology information exploiting the Network Exposure Function (NEF). Using the QoS requirements of the application and the 5G logical bridge-related information, the Centralized Network Configuration (CNC) can then determine suitable configuration parameters and schedules for all (virtual) TSN switches for each end-to-end stream. The second role of the TSN AF is to map the TSN QoS profiles provided by the CNC to 5G specific parameters. A suitable 5G QoS flow configuration is generated and negotiated over the TSN AF with the Policy Control Function (PCF). The configuration is deployed over the Access Management Function (AMF) and Session Management Function (SMF) to the respective flow. In this way, an end-to-end 5G QoS flow from the respective user equipment (UE) over the gNodeB (gNB) toward the User Plane Function (UPF) can be configured according to the TSN stream requirements.

4 Challenges from Practical Deployments

The previous analysis has shown that both technologies, TSN and 5G, have a lot of potential to pave the way for innovative factory applications. Mechanisms for seamless vertical integration, distributed and time-aware systems along with reliable, industrial-grade wireless connectivity, are the cornerstones for the envisioned Factory of the Future. However, to successfully bring the new concepts to practice and achieve a broad acceptance of the new technologies in a well-established domain, several considerations have to be taken into account. In the following, we shed some light on different aspects that may be relevant in this regard.

Quality of Service In a converged network, as exemplary depicted in Fig. 1, there are diverse end-to-end QoS requirements that need to be met. Depending on the combination of given requirements (e.g., cycle time, latency, data rate, etc) and the actual topology and size of the network, computation (and optimization) of a feasible configuration and resource allocation can become very complex. While there has been quite some research on efficient scheduling and resource allocation algorithms, practical implementations still need to be found that are capable of handling realistic network deployments efficiently. An unfavorable choice of cycle times and data lengths on a single port can easily result in inefficient resource utilization; hence optimized solutions are required that usually result in long computing delays for configuration. Especially in dynamic scenarios, no matter if application demands or network capabilities like channel conditions are frequently changing, this can result in long down-times whenever the system needs to be adjusted to new requirements and constraints. Clearly, more efficient ways for (partial) re-configuration are needed in order to keep costly production downtimes at a minimum and also to guarantee continuous availability of coexisting production processes and applications. If, for instance, the changing requirements of an application *A* necessitate an adaptation of a network segment, this should not disturb the operation of another application *B* in the same infrastructure. Even though the amendment SRP Enhancements and Performance Improvements in IEEE 802.1Qcc describes mechanisms to roll out new schedules during operation, it is challenging or even unattainable to calculate such schedules for a partial re-configuration that prevents application *B* from experiencing a (temporal) loss of connectivity.

The ongoing trend to decentralized architectures, virtualized functions, and the utilization of cloud technologies poses further challenges on the QoS provision in the converged network. Particularly, the idea of running time-critical applications not directly on the machine locally, but on some shared edge or on-premise cloud in the network, introduces new challenges. For instance, the edge server may be required to run real-time capable hypervisors and network protocol stacks. Depending on the specific application requirements, the provision of time synchronization and sufficient QoS guarantees across multiple (time) domains might become necessary.

Reliability Reliability and availability are essential factors for industrial automation to guarantee the continuous operation of production processes. Following the trend of Industry 4.0 to implement control functions on a shared, centralized resource such as a cloud server raises the issue of a single point of failure (SPOF). If, by any reason, the server itself or the connection to it gets unavailable, there is the risk that the whole production or at least some part of it breaks down. Clearly, this needs to be addressed, e.g. through appropriate resilience concepts and redundant network paths. Standards like IEEE 802.1CB already provide feasible redundancy mechanisms on the network level. But the need for resilience may also exist for the network management plane: Following the centralized configuration approach described in IEEE 802.1Qcc turns the CNC and Centralized User Configuration (CUC) into potential SPOFs as well, as long as their implementations do not provide appropriate measures of redundancy. But also, the network links need to provide the required minimum level of reliable data transport, which is challenging when it comes to wireless transmission. While the deployment of mobile devices in an industrial environment is nothing new, the idea of seamlessly integrating them into coordinated automation processes needs much more stringent limitations of latency and delivery guarantee. TSN over 5G is a promising concept to fulfill these requirements. Still, it needs to be analyzed how reservation-based communication in the presence of highly varying radio channels can be realized to meet the expected end-to-end guarantees. Additionally, mobility introduces even greater challenge, since uninterrupted QoS must be guaranteed during handover processes between multiple wireless access points. It remains unclear how this can be achieved with the required reliability of industrial applications. Possible approaches are to consider a cooperative resource management of base stations taking mobility patterns and channel prediction of mobile users into account.

Network Integration and Deployment Due to several reasons, e.g. high investments, well-established reliable technologies, or vendor lock-ins, the life cycles of today's factory automation systems are typically very long, lasting 30 years and beyond. As a consequence, the landscape of underlying technologies that need to coexist in a facility can become very broad. Obviously, an entire replacement of legacy systems by TSN- and 5G-capable devices is too costly, making integration capabilities for legacy technologies in the new infrastructure indispensable. This soft migration could be realized, for instance, based on tunneling or appropriate gateways to translate one protocol into another. An approach to how this can be done for Sercos III and TSN was described in [9]. There should also be considerations for monitoring heterogeneous networks and enabling configuration of end-to-end streams beyond the boundaries of a technology domain. A potential technology-agnostic description format for that purpose was proposed in [10]. Furthermore, while standards usually allow a certain amount of optional configuration settings, interoperability needs to be assured not just between heterogeneous technologies but also between devices using a common networking technology yet with vendor-specific parameterization. Ongoing activities, such as in IEC/IEEE 60802, are approaching this dilemma by defining profiles and conformity classes feasible for specific application domains. Further questions regarding different configuration concepts of network segments within a shared network infrastructures, referred to as TSN domains, remain open. Depending on the arrangement of these TSN domains, there may be overlaps so that network devices need to logically split their resources and let them be managed as part of different TSN domains. Especially in the case of inter-technology, i.e., TSN and TSN over 5G, and inter-vendor operation, this may lead to yet inconclusive questions, e.g., regarding the technical implementation or liability in case of failures.

Security One major driver for the development of TSN was to get a basis for converged networks and to allow vertical integration from a cloud down to sensors or other field-level devices. Nevertheless, this vision of connected industry strongly contradicts the network architectures and regulations that are typically in place today for most larger enterprise networks. To protect these networks from critical security threats, one premise is, besides other measures, to divide a network into various segments. Depending on their individual criticality, every segment has its own rules for physical or remote access, data forwarding, etc. This enables an improved maintainability, rules, and rights management and even allows to shut down single segments in case of cyber attacks without affecting the remaining infrastructure. Hence, the need for delivery of data across different administration levels requires fundamental changes in existing network deployments and business processes and regulations. A completely transparent vertical integration is hardly possible, as long as no other appropriate security mechanisms, e.g. service-oriented paradigms, are in place. Please refer to the chapter "Security Challenges in Industrial IoT Networks" for further details.

Engineering The previously mentioned soft migration from legacy toward newly arriving technology and the involved brownfield scenarios do not only play an important role in regard to network technologies but also for related engineering processes. Handling the set of configuration parameters from multiple coexisting technologies can lead to a level of complexity that can negatively affect user experience. This needs to be addressed by extending existing engineering processes and tools in a suitable way, for example, through semi-automated or self-aware configuration processes. Of course, this kind of mechanisms requires an appropriate conformity certification. Similar to the requirement of defining conformity classes for network components, this kind of certification is also important for engineering and network management tools if cross-vendor interoperability shall be achieved. One example is the CNC, where no non-proprietary implementation is available on the market yet. In general, there should be a generalized information and device description model to allow generic identification and configuration of endpoints and network infrastructure elements. While the ongoing standardization of YANG models as part of the IEEE TSN TG seems to be promising, equivalent activities would be required for all related interfaces needed for the interoperation across different vendors and technologies. For example, there is still the need to further define the configuration interface between 5G and TSN in order to deploy TSN over 5G. OPC Unified Architecture (OPC UA) specifies a widely recognized communication protocol along with appropriate information models. This is being extended since 2018, when some of the leading players in industrial automation joined forces by following the Object Linking and Embedding for Process Control (OPC) foundation's Field Level Communication (FLC) initiative to pursue their vision of an open and unified solution. It is supposed to enable industrial-grade communication from field-level devices to cloud services. This protocol is a promising approach to attain convergence on the higher layers of the networking technologies TSN and 5G and to support a unified solution for deterministic data transfer across different domains [11].

5 Conclusion

Industry 4.0 poses greater challenges on the network infrastructure than ever before. Currently, most common industrial network deployments are configured statically within a strict network hierarchy. However, our use case analysis has shown that future industrial applications require networks that are highly flexible with support of diverse QoS requirements. To fully support the new manufacturing paradigms involving modular and distributed control systems or mobile devices, new network technologies are needed. Our comparison between use case requirements and the emerging technologies TSN and 5G have shown that many aspects of Industry 4.0 are addressed by these communication standards. Both technologies are able to support multiple QoS classes simultaneously with service guarantees over heterogeneous network infrastructures with different forwarding and shaping mechanisms. End-to-end configuration of TSN streams and 5G QoS flows allows flexible provision of service guarantees to devices across the network. Most importantly, with the integration of TSN into 5G, many functionalities of TSN are supported across both technologies. This includes time synchronization, mapping of domain-specific QoS streams, and provision of common management interfaces.

However, from a practical point of view, many open questions remain on how a TSN or TSN over 5G network can be operated feasibly. A converged network with flat hierarchies enabling vertical integration creates vast possibilities for new use cases but poses new challenges on the network configuration and engineering. In TSN, to guarantee services for each user, the respective resources across the entire network must be reserved for each stream. To support an entire manufacturing environment, highly scalable and efficient methods to calculate and roll out schedules dynamically without influencing ongoing production and network streams are indispensable. Supporting on the one hand a highly flexible network and guaranteeing on the other hand extremely high reliability – to mitigate any production halt that usually results in high costs – are a major challenge. This is especially the case for the 5G network, which suffers from unreliable communication over the wireless link compared to wired technologies. These issues need to be further addressed before TSN and 5G can be integrated into or replace existing industrial network infrastructures to pave the way for the Factory of the Future.

References

1. Belliardi R. Use cases IEC/IEEE 60802, version 1.3. Accessed 19 Feb 2019. [Online]. Available: http://www.ieee802.org/1/files/public/docs2018/60802-industrial-use-cases-0918-v13.pdf
2. 3GPP, TR 22.804 study on communication for automation and vertical domains, version 16.2.0. Accessed 21 Oct 2019. [Online]. Available: https://www.3gpp.org/ftp/Specs/archive/22_series/22.804/
3. IEEE, IEEE std 802.3-2015 standard for ethernet. Accessed 21 Oct 2019. [Online]. Available: https://ieeexplore.ieee.org/document/7428776
4. IEEE, IEEE std 802.1Q-2018 standard for local and metropolitan area networks – bridges and bridged networks. Accessed 21 Oct 2019. [Online]. Available: https://ieeexplore.ieee.org/document/8403927
5. IEEE, IEEE std 802.1AS-2011 standard for local and metropolitan area networks–timing and synchronization for time-sensitive applications in bridged local area networks. Accessed 21 Oct 2019. [Online]. Available: https://ieeexplore.ieee.org/document/5741898
6. IEEE, IEEE std 802.1CB-2017 standard for local and metropolitan area networks–frame replication and elimination for reliability. Accessed 21 Oct 2019. [Online]. Available: https://ieeexplore.ieee.org/document/8091139
7. 3GPP, TS 23.501 system architecture for the 5G system (5GS), version 16.2.0. Accessed 21 Oct 2019. [Online]. Available: https://www.3gpp.org/ftp/Specs/archive/23_series/23.501/
8. Cavalcanti D et al (2019) Extending accurate time distribution and timeliness capabilities over the air to enable future wireless industrial automation systems. Proc IEEE 107/66:1132–1152

9. Nsaibi S, Leurs L, Schotten HD (2017) Formal and simulation-based timing analysis of industrial-ethernet sercos III over TSN. In: 2017 IEEE/ACM 21st International Symposium on Distributed Simulation and Real Time Applications (DS-RT)
10. Olaya SSP et al (2018) Communication abstraction supports network resource virtualisation in automation. In: 2018 IEEE 27th International Symposium on Industrial Electronics (ISIE)
11. OPC Foundation, Initiative: field level communications (FLC). Accessed 4 Feb 2020. [Online]. Available: https://opcfoundation.org/flc-pdf

Random Access Protocols for Industrial Internet of Things: Enablers, Challenges, and Research Directions

Mikhail Vilgelm, H. Murat Gürsu, and Wolfgang Kellerer

1 Introduction

Industrial Internet of Things (IIoT) comprises a variety of applications related to automation in different domains: smart power grids, smart cities, vehicle-to-X, and factory automation. IIoT applications typically involve sensors and actuators exchanging critical monitoring and control commands. An exemplary application is monitoring production lines and controlling tasks in the manufacturing process. IIoT applications can greatly benefit from wireless communication, as it reduces installation and maintenance costs and device enables mobility. On the other hand, with the stagnating revenue growth from conventional users, telecommunication providers are looking for new sources of revenue. To that end, IIoT became one of the driving use cases for 5G wireless communications.

In 3rd Generation Partnership Project (3GPP) New Radio (NR), supporting challenging IIoT requirements has been one of the major focuses for Release 15 and for the ongoing Release 16 (5G phase 2) standardization. While the envisioned data rates of IIoT devices are typically low, IIoT applications dictate strict Quality of Service (QoS) requirements with respect to latency and reliability. Reference use cases defined by 3GPP require down to 1 ms latency with 10^{-5} reliability [1]. Moreover, the key parameter for IIoT communication is its *predictable performance*. The requirements of IIoT are expected to be addressed by *Ultra-Reliable Low-Latency Communications (URLLC)* mode of NR. In an effort to enable URLLC, 3GPP has introduced some notable features, such as mini-slot scheduling and aggressive timing capabilities for some procedures (e.g., grant processing and re-transmissions), all together greatly reducing data plane latency. However, radio

M. Vilgelm (✉) · H. Murat Gürsu · W. Kellerer
Chair of Communication Networks, Technical University of Munich, München, Germany
e-mail: mikhail.vilgelm@tum.de

N. H. Mahmood et al. (eds.), *Wireless Networks and Industrial IoT*,
https://doi.org/10.1007/978-3-030-51473-0_4

resource management procedures to obtain and maintain a connection are still derived from legacy LTE networks and thus constitute a latency bottleneck in the control plane latency.

The de facto standard resource management principle for applications with QoS requirements has been *dynamic grant-based scheduling*. Dynamic scheduling allows fine granular prioritization of User Equipments (UEs), as well as exploits variations of channel quality in time and space to maximize spectral efficiency. Dynamic scheduling found its success for applications requiring QoS in terms of the *guaranteed data rate*. These applications, with a notable example of video streaming, are inherently bursty and datarate-hungry, meaning that overhead and latency introduced by the dynamic grant acquisition is negligible. The situation is different for IIoT applications: Users typically only have small amounts of data to be transmitted and thus do not pose high data rate requirements. In addition to that, unlike heavy-tailed video streaming or strictly periodic Voice-over-IP traffic, industrial applications tend to transmit their data *sporadically*. Such transmission patterns do not allow to neglect grant acquisition procedure, which add to latency of every transmitted packet. Hence, to make NR control plane IIoT ready, there is a need to improve the *grant acquisition procedure* or to adopt *grant-free access* as an alternative.

To that end, in this chapter, we revisit *random access* as an underlying resource management principle behind grant acquisition and grant-free access. On one hand, probabilistic broadcast-based access coordination makes random access efficient and allows low-latency access for sporadic IIoT applications. On the other hand, its stochastic nature means that novel techniques have to be introduced to efficiently handle collisions and, more importantly, to *provide predictable protocol performance*, which is the main requirement for IIoT. Here, we give an overview of advances in random access enabling its application in IIoT scenarios. We present our view on the standing challenges in random access and outline early works to solve these challenges and toward further research directions.

The remainder of the chapter is structured as follows. In Sect. 2, we review the basics of random access protocols: system model assumptions, protocol properties, and its important applications. In Sect. 3, we briefly introduce the reader to novel techniques which serve as enablers for IIoT use cases. After that, in Sect. 4, we explore the challenges on way of random access for IIoT. Finally, we conclude with a summary in Sect. 5.

2 Random Access Overview

In this section, we introduce the reader to random access protocols. We start by motivating the use of random access protocols and explain its application areas (Sect. 2.1). Then, we introduce the basic definitions and typical modeling assumptions behind random access (Sect. 2.2). Finally, we present a holistic view on a random access protocol and explain its building blocks (Sect. 2.3).

2.1 *Random Access in Industrial IoT Communications*

The protocols for coordinating access to radio resources can be classified into two main categories: schedule-based and random-access-based. Schedule-based access implies that a Base Station (BS) assigns dedicated radio resources to every UE. The assignment can be predefined statically, e.g., as in Time Division Multiple Duplex (TDMA) protocols, or performed via dynamic assignments of *individual grants* as in LTE or NR. In contrast to the deterministic operation of schedule-based access, random access protocols assign resources to UEs in a *stochastic fashion*. BS indicates available resources and access parameters via broadcast messages instead of individual transmission grants. UEs decide to access the available resources or not probabilistically based on their *activity and contention parameters* (back-off window, barring probability, etc.). The possibility of interference between multiple UEs, if they decide to access the same resources at the same time, is explicitly accepted by the protocol. This interference can result in a reliability penalty and thus has to be carefully managed for IIoT applications.

The answer to the question whether to use schedule-based or random-access-based protocols is fully determined by the application requirements, traffic pattern, and its deployment scenario. Schedule-based access allows to avoid or minimize interference, and its dynamic version additionally allows to exploit time and frequency diversity of UEs' channels. However, it comes with a drawback of signaling overhead for grant acquisition, which causes efficiency loss and signaling delay. These drawbacks are especially relevant for IIoT applications with their dominantly sporadic traffic patterns. In contrast to schedule-based access, random access protocols allow to avoid the overhead of grant acquisition, yet they introduce uncertainty in delay and reliability due to the stochastic access coordination. In the next sections, we will present enabling state-of-the-art methods to control this uncertainty to provide reliable communication for IIoT. But this, in this section, we review common scenarios where IIoT communication can use random access protocols.

Connection Establishment

In practice, many technologies are combining random access and schedule-based access protocols. Even though LTE and 5G NR are designed for grant-based operation, their connection establishment procedure is inherently relying on random access, since the arrivals of UEs' connection requests cannot be deterministically predicted. The connection establishment, known as *random access (RA) procedure*, is triggered whenever UE transits from RRC-IDLE (or RRC-INACTIVE in NR) to RRC-CONNECTED states. This transition occurs not only when a UE joins the network but also after prolonged periods of inactivity (in the order of tens of seconds) and lost synchronization.

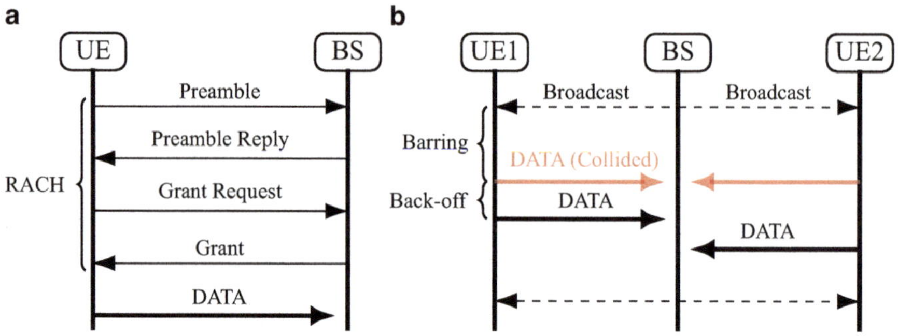

Fig. 1 (**a**) Exemplary schedule-based protocol (RACH): dynamic grant acquisition with a four-way handshake (preamble-based). (**b**) Exemplary random access protocol: grant-free operation with shared resources, using barring and back-off. The figure (**b**) illustrates a scenario where UE1 and UE2 are collided in the first attempt, but the collision is resolved after a random back-off

The RA procedure is implemented as a 4-step handshake, as illustrated in the timeline in Fig. 1a. It utilizes a dedicated Physical RA CHannel (PRACH) and its preambles as contention resources. While the preambles are efficient due to code domain multiplexing, they also add extra latency for grant acquisition. Moreover, predominantly sporadic IoT traffic patterns create overload in the channel. To control RACH load, access class barring (ACB) and its extended version are standardized by 3GPP [29]. The mechanisms are probabilistically regulating access to PRACH by broadcasting access barring factor to all UEs via system information blocks.

LTE and NR RACH have received a lot of research attention as a part of the effort to enable cellular *massive* IoT. Various improvements to preamble allocation, barring and back-off mechanisms, and physical layer design have been suggested in the literature [47] to improve the throughput of the channel. Since IIoT devices have sporadic activity and small amounts of data, RACH is expected to be triggered very often, introducing extra latency and compromising reliability in case of an overload. Thus, RACH must be accounted in *reliability and latency* analysis of NR. We will illustrate an approach RACH reliability analysis later in Sect. 4.1.

Grant-Free Access

Since RACH has a negative impact on reliability and latency, a natural solution is to avoid it by pre-allocating the resources for *grant-free access*. Grant-free access can be contention-free, where the users are uniquely assigned to resources, and contention-based, where users can share some resources. An example of contention-free grant-free access is semi-persistent scheduling in LTE, which is used for periodic applications, such as Voice-over-IP. However, contention-free resource assignments are wasteful for applications with sporadic activity; hence, 5G NR supports contention-based grant-free access to achieve statistical multiplexing gains. The mechanism to allocate grant-free resource in NR is referred to as semi-static scheduling [11]. The BS determines the amount of resources to be allocated and communicates it to UEs individually via Radio Resource Control (RRC) procedures. The resources might be also individually "activated" for the UEs using downlink control channel. The standard does not determine how many resources are to be allocated and how do UEs access them, leaving the options to be implementation specific.

In academic research, grant-free access is often abstracted from its LTE and NR implementation. A schematic example of grant-based vs grant-free protocol timeline is depicted in Fig. 1. The papers treat it as a generic random access protocol with some degree of coordination using barring, back-off [42], or resource allocation techniques [24]. The research has been focused on the various aspects of performance analysis [42], decoding techniques [3], and integration with modern interference cancellation techniques [21].

Millimeter Wave Spectrum

A prominent feature of 5G systems is the use of millimeter wave (mmWave) spectrum, also referred to as Frequency Range 2 (FR2) in 3GPP. MmWave can be considered an enabler for Industrial IoT applications due to its potential for ultra-low-latency links and "built-in" security due to the need for line-of-sight. However, efficient mmWave operation relies on beam management procedures to utilize the gains of directional communication. Beam management, i.e., establishment of the beam pair between UE and BS, adds complexity and new challenges to control plane procedures, in particular, to the random access procedure [17, 40].

As we will see in later sections, random access protocols are typically treated as interference-limited systems. These assumptions have to be revised for mmWave spectrum since directional communication significantly reduces interference between users. Instead, mmWave spectrum suffers from blockage and deafness effects. In both cases, many classical random access overload control techniques are rendered useless. E.g., consider a scenario where UE does not receive any feedback from BS after a grant request (see Fig. 1). Typically, after a timeout, UE assumes either a collision or low SNR. If a collision is assumed, UE continues waiting a certain time for back-off. In mmWave spectrum, back-off often introduces addition delay without a payoff, since mmWave networks are less prone to interference (and hence collisions) due to high directionality of communication [41]. If low SNR is assumed, UEs typically re-try with power ramping. However, in a deafness scenario, increase of transmission power is not likely to solve the problem.

Satellite Communications

For completeness, satellite communication must be mentioned here. It is often preferred whenever communication over large areas is required, such as in logistics, agriculture, or military applications, i.e., for industrial applications in a wider sense. There are also many start-ups launching their own CubeSats in order to provide a satellite-based ubiquitous IoT support [12, 38]. High propagation delay makes handshaking and grant acquisitions very costly in satellite scenarios; therefore this industry was one of the early adopters of random access protocols. It has been also a driving force behind the development of more advanced random access techniques with interference cancellation, which we will discuss in the next chapters.

2.2 Performance Model

In this subsection, we introduce a basic MAC layer model of random access [47]. Let us assume that a certain number of resources (i.e., resource pool) is allocated with a given periodicity. This periodicity is thus defining the duration of a single *contention round*. The resource pool is subdivided into *transmission opportunities (TO)*: smallest amount of resources needed to transmit UE's data. TOs can be allocated both in time, frequency, or code domain, e.g., Random Access Opportunity in NR is six resource blocks × one sub-frame × one preamble. Let us further consider *a single TO*, and a total population of n users, and let us define variable α_i indicating whether UE i is accessing the TO or not.

Given that UEs using the same TO might interfere with each other, let us apply a Signal-to-Interference-to-Noise Ratio (SINR) threshold model to determine whether the data of a particular active UE j in a given TO is successfully decoded:

$$\gamma_j = \frac{|h_j|^2 P_{\text{tx},j}}{\sum_{i\in\{1\ldots n\}\setminus j} \alpha_i |h_i|^2 P_{\text{tx},i} + \eta} \geq \gamma_{\min}, \tag{1}$$

where η is the noise at the receiver; $P_{\text{tx},i}$, $P_{\text{tx},j}$ are transmission powers of the ith and jth UE, respectively; and h_i, h_j are the channel coefficients from users i, j to the receiver, respectively. In other words, SINR γ_j at the receiver must be beyond a certain threshold $\gamma_{\min}$ for the data to be successfully decoded.

A common analysis approach is a *collision channel model*. The standard 0/1 collision channel assumes that following two conditions are satisfied:

$$\frac{|h_j|^2 P_{\text{tx},j}}{\eta} \geq \gamma_{\min}, \quad \forall j \in \{1\ldots n\}. \tag{2a}$$

$$\frac{|h_j|^2 P_{\text{tx},j}}{|h_i|^2 P_{\text{tx},i} + \eta} < \gamma_{\min}, \quad \forall i, j, \quad i \neq j. \tag{2b}$$

Condition (2a) assumes *high Signal-to-Noise Ratio (SNR) regime*, i.e., any individual UE's SNR is always greater than the threshold. Condition (2b) indicates that if there is more than one UE using a TO, the interference is fully destructive. Under the 0/1 collision model, the probability p_s of a given TO to have a successfully decoded transmission is described as:

$$p_s = \begin{cases} 1 & \text{if} \sum_{i \in 1 \ldots n} \alpha_i = 1, \\ 0 & \text{otherwise.} \end{cases} \tag{3}$$

This performance model can be readily extended to analyze time-varying behavior for multiple TO. It provides an abstraction to be used for queuing-theoretic protocol analysis. In some scenarios, one or both conditions (2a)–(2b) do not hold, leading to such effects as *detection error* (condition (2a) violated), or *capture* (condition (2b) violated). In such cases, variations of collision model are often studied, e.g., channels with multi-packet reception [19]. In other cases, the time diversity can be utilized to cancel the interference between UEs, which is the premise for Successive Interference Cancellation (SIC) techniques. We discuss these effects later in Sects. 3.1 and 3.2.

The activity indicator α_i depends on the *traffic pattern* of every application and on the *contention resolution parameters*. If a collision occurs, it has to be resolved by means of a *collision resolution procedure*. An ideal procedure should introduce as less delay as possible and distribute UEs to resources in the way to maximize their decoding probability. Historically, collision resolution is performed in time domain by means of a *random back-off*. If high load is anticipated prior to transmission, or if prioritization needs to be applied, contention resolution process can start before the actual transmission, i.e., by means of probabilistic access barring. Access barring and back-off procedures are illustrated in Fig. 1. In addition to back-off, UEs might deploy power ramping, where the transmission power is increased after every failed attempt. Power ramping is beneficial if high SNR regime cannot be assumed, but it can also serve as an implicit prioritization mechanism.

Finally, it is important to distinguish two different flavors of random access protocols depending on the synchronization between UEs and the BS: time-slotted and un-slotted. Simplest examples date back to the works of Abramson [2], ALOHA and slotted ALOHA protocols. Time-slotted random access introduces additional overhead but allows to reduce collisions probability down to half, if no additional techniques [7] to recover collided packets are applied. In modern cellular systems, where BS is a single common received for all UEs in the cell, time synchronization can be achieved with periodic low overhead broadcasts; therefore slotted random access is more common.

2.3 Holistic View on Random Access Protocols

In this subsection, we present building blocks of a typical random access protocol. For illustrative purposes, we consider a simple case where the protocol aims to optimize its performance on a *contention-round basis*. The goal of a well-designed protocol is to derive a set of optimal contention parameters $\mathcal{P}$ to maximize a certain utility function U under a set of constraints C in a given contention round. For example, a utility function can be throughput: The number of successfully decoded packets during the contention round. With the set of contention parameters, BS controls how UEs access the TOs; hence, it is a random access counterpart of the deterministic resource allocation in grant-based access. The contention parameters might include barring probability, back-off, limit on transmission attempts, power ramping parameters, etc. The amount of allocated resources and their split can also be advertised with contention parameters.

Let us define the activity vector $\boldsymbol{\alpha} = [\alpha_0, \ldots, \alpha_n]^\mathsf{T}$, where α_i denotes that UE i is active in a given contention round,[1] and channel coefficients vector $\boldsymbol{h} = [h_0, \ldots, h_n]^\mathsf{T}$. Using a performance model, activity, and channel information, the protocol derives:

$$\mathcal{P}^\star = \arg\max_{\mathcal{P}} U(\boldsymbol{h}, \boldsymbol{\alpha}, \mathcal{P}), \quad \text{s.t.} C \tag{4}$$

Unlike in grant-based scheduling, BS typically cannot obtain all activity and channel information to derive optimal contention parameters directly from UEs. Instead, it has to rely on *contention outcome observations* $\mathcal{O}$ and, if available, on a priori information, such as activity pattern or its model. The outcome observation can include number of successfully decoded UEs, number of idle TOs, and number of collided TOs. Moreover, both channel and activity are random variable: The two factors influencing activity pattern – application traffic profile and contention resolution process – are stochastic, and thus $\boldsymbol{\alpha}$ is generally unknown. Thus, this uncertainty has to be resolved via *user activity estimation* and channel estimation.[2] Thus, the protocol has to include estimation blocks. The full building blocks of the protocol are illustrated in Fig. 2.

In practice, these building blocks are either implemented implicitly or neglected. However, strict requirements of IIoT applications require careful re-considerations of all protocol parts and introduce novel challenges:

- The performance of random access protocols is limited by destructive interference between UEs, i.e., collisions. Hence, performance enhancing techniques like interference cancellation and multi-packet reception must be carefully studied for their improvement potential.

[1] Note that we have overloaded earlier definition of activity indication per TO.

[2] Assuming fixed location of IIoT users, channel uncertainty can be partly mitigated by either allocating separate estimation resources or by pre-estimating the channel in advance.

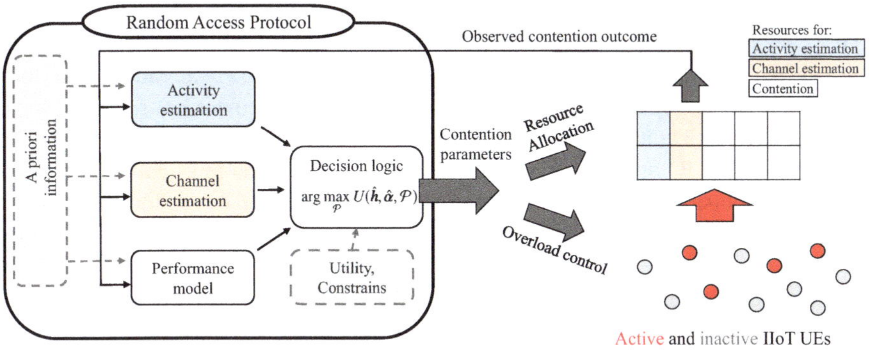

Fig. 2 A random access protocol has to solve three main tasks: (1) estimate active UEs $\hat{\boldsymbol{\alpha}}$, (2) estimate channels $\hat{\boldsymbol{h}}$, and (3) derive optimal contention parameters $\mathcal{P}$. The history of observed contention outcomes and the history of contention parameters serve as input information to the estimators. Derived optimal contention parameters are then broadcast to UEs

- The classical choice of the utility function, i.e., normalized throughput, has to be revised and, at the very least, aided with latency and reliability constraints [16].
- The estimation building blocks should consider reliability constraints and to include probabilistic performance characterization [46].
- A priori information has to be tailored to the IIoT scenario, that is, it has to provide reliable information necessary to enhance the estimation performance. In addition, a priori information can include *IIoT application-specific* information [47], which can help prioritize users depending on the urgency.

In the next chapters, we outline main performance enhancing enablers for random access (Sect. 3) and describe in detail these design challenges for IIoT random access (Sect. 4).

3 Selected Enablers for IIoT Random Access

In this section, we present the enabling techniques which allow to boost random access performance by reducing the impact of destructive interference and allowing additional coordination: interference cancellation, multi-packet reception, and feedback.

3.1 Interference Cancellation

Due to the lack of coordination, classical random access protocols do not allow deterministic user separation in time or closed-loop power control to reduce interference. Instead, advanced decoding techniques for interference cancellation (IC) can be deployed to recover collisions and thus increase utilization of contention resources. While interference cancellation had been known already for a long time [26], in the recent years, the increased computation power has rendered more complex decoding techniques real-time capable. In this section, we present the implications of IC on random access protocols, whereas the physical layer aspects of IC are outside of our scope. Interference cancellation in random access can be classified into Inter-slot Interference Cancellation (IeIC), Intra-slot Interference Cancellation (IaIC), and Asynchronous Interference Cancellation (AIC). AIC can be deployed in unslotted random access, yet it is complex to implement as one has to search a continuous space for the IC process to start [7]. Here, we focus on synchronous IC techniques.

Protocols with IeIC, usually referred to as successive interference cancellation (SIC), exploit signal replicas and time diversity to recover collided packets. Instead of transmitting a signal in only one TO, every UE generates a certain number of replicas, each to be transmitted in a different TO. Additionally, each replica contains pointers to TOs with other replicas, such that even if only one replica is successfully decoded, the time-frequency location of all others is known. The iterative decoding process usually starts with a *singleton TO*, where no collision has occurred. If a singleton signal is successfully decoded, its replicas are found using pointers, and the interference of its replicas is canceled from the respective TOs. The process is then repeated until there are no singleton TOs remaining.

IeIC is illustrated in Fig. 3 with two examples. Consider a contention round consisting of three TO and three UEs (A, B, C). First, consider Example I, where UE A sends three replicas of a packet and UE B sends two replicas. The protocol

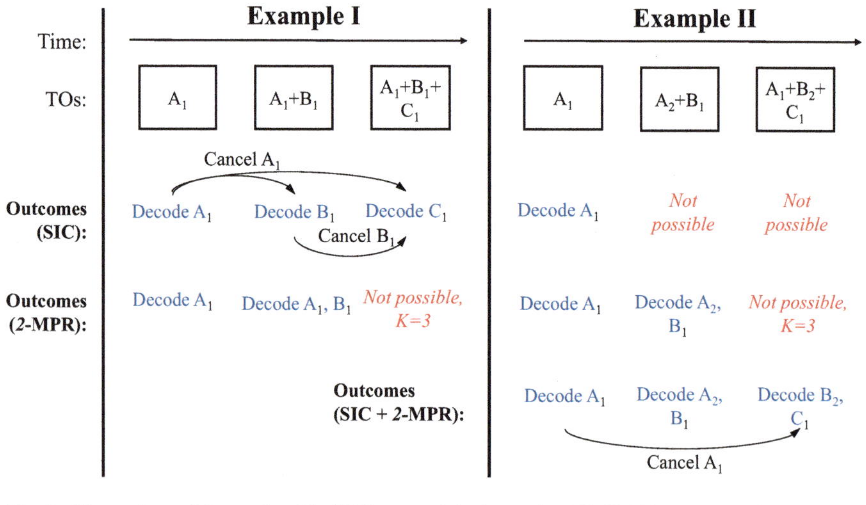

Fig. 3 Illustration of inter-slot IC (SIC example), intra-slot IC (2-MPR example), and their hypothetical combination

thus starts with the first TO (singleton), decodes packet A_1, and then cancels the interference caused by A_1 from TO 2 and 3. It follows with decoding B_1, since it is now in a singleton TO, and then canceling its interference from third TO. Following that, the protocol completes the process by decoding the remaining signal C_1 in the third TO. Hence, in Example I, SIC is able to recover all packets despite collisions, resulting in normalized per-TO throughput 1. Now consider Example II, where UEs only partially use replicas and UE A sends different packets in the first and the second TO. Here, the protocol is only able to decode A_1 in the first TO, since no interference cancellation is possible without replicas. Thus, its throughput is reduced to $1/3$. Most of the known examples of protocols with SIC stem from satellite communications, where long propagation delays led to early adoption of random access. E.g., contention resolution diversity slotted ALOHA [5] suggests to use a fixed number of replicas and send them in randomly selected TOs within a frame, whereas authors in [31] demonstrate that higher throughput can be achieved by choosing the number of replicas from a given probability distribution.

3.2 Multi-packet Reception

Unlike IeIC protocols, IaIC does not rely on packet replicas and allows to recover multiple packets from the same TO without the packet's replicas. The capability to perform Intra-slot IC is often referred to as *multi-packet reception (MPR)*. A special case here is a capability to recover only the strongest signal despite the interference, denoted as *capture effect*. Capture effect occurs in a variety of the conditions, but it is more common if the difference between the received signal strengths is high. Capture effect is also more prominent in unslotted random access, where the collided signals overlap only partially.

While capture allows only to recover the strongest signal, Non-Orthogonal Multiple Access (NOMA) techniques allow to recover multiple users' signals within the same TO using multi-user detection algorithms on the receiver side. In a variety of works, NOMA has been shown to outperform orthogonal access in terms of its spectral efficiency [9]. To enable multi-user detection, *additional diversity* between the users within the same TO is created, e.g., in code domain (spreading sequences) or in power domain (received signal strength differences). The users are then iteratively decoded, typically starting with the strongest signal. Another group of NOMA protocols based on *Compressed Sensing (CS)* [10] has recently gained attention for massive IoT communications [13, 22]. In its essence, CS allows to recover a sparse signal from a small number of observations by solving an under-determined linear system. The sporadic activity of IoT users allows a convenient mapping to a sparse optimization problem. CS techniques is to be used both for user activity detection [13] and for joint activity-detection-channel estimation-data-decoding [49]. It has been shown that CS-based random access can greatly benefit from power control [13] and massive Multiple-Input-Multiple-Output (MIMO) [30].

While NOMA protocols are primarily studied in the context of grant-based communication, they can be efficiently applied to random access and grant-free communication [48]. From MAC layer perspective, these protocols are often modeled as K-MPR random access [44]. The ability to decode multiple packets is typically limited to a certain number of iterations K due to a variety of factors, e.g., residual noise. K-MPR model is thus an extension of a $0/1$ collision model: If more than K users transmit in the same TO, then all of them are declared undecodable (i.e., *hard collision* occurs); otherwise all users are assumed to be successfully decoded. We illustrate the difference of SIC and K-MPR random access with the exemplary $K = 2$ in Fig. 3. We have seen earlier that SIC is able to decode all packets in Example I. In contrast to that, 2-MPR protocol can only decode first (no interference) and second TO, where A_1 and B_1 are decoded iteratively. Note that it does not use the signals decoded in the third TO; and, thus, it is not able to decode any signal there, since there are three users colliding. As a result, 2-MPR protocol yields normalized throughput $2/3$ (in contrast to SIC with throughput 1). Now consider the second example, where the advantage of 2-MPR becomes clear. Here, the protocol is also able to decode only first and second TO; however, since no replicas were in use, 2-MPR can recover two different packets from user A. Hence, a total of three distinct packets have been recovered, resulting in normalized throughput 1 (in contrast to SIC delivering throughput $1/3$). In principle, SIC and MPR can be combined together, resulting in throughput $5/3$ as illustrated in the Example II.

The deployment of SIC and MPR in IIoT random access, while potentially very promising to increase throughput and reliability, also brings additional design challenges. Both SIC and MPR require more complex procedures to manage the resources for contention and estimation. SIC protocols are designed and optimized for time-domain diversity and thus need adaptation to multi-channel systems with orthogonal TOs, such as NR [21]. For MPR protocols, the additional diversity parameters must be allocated and indicated to be UEs in advance. Moreover, MPR can be greatly enhanced with channel and activity information; thus the trade-offs in allocating additional resources (e.g., pilots) for estimation must be evaluated [20]. Additionally, all classical overload control and hard collision handling methods (access class barring, back-off, etc.) must be revised to account for interference cancellation capabilities.

Traditionally, both SIC and NOMA-based access protocols are designed to maximize the throughput. However, throughput-driven optimization often contradicts with IIoT design goals. For example, in CS-based random access, total throughput is often increased at the expense of decreased detection probability of individual users [22]. In another example, many SIC protocols deliver optimal throughput only with infinitely large contention rounds, which obviously can lead to latency constraints violations. Therefore, throughput-maximization techniques cannot be applied to IIoT use cases without a careful study of their behavior under latency and reliability constraints.

The integration of SIC and MPR also attracts high theoretical research attention [18, 43]. Even though performance gains of their combination are clear, to the best of our knowledge, there are no existing practical implementations or prototypes due to high complexity of a combined protocol.

3.3 Feedback

Random access protocols can operate with or without contention feedback to users. In many systems, such as satellite, feedback is costly due to additional delay, and thus it is avoided. However, the feedback has been shown to be strongly beneficial in wireless communications. It has been theoretically shown that feedback improves the channel capacity [28]. Additionally, it enables hypothesis testing and the estimation of the stochastic processes, i.e., in context of random access, of the user activity $\boldsymbol{\alpha}$. The first proposal of using NACK for random access is the *tree algorithms* by Capetanakis [4], where it has been shown how feedback can improve stability and increase the throughput. The NACK can be explicit, such as in RFID communications in the form of a specific message sent by the receiver. Alternatively, it can be implicit: If no signal is received from the receiver until a certain time, then NACK is assumed, e.g., in LTE RACH. Explicit feedback, such as in some variations of tree algorithms, can be more powerful as it carries more tailored information and allows for faster reaction [4, 35].

Feedback acts as an input for estimating two stochastic processes of wireless communication. First, the channel realization is a stochastic process that the random access algorithms abstract away as it is assumed that the physical layer is dealing with that process. Secondly, the user activity is another stochastic process that the random access algorithm can choose to react or not. The feedback may be unfiltered such that it may carry information related to both of these stochastic processes. The algorithm has to include a post-processing of the feedback before feeding it in either of the two estimations. For instance, with the user activity, the failures due to transmission of the users have to be separated from channel failures. In case where UEs are aware of the other UE's decision, they can filter the feedback with respect to the other UE's decisions. This would be a distributed filtering. In case the BS does such a filtering for the UEs, as more processing capability is available there, the BS can transmit separate feedback for channel failure and user activity-based failure.

4 Challenges and Research Directions

Even with the promising enables, there exist a number of standing challenges on the way to applying RA protocols for industrial IoT scenarios. These challenges lie in the details of implementing individual building blocks of random access protocols (see Fig. 2). In this section, we first go into the problem of designing utility function for IIoT random access (Sect. 4.1). Then, we discuss the estimation (Sect. 4.2) and channel modeling problems (Sect. 4.3) in the context of IIoT. Finally, we show how application domain knowledge can contribute to all the building blocks (Sect. 4.4).

4.1 Probabilistic Performance Characterization

Since the early works on random access, its *expected* performance has been used as utility function, e.g., contention parameters have been chosen to maximize average throughput or decrease average delay. Thus, performance modeling for random access has been limited to the models assessing average performance. However, for applications in the IIoT domain, assessing the average performance is insufficient. Instead, it is important to characterize higher-order statistics or determine tail of latency distribution. For random access protocols, this means that their *stochastic performance bounds* must be evaluated.

One way to describe the stochastic performance is by using a reliability-latency curve, i.e., the probability to obtain a certain latency. Such performance characterization is important for network dimensioning, e.g., determining how many UEs can be supported in the network for a given target reliability-latency level. One can further differentiate between reliability-latency requirements of *individual* UEs and *system-level* reliability-latency requirements, e.g., a requirement on the time it takes to connect all system elements to the network [46]. Some early research efforts have emerged in this direction in the recent years. For individual UEs, a method to characterize reliability-latency of multi-channel tree resolution algorithms has been introduced in [14]. The distribution of random access delay has been derived in [23], and reliability of grant-free access protocols has been investigated in [42].

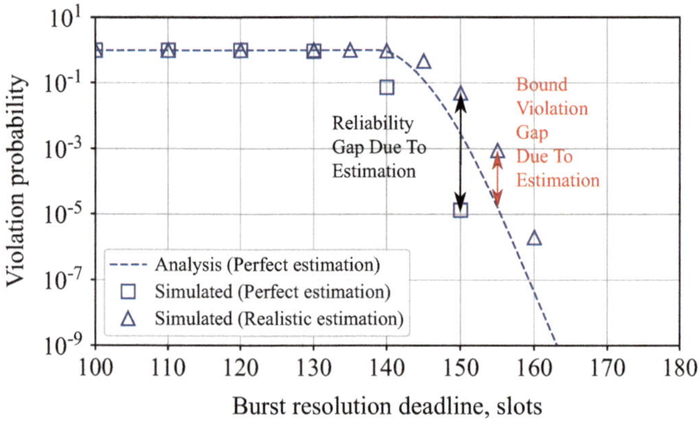

Fig. 4 Exemplary reliability-latency performance curve for a burst resolution process [46]. Illustrated are analytical bound with perfect estimation assumption and simulation results with and without perfect estimation assumption

For the system-level scenarios, the burst resolution delay, corresponding to a variety of emergency scenarios in industrial automation, such as blackout recovery or system reboot, has been studied by the authors in [46]. The authors have illustrated that the reliability assessment is additionally complicated by the uncertainty of user activity; see Fig. 4. That is, even if the performance bounds on protocol behavior are characterized, they are conditioned on the user activity profile. Thus, the user activity profile either has to be enforced by means of *traffic shaping tools* or has to be *estimated with sufficient precision*.

The next step after evaluating stochastic performance is to incorporate reliability-latency into the optimization, either as objective or constraints. Recent work in this direction includes [15, 32]. The authors in [32] have introduced frameless ALOHA, providing guaranteed reliability-latency performance by means of SIC. The protocol operates in a way that adds time slots on-the-fly in order to meet a certain performance level. Latency-constrained IeIC with feedback is investigated in [15], where users are allocated specific identities to provide guaranteed random access performance.

4.2 Reliable Activity Estimation

Providing precise estimation of user activity is a prerequisite for reliable access protocols. Three activity estimation problems are common in random access: (1) total population of users, (2) active users, and (3) collision multiplicity estimation. For IIoT communication, since the sensors are deployed in a controlled well-dimensioned environment, it is typical to assume that the total number of users is known. An estimate on the number of active users is needed in order to allocate appropriate number of resources and devise overload control measures (e.g., barring factor). Estimating the collision multiplicity allows to deploy more advanced protocols for faster and more efficient collision resolution.

The estimation can be performed either under assumption of a known prior distribution or without any prior assumptions. A prior typically comes for a model of user behavior; thus it requires knowledge of the users' activity profile, which can be hard to obtain if user's activity is comprised of many different applications. Without a known prior, it has been a common approach to assume Poisson distribution and iteratively update the estimate [45]. The approach might work well in case of massive IoT, where, due to a large number of independent users, the Poisson limit theorem can be assumed to hold. This is however not the case for IIoT, where due to tight reliability requirements much less users per BS can be supported. If the total number of independent users is low, Poisson approximation does not hold anymore. Furthermore, Poisson approximation assumes independence of users' activity, thus neglecting a possibility of correlations. The correlation arises in exemplary cases where multiple sensors monitor the same process or in emergency scenarios, creating burst arrivals [29] and highly impacting random access performance [25].

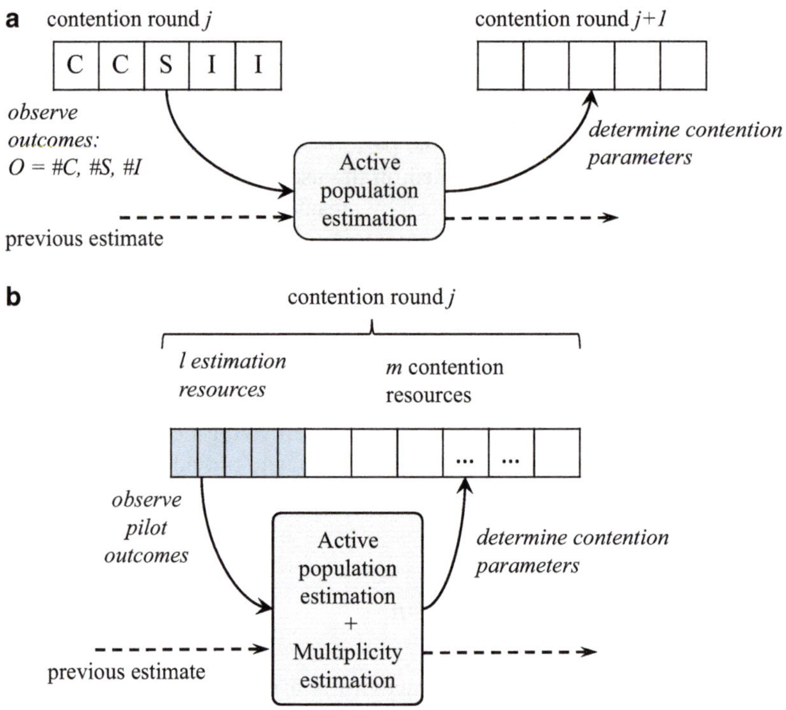

Fig. 5 Two exemplary options how to implement activity estimation: (**a**) Observe outcome in contention resources (e.g., number of collided, successful, and idle TOs), and use the estimate to derive contention parameters for the next contention round; and (**b**) separate contention round in estimation and contention phase, observe outcome of pilot transmissions in estimation resources, and use the estimate to determine contention parameters for the contention phase. Previous estimate can be used to enhance the precision [39]

User activity can be estimated by *observing* the outcomes in the allocated resources. Here, we are interested in estimation on the BS side. As the number of resources is scarce, intuitive approach would be to directly use the *contention outcomes* O as an observation (see Fig. 5a). In that case, no additional resources need to be spent on estimation. There is a number of seminal works investigating estimators using the contention outcomes [8, 39] given the number of users and their activity pattern. Typically, the expected number of active UEs is obtained with this method.

An alternative approach is to deploy additional resources specifically for the estimation of the user activity (see Fig. 5b). Using separate resources for estimation provides multiple advantages. First, one can use pilots instead of actual data packets, thus potentially allowing smaller transmission slots. Second, the more resources are used for estimation, the better precision can be achieved. With more resources, it is possible to obtain a multiplicity estimate [8] on the instantaneous number of UEs per TO. The multiplicity estimate can be used to guide and dimension collision

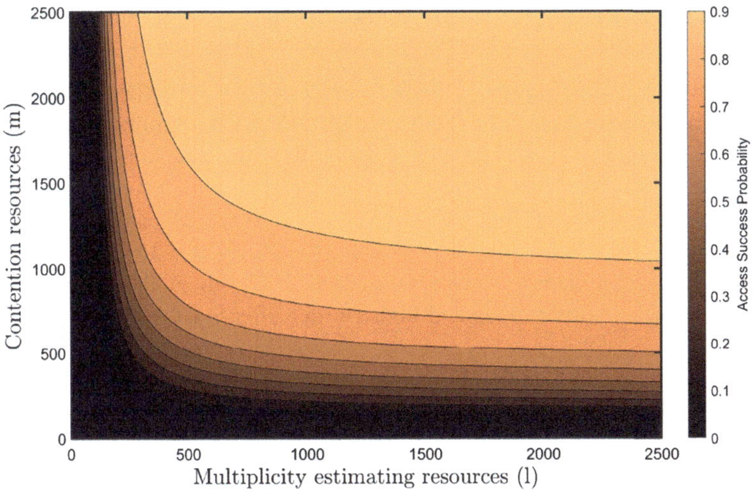

Fig. 6 Illustration of the trade-off between allocating estimation and contention resources [20]. Success probability is depicted as a function of (m, l) configuration of a contention round resources: m contention resources and l estimation resources

resolution techniques, such as tree resolution algorithms, which in turn can provide tighter performance guarantees. It is an ongoing research to assess the trade-off between allocation of estimation resources and the estimation precision [20, 33]. The balance between allocation of a contention resources and estimation resources is evaluated from MAC layer perspective in [20]. For a fixed reliability requirement and total number of active users, the authors evaluate different allocations of l resource units for multiplicity estimation and m resource units for contention resources. This trade-off is illustrated in Fig. 6. We observe that a combination of $l, m = (400, 2500)$ resources achieves the same success probability, 0.9, as $l, m = (2500, 1100)$ combination, whereas it spends less total amount of resource units (2900 vs 3600). This illustrated that, in some cases, allocation of additional estimation resources can increase the resource efficiency.

In addition to the centralized activity estimation on the BS, there is an option to apply a decentralized approach and delegate the estimation task to UEs. However, typically, UE can only observe contention outcome in the TO where it transmits itself, making the resulting estimation very coarse. Whether the sensor has access to outcome of other TOs depends on whether it has access to the feedback to other sensors (e.g., if feedback from BS is transmitted as a broadcast). Continuous listening to the feedback introduces additional power consumption. This perspective is investigated in the following work [50].

4.3 Channel and Performance Modeling

As we have explained earlier in Sect. 2, random access protocols are typically analyzed under the assumptions of collision channel model. Unlike its information-theoretic counterparts, such as Gaussian MAC [36], collision channel is less complex and thus applicable for queuing theoretic analysis. Low complexity allows to use the analysis in real time to optimize the resource allocation and overload control. However, while it provides a powerful abstraction to analyze *expected* behavior, for high reliability scenarios such as IIoT, tail of the distribution also matters; therefore, corner cases and higher-order effects must be captured by the model.

Additionally, novel techniques such as SIC or MPR allow better performance than that predicted by collision channel model. Therefore, research community is currently investigating generalizations of collision channel models. Prominently, the K-MPR channel model [19] has been revisited. Its probabilistic version [42] allows more precise modeling of higher-order effects. However, K-MPR model is typically assuming that MPR capabilities are static and can be determined in advance. This assumption often does not hold, since MPR capabilities depend heavily on the deployment scenarios, power control, channel coefficients, and even on the contention parameters. More research in this direction is needed to find suitable models to characterize the high reliability region of performance.

4.4 Application Awareness and Cross-Layer Design

A promising way to improve the performance of random access protocols is to provide more information from the application layer. This approach is often referred to as cross-layer design [6], and more recently, it has been coined as semantic networking [37]. Multiple building blocks of a random access protocol can benefit from application domain knowledge, as illustrated in Fig. 7. It can reduce uncertainly in the user activation and provide semantic information to increase efficiency and allow dynamic prioritization.

Let us first consider the use of application domain knowledge for activity estimation. It has been a common assumption that random access does not cope well with synchronized or correlated arrivals [29]. However, as it has been demonstrated in recent work [25], anticipating the correlations can also boost the performance beyond slotted ALOHA throughput. Known correlations can be exploited by adjusting contention parameters or allocating appropriate amount of resources. Moreover, the activity of IIoT devices often tends to be correlated, e.g., consider redundant sensors monitoring environment conditions or cascading failure events propagating in a facility. Exploiting *correlation in the transmitted information* can further boost the performance, leading to the information-centric networking

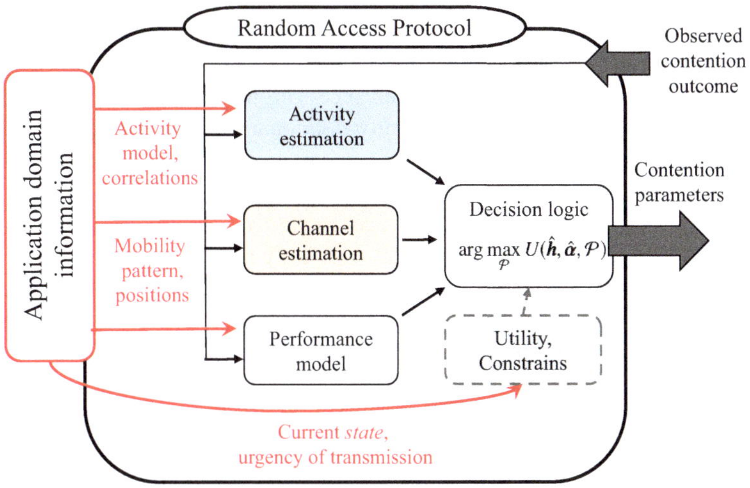

Fig. 7 Incorporating application domain knowledge into random access: Information about the current state and correlation between users' activity can be used for more precise activity estimation; application-specific utility functions and current state of the application can be used to improve performance estimation

approach, where only the *information itself* matters and not the identity of its transmitting UE [27].

Additionally, the application domain can provide dynamic input to the utility function and decision logic. Utility function can take the current *state of an application* into consideration, and the random access protocol can dynamically react upon it. Current state can determine the *urgency* of data transmission, which helps to prioritize UEs and thus improve the efficiency of the protocols, when the data is only transmitted when it is really needed. Let us take an example of networked control systems, a general class of IIoT applications, where the feedback control loop is closed over a wireless channel [34]. Here, the state of an application can be expressed as the *network-induced error*, i.e., the deviation of the state of the plant from the desired operating point. The authors [34] demonstrate that current state of the application can be dynamically used to locally determine access probability, which reduces collision rates and improves application performance.

In order to fully benefit from application domain knowledge, further work is required both for theoretical foundations and performance limits and for cross-layer architecture.

5 Summary and Conclusions

In this chapter, we have presented random access protocols for IIoT communication as an alternative approach to the currently dominant dynamic grant-based access. Despite its stochastic nature and notorious uncertainty, random access has potential to achieve high efficiency and lower latency for sporadic IIoT users by avoiding excessive signaling overhead. We have presented interference cancellation and feedback as enabling techniques to boost efficiency and control the uncertainty of random access protocols. We have indicated the implications of IIoT requirements on the building blocks of a random access protocols: activity estimation, performance modeling, and utility function choice. We have shown that all the building blocks of a random access protocol must be carefully designed to fulfill IIoT requirements for predictable performance. Finally, we argued the random access protocols can benefit from application layer information.

References

1. 3GPP (2019) TR 38.825 V16.0.0-Technical Specification Group Radio Access Network; Study on NR industrial Internet of Things (IoT); (Release 16) (2019)
2. Abramson N (1977) The throughput of packet broadcasting channels. IEEE Trans Commun 25(1):117–128
3. Bayesteh A, Yi E, Nikopour H, Baligh H (2014) Blind detection of SCMA for uplink grant-free multiple-access. In: 2014 11th international symposium on wireless communications systems (ISWCS), pp 853–857. https://doi.org/10.1109/ISWCS.2014.6933472
4. Capetanakis J (1979) Tree algorithms for packet broadcast channels. IEEE Trans Inf Theory 25:505–515
5. Casini E, De Gaudenzi R, Del Rio Herrero O (2007) Contention Resolution Diversity Slotted ALOHA (CRDSA): an enhanced random access schemefor satellite access packet networks. IEEE Trans Wirel Commun 6(4):1408–1419
6. Chiang M, Low SH, Calderbank AR, Doyle JC (2007) Layering as optimization decomposition: a mathematical theory of network architectures. Proc IEEE 95(1):255–312
7. Clazzer F, Munari A, Giorgi F (2017) Asynchronous random access schemes for the VDES satellite uplink. In: OCEANS 2017 – Aberdeen, pp 1–7
8. Cidon I, Sidi M (1988) Conflict multiplicity estimation and batch resolution algorithms. IEEE Trans Inf Theory 34(1):101–110
9. Ding Z, Liu Y, Choi J, Sun Q, Elkashlan M et al (2017) Application of non-orthogonal multiple access in LTE and 5G networks. IEEE Commun Mag 55(2):185–191
10. Donoho DL (2006) Compressed sensing. IEEE Trans Inf Theory 52(4):1289–1306
11. Dahlman E, Parkvall S, Skold J (2018) 5G NR: the next generation wireless access technology. Amsterdam: Academic
12. De Sanctis M, Cianca E, Araniti G, Bisio I, Prasad R (2015) Satellite communications supporting internet of remote things. IEEE Internet Things J 3(1):113–123
13. Fletcher AK, Rangan S, Goyal VK (2009) On-off random access channels: a compressed sensing framework. arXiv preprint: 0903.1022
14. Gürsu HM, Alba AM, Kellerer W (2017) Delay analysis of multichannel parallel contention tree algorithms (MP-CTA). arXiv preprint: 1707.09754

15. Gürsu HM, Guan F, Kellerer W (2019) Hard latency-constraints for high-throughput random access: SICQTA. In: IEEE international conference on communications (ICC)
16. Gürsu HM, Kellerer W, Stefanović C (2019) On throughput maximization of grant-free access with reliability-latency constraints. In: Proceedings of IEEE International Conference on Communications (ICC)
17. Giordani M, Mezzavilla M, Zorzi M (2016) Initial access in 5G mmWave cellular networks. IEEE Commun Mag 54(11):40–47
18. Goseling J, Stefanović Č, Popovski P (2018) Sign-compute-resolve for tree splitting random access. IEEE Trans Inf Theory 64(7):5261–5276
19. Ghez S, Verdu S, Schwartz SC (1988) Stability properties of slotted Aloha with multipacket reception capability. IEEE Trans Autom Control 33(7):640–649. ISSN: 2334-3303
20. Gürsu HM, Köprü B, Ergen SC, Kellerer W (2018) Multiplicity estimating random access protocol for resource efficiency in contention based NOMA. In: Proceedings of IEEE international symposium on personal, indoor and mobile radio communications (PIMRC). IEEE, pp 817–823
21. Gürsu HM, Moroglu C, Vilgelm M, Clazzer F, Kellerer W (2019) System level integration of irregular repetition slotted ALOHA for industrial IoT in 5G new radio. In: Proceedings of IEEE International Symposium on Personal, Indoor and Mobile Radio Communications (PIMRC), Oct 2019
22. Hong J, Choi W, Rao BD (2015) Sparsity controlled random multiple access with compressed sensing. IEEE Trans Wirel Commun 14(2):998–1010. ISSN: 1558-2248
23. Jian X, Liu Y, Wei Y, Zeng X, Tan X (2016) Random access delay distribution of multichannel slotted ALOHA with its applications for machine type communications. IEEE Internet Things J 4(1):21–28
24. Jacquelin A, Vilgelm M, Kellerer W (2019) Grant-Free access with multipacket reception: analysis and reinforcement learning optimization. In: Proceedings of IEEE/IFIP Wireless On-demand Network Systems and Services Conference (WONS). IEEE, Jan 2019, pp 1–8
25. Kalor AE, Hanna OA, Popovski P (2018) Random access schemes in wireless systems with correlated user activity. In: Proceedings of IEEE International Workshop on Signal Processing Advances in Wireless Communications (SPAWC), pp 1–5
26. Kaitsuka T, Inoue T (1984) Interference cancellation system for satellite communication earth station. IEEE Trans Commun 32(7):796–803
27. Kassab R, Simeone O, Popovski P (2019) Information-centric grant-free access for iot fog networks: edge vs cloud detection and learning. arXiv preprint: 1907.05182
28. Kramer G (2002) Feedback strategies for white Gaussian interference networks. IEEE Trans Inf Theory 48(6):1423–1438. https://doi.org/10.1109/TIT.2002.1003831
29. Laya A, Alonso L, Alonso-Zarate J (2014) Is the random access channel of LTE and LTE-A suitable for M2M communications? A survey of alternatives. IEEE Commun Surv Tutorials 16(1):4–16. ISSN: 1553-877X
30. Liu L, Larsson EG, Yu W, Popovski P, Stefanovic C et al (2018) Sparse signal processing for grant-free massive connectivity: a future paradigm for random access protocols in the Internet of Things. IEEE Signal Process Mag 35(5):88–99
31. Liva G (2010) Graph-based analysis and optimization of contention resolution diversity slotted ALOHA. IEEE Trans Commun 59(2):477–487
32. Lázaro F, Stefanović Č, Popovski P (2019) Reliability-latency performance of frameless aloha with and without feedback. arXiv preprint: 1907.06383
33. Magrin D, Pielli C, Stefanovic C, Zorzi M (2018) Enabling LTE RACH collision multiplicity detection via machine learning. arXiv preprint: 1805.11482
34. Mamduhi M, Vilgelm M, Kellerer W, Hirche S (2017) Prioritized contention resolution for random access networked control systems. In: Proceedings of IEEE Conference on Decision and Control (CDC), Dec 2017. https://doi.org/10.1109/CDC.2017.8264667
35. Madueño GC, Stefanović Č, Popovski P (2015) Reliable and efficient access for alarm-initiated and regular M2M traffic in IEEE 802.11 ah systems. IEEE Internet Things J 3(5):673–682

36. Polyanskiy Y (2017) A perspective on massive random-access. In: 2017 IEEE International Symposium on Information Theory (ISIT). IEEE, pp 2523–2527
37. Popovski P, Simeone O, Boccardi F, Gunduz D, Sahin O (2019) Semantic-effectiveness filtering and control for post-5G wireless connectivity. arXiv preprint: 1907.02441
38. Qu Z, Zhang G, Cao H, Xie J (2017) LEO satellite constellation for Internet of Things. IEEE Access 5:18391–18401
39. Rivest R (1987) Network control by Bayesian broadcast. IEEE Trans Inf Theory 33(3):323–328
40. Shokri-Ghadikolaei H, Fischione C, Fodor G, Popovski P, Zorzi M (2015) Millimeter wave cellular networks: a MAC layer perspective. IEEE Trans Commun 63(10):3437–3458
41. Shokri-Ghadikolaei H, Fischione C (2015) The transitional behavior of interference in millimeter wave networks and its impact on medium access control. IEEE Trans Commun 64(2):723–740
42. Singh B, Tirkkonen O, Li Z, Uusitalo MA (2017) Contention-based access for ultra-reliable low latency uplink transmissions. IEEE Wirel Commun Lett 7(2):182–185
43. Stefanović Č, Paolini E, Liva G (2017) Asymptotic performance of coded slotted aloha with multipacket reception. IEEE Commun Lett 22(1):105–108
44. Stefanović Č, Paolini E, Liva G (2018) Asymptotic performance of coded slotted ALOHA with multipacket reception. IEEE Commun Lett 22(1):105–108
45. Tebaldi C, West M (1998) Bayesian inference on network traffic using link count data. J Am Stat Assoc 93(442):557–573
46. Vilgelm M, Schiessl S, Al-Zubaidy H, Kellerer W, Gross J (2018) On the reliability of LTE random access: performance bounds for machine-to-machine burst resolution time . In: Proceedings of IEEE International Conference on Communications (ICC), May 2018. https://doi.org/10.1109/ICC.2018.8422323
47. Vilgelm M (2019) Random access protocols for massive and reliable machine-to-machine communication. Ph.D. thesis. Technical University of Munich
48. Wu Z, Lu K, Jiang C, Shao X (2018) Comprehensive study and comparison on 5G NOMAschemes. IEEE Access 6:18511–18519
49. Wunder G, Stefanović Č, Popovski P, Thiele L (2015) Compressive coded random access for massive MTC traffic in 5G systems. In: 2015 49th Asilomar Conference on Signals, Systems and Computers. IEEE, pp 13–17
50. Zhu H, Li X, Xu Y, Li X, Liu Y (2012) An energy-efficient link quality monitoring scheme for wireless networks. Wirel Commun Mob Comput 12(4):333–344

Part II
Low-Power Wide Area Networks for Massive IoT

Wireless Communications for Industrial Internet of Things: The LPWAN Solutions

Emiliano Sisinni and Aamir Mahmood

1 The LPWAN Umbrella: A Snapshot of Common Features

The burgeoning *Industry 4.0* paradigm, introduced in 2011 as a German government initiative to improve efficiency in the industry, aims at exchanging and collecting information along the whole life cycle of a product. This information is stored in a repository to create a virtual "digital twin" of the product itself. The digital twin can exploit the valuable information collected all along the product value chain to implement forecasting models, minimizing losses against unexpected events, thus improving the overall business process. Generally, the ensemble of both the physical and cyber components is referred to as a cyber-physical system (CPS). In particular, the operation phase is where *Industry 4.0* meets the Internet of things (IoT), leading to Industrial IoT (IIoT).

For this reason, IIoT is considered an evolution rather than a revolution as for the consumer counterpart. Indeed, low-cost, low-power consumption, battery-powered wireless devices were already available, in the late 2000s, with the advent of the so-called wireless mesh network protocols. In particular, the availability of reliable IEEE802.15.4 radios paved the way to international standards devoted to process automation and utility networks, like the well-known IEC62591 (a.k.a. WirelessHART) and IEC62734 (a.k.a. ISA100.11a). On the other side, wireless in factory automation is still mainly based on customary solutions derived from consumer market technologies such as WiFi (IEEE802.11) and Bluetooth (IEEE802.15.1), due to stringent time constraints.

E. Sisinni (✉)
University of Brescia, Brescia, Italy
e-mail: emiliano.sisinni@unibs.it

A. Mahmood
Mid Sweden University, Sundsvall, Sweden
e-mail: aamir.mahmood@miun.se

N. H. Mahmood et al. (eds.), *Wireless Networks and Industrial IoT*,
https://doi.org/10.1007/978-3-030-51473-0_5

This consolidated scenario started to change recently, due to the widespread diffusion of protocols originally designed for the IoT and the smart things, that contaminated the industrial automation as well. New actors came into play, allowing for multi-km network coverage by mimicking the mobile systems approach. The term coined for addressing such technologies is low-power wide-area network (LPWAN). A survey, carried out by the ON World research firm in 2018 on IIoT topics [1], confirms the interest toward these technologies. In the survey, more than 100 industrial automation vendors, end-users, systems integrators, and service providers were contacted, and 19% of respondents affirmed they were planning the development of an LPWAN network.

It is worth stressing that the main design goals of LPWAN technologies are wide-coverage and low-power operation, resulting in low data rate (usually in the order of few kilobits per second) and high latency (usually in the order of seconds or minutes). Accordingly, LPWANs are well-suited for niche areas, including delay-tolerant machine-type communication (MTC), which typically emphasizes low-power consumption for low-cost devices. Such applications are addressed with the term massive MTC (or mMTC), in contrast to the critical MTC (or cMTC) applications [2], with the latter needing ultra-low latency and ultra-high reliability.

Although the LPWAN family includes several members with different characteristics, the most common features are: *(a)* large area coverage, *(b)* low-power consumption, *(c)* low-cost and, *(d)* scalability [3]. In Fig. 1, a generalized diagram of an LPWAN system is sketched. The network architecture mimics cellular networks; one or more base stations (BS) are the centers of a wireless star network (single-hop topology) and provide connectivity to backend servers using a backhaul network. As better explained in the following, this configuration allows to move most of the complexity in the cloud, where higher computational resources are available, and

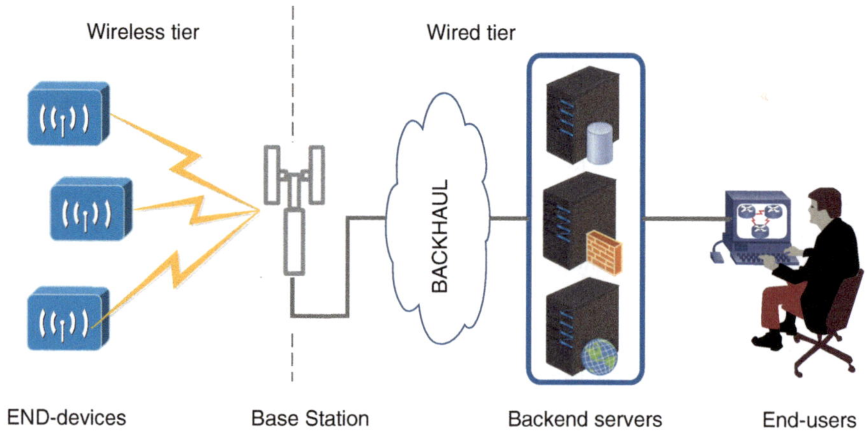

Fig. 1 A generic LPWAN network architecture

permits to reduce energy consumption (wireless routing is not required) and cost (due to the centralized network management).

1.1 Large Area Coverage

LPWANs are generally based on single-hop wireless links; consequently, wide coverage in possible hybrid indoor/outdoor scenarios must be ensured by means of very good receiver sensitivity. For this reason, excluding very few examples (notably, the solutions derived from mobile communications, as NB-IoT, and proprietary solutions, as Ingenu), most of LPWAN technologies operate in the license-free sub-GHz region of the spectrum. One of the advantages is that, compared to other license-free bands, both the attenuation and the multipath fading are lower. Additionally, the radio environment is generally less crowded, resulting in a reduced number of interferences.

The price to pay is the reduced available bandwidth, which limits the overall throughput. When possible, the limited number of physical channels is increased through virtual channels, exploiting some kind of message orthogonalization permitted by advanced digital modulation techniques. The good news is that when narrowband modulations are adopted, most of the noise and interferences can be filtered out, which greatly improves the power budget. The power budget is generally more than 150 dB (even though maximum transmitting power is necessarily reduced for limiting the energy consumption), thus allowing for multi-km links.

1.2 Low-Power Consumption

Smart things (i.e., IIoT end nodes) are typically battery-powered; thus, reasonably long lifetime (in the order of several years for many real-world applications) must be ensured by very-low-power consumption. The wireless mesh approach, underlying WirelessHART and ISA100.11a, is not a viable solution since wasting precious energy in overhearing to other devices and relaying their traffic is not tolerable. For this reason, borrowing the approach pursued in mobile networks, the preferred choice is a star topology. Additionally, the BS is generally connected to a wired infrastructure, and it can always be on, thus easily allowing for event-based node communications (since uplink is of main interest in most of the applications).

Another important knob for controlling energy consumption is the careful choice of the medium access control (MAC) strategy, which must be simple enough to require minimal computational resources (for reducing the cost) but must minimize the collisions as well. For all these reasons, based on the assumption that the nodes transmit sporadically, a common solution is the use of pure ALOHA, a distributed random access MAC protocol in which end devices transmit without performing any carrier sensing. Note that the radios are already equipped with carrier sensing capabilities, so the listen-before-talk (LBT) mechanism (as the carrier sense multiple access with collision avoidance–CSMA/CA–used in most of the short-range wireless networks) can be implemented as well. However, LBT is not effective in dense radio environments [4], thus discouraging its adoption.

An additional mechanism for reducing the consumption is duty cycle limitation, i.e., forcing quite periods among consecutive transmissions. Very often, such a constraint is mandatory to fulfill regional regulatory requirements related to license-free bands.

1.3 Low Cost

The (I)IoT idea is effective only if the cost per device is very low, in the order of a few USD, including the cost for connectivity. For this reason, LPWANs successfully compete with legacy licensed solutions, as the cellular-like communications. LPWAN networks permit new players to implement a cellular network, without the need of a subscription to a third-party provider. Accordingly, the backend infrastructure must be simple as well. The use of the Internet as the backbone, similarly to all the other IoT approaches, goes in the right direction. Another common approach is the use of transparent BSs (often named gateways), for which data coming from (and going to) the field via the wireless link are opaque. By this, their complexity is greatly reduced, and security-related operations for authentication and encryption are moved into the backend.

1.4 Scalability

As previously stated, target applications include mMTC; hence, one of the key requirements of all LPWANs is supporting a massive number of devices sending small chunks of data. Thus, network scalability is a challenge. The spectrum scarcity, as mentioned earlier, must be compensated by using different diversity techniques, including efficient modulations and some form of orthogonal multiplexing.

On one side, centralized backend, able to check the overall network status in real-time, helps in implementing an adaptive selection of channels and data rates. On the other, recall that most of the communications occur from the field toward the BS. Therefore, it is generally preferred to rely on simple distributed mechanisms, e.g., simple message retransmission or a random selection of channel and data rate.

2 Different LPWAN Flavors

Many different technologies are covered by the LPWAN umbrella, including both proprietary and standardized approaches. The solutions that attracted the highest interest are probably NB-IoT, SigFox, and LoRa/LoRaWAN. Some works exist that try to compare these technologies, all sharing similar design goals (e.g., as reported in [5]). In the following sections, basic characteristics of NB-IoT, SigFox, and LoRa are outlined, whereas LoRaWAN solution is discussed in detail in Sect. 3. Although NB-IoT seems to offer the best coverage, the most diffused LPWAN, well-accepted by both the academia and the industrial world, is LoRaWAN, for its flexible backend implementation and for permitting to manage very diverse applications.

2.1 NB-IoT

In the past, the 3rd Generation Partnership Project (3GPP) group proposed some amendments in mobile networks to encompass requirements of low-power and wide-area networking applications. MTC has been supported in Long-Term Evolution (LTE) since Release 10, in which the 3GPP defined a new profile, called CAT-0, which reduced system complexity offering a maximum data rate of 1 Mbps, but keeping the same 20 MHz maximum system bandwidth of regular services. In Release 13, two new categories, CAT-M1 and Narrowband IoT (NB-IoT), have been defined, which lowered the available bandwidth to 1.4 MHz and 200 kHz, respectively. NB-IoT communication coexists with LTE, operates in licensed frequency bands (e.g., 700, 800, and 900 MHz), and occupies a single resource block of regular LTE transmission.

In particular, NB-IoT specifications have been developed according to the following main design goals [6]:

- enhanced indoor coverage, obtained by a maximum coupling loss (MCL, the metric chosen by the 3GPP to evaluate the radio coverage and defined as the ratio of the power at the transmitter to the sensitivity at the receiver) in the order of 164 dB,
- support for a large number of low-demanding nodes per single BS. The goal was to manage smart metering applications, considering the estimated household density in London, where there are 1517 households per square km, and Tokyo, where there are 2316 households per square km. In both cases, the inter-cell distance is in the order of 1.7 km,
- support for low-throughput nodes; indeed, the MCL of 164 dB, should ensure an end-user data rate of 160 bps for both the uplink and downlink exchanges,
- support for low-complexity and low-cost implementation of devices, as required by typical IoT applications; in turn, better power efficiency is achieved, thus extending the lifetime of battery-powered devices,

- capability of delivering exception reports in less than 10 s for at least 99% of the devices.

NB-IoT defines three different usages: stand-alone (reusing GSM frequencies bands), guard-band operation (leveraging on unused resource blocks within an LTE carrier's guard band), and in-band operation (leveraging on resource blocks within an LTE carrier). The NB-IoT amendment can be considered as a new air interface tailored for simple devices that can be connected to a network operator/provider exploiting the well-established LTE infrastructure. It uses the single-carrier frequency division multiple access (FDMA) in the uplink and orthogonal FDMA (OFDMA) in the downlink, and the adopted modulation is the quadrature phase-shift keying (QPSK) [7]. The data rate is limited to 200 kbps for the downlink and 20 kbps for the uplink, with a maximum payload size of 1600-byte per message. Standardization of NB-IoT continued in Release 14 and 15 to include localization methods based on observed time difference of arrival, multicast services (e.g., for over-the-air update), as well as latency and power consumption reduction [8].

The NB-IoT protocol stack is split into two planes, the control and the user one. The access stratum (AS) layer occupies the level 2 of the stack, and it is in charge of transporting data over the wireless connection and managing radio resources. The data transportation is offered via the packet data convergence protocol (PDCP), which determines the 1600-byte payload limit, whereas the radio resource management provided by the radio resource control (RRC) protocol aims at minimizing signaling by suspending/resuming the operation of the user plane. Above AS, there is non-access stratum (NAS), which conveys non-radio signals exchanged among the user equipment (UE) and the core network. NAS handles authentication procedures, performs security controls, and manages mobility and bearer, while the routing is performed using Internet protocol (IP). The UEs access the medium according to the contention-based random access channel (RACH) procedure. First, a preamble is transmitted; if the transmission fails, the preamble is re-transmitted for a maximum number of retries, which depends on the desired coverage enhancement (CE) level; subsequently, the next CE level is adopted. Once the preamble is correctly received, the associated random access response is received by the UE. Finally, the contention resolution process is started by the transmission of a scheduled message, and it is concluded when the user equipment receives the associated contention resolution message.

2.2 SigFox

NB-IoT offers relatively high data rates, high-power BSs, and advanced features for routing and multicasting. On the opposite side, we find SigFox proprietary solution, which exploits so-called ultra-narrowband (UNB) communication. Although SigFox operates in unlicensed ISM bands (e.g., 868 MHz in Europe, 915 MHz in North America, and 433 MHz in Asia), its business model is similar to the one used by mobile operators. SigFox deploys its BSs equipped with proprietary software-defined radios (SDRs). An IP-based network is used to connect the BSs to the backend servers, representing Sigfox Cloud. End-users must integrate their applications with the SigFox Cloud, by means of the two available methods: the "Callback API" and the "REST API." In particular, Callbacks are HTTP requests implementing one-way only notification messages, thus permitting to retrieve messages from the devices. On the other hand, the REST APIs implement bi-directional HTTP data flows used for administrating the behavior of the devices and implementing related services.

The SigFox protocol stack includes the radio-frequency layer, implementing the modulation scheme and providing services to the data link layer performing the medium access control and error detection. Finally, the users' requirements and specifications are managed at the application layer. The adopted modulation is binary phase-shift keying (BPSK) with very limited bandwidth (100 Hz). When devices operate in the EU, the unlicensed region between 868.180 and 868.220 MHz is divided into 400 different channels (including 40 reserved). Unfortunately, the maximum throughput is 100 bps. This bandwidth choice has been dictated by the needs of minimizing the noise level (thus obtaining exceptional sensitivity), the end device cost, and the power consumption. End-device design cost is reduced since precise and stable oscillators are not required [9], thanks to processing carried out in the BSs, as better explained in the following.

Sigfox uses a random frequency and time division multiple access (RFTDMA) strategy to transmit frames. RFTDMA allows the nodes to access the medium randomly both in time and frequency, without performing any contention-based mechanisms. Although it can be related to the pure ALOHA protocol, it allows choosing the carrier frequencies from a well-defined continuous interval (let's call it B), and not from a predefined discrete one. At the BS side, the SDR listens at the full available bandwidth B instead of recognizing the actual carrier used by the transmitting device. Thus, message preambles are continuously searched within the whole B band, and, once recognized, the whole message frames are demodulated/decoded. Such an access scheme, on the one hand, limits the power consumption but possibly introduces interference among active nodes, especially if other co-located wireless systems are in the field. For instance, in [9], it is reported that in order "to ensure a high level of performance (e.g., PER below 10%), the highest number of nodes which can communicate at the same time is approximately 100, if the number of available channels is 360. At the same time, with the increase in the number of sensors, an avalanche effect is triggered, which drastically lowers the performance level."

A SigFox frame consists of a 4-byte preamble and a 2-byte frame synchronization field, followed by a 4-byte device identifier, up to a 12-byte payload, a variable length Hash code for authentication, and a 2-byte cyclic redundancy check (CRC) field for error detection.

Initially, uplink-only communication was supported, but later asymmetric bidirectional communications have been permitted. Subsequently, two-way communication has also been permitted, in which a downlink payload can be transferred after a node explicitly sets a request flag within an uplink. The communication is always started by nodes, thus minimizing the time spent in listening the medium. The maximum number of uplink messages is 140 messages per day, while the message payload length for each uplink is limited to 12 bytes, as previously stated. Moreover, up to four downlink messages per day with a maximum user payload of 8-byte are permitted (and obviously, acknowledgment of every uplink message is not supported).

Communications reliability is provided by retransmissions, featuring time and frequency diversity. Each end-device message is transmitted three times by default over different frequency channels. The SDR approach at the BS allows the simultaneous reception of messages in multiple channels so that transmission frequency is randomly chosen.

2.3 *LoRa*

The LoRa radio technology has been originally introduced (and patented) by Semtech. LoRa adopts chirp spread spectrum (CSS) modulation in order to code, with a single chirp frequency trajectory, SF (spreading factor) bits. Consequently, each transmitted symbol is SF bits long. The choice of CSS has been motivated by the very good correlation properties shown by chirp symbols; virtual channels and adaptive data rate strategy can thus be easily achieved by transmitting messages with different SF values. In detail, the chirp bandwidth BW is fixed (BW – 125, 250, or 500 kHz), while the chirp duration can be calculated as $T_C = 2^{SF}/BW$. Accordingly, a higher SF means a lower data rate but a better noise immunity (thanks to additional processing gain). Forward error correction mechanism has been considered to increase robustness again noise and interference, but the price to pay is a reduced throughput; however, several coding rates (CR) can be specified in the range from $CR = 4/5$ to $CR = 4/8$.

3 The LoRaWAN Solution

A complete communication solution has been designed around the LoRa physical layer, which is known as LoRaWAN. Regarding the physical layer (i.e., the previously described LoRa), LoRaWAN specifications add other constraints that depend on the regional parameters. For instance, when operated in Europe, the allowed bandwidth is $BW \in [125, 250]$ kHz and $SF \in [7..12]$. Accordingly, a single physical channel can host up to six quasi-orthogonal virtual links (one per SF value [10]). Unfortunately, due to the different symbol duration, the actual data rate ranges from about 300 bps to 11 kbps. The message payload has a maximum length in the range of 51-byte (at $SF = 12$) to 242-byte (at $SF = 7$).

3.1 The LoRaWAN Architecture

LoRaWAN specifications are drafted by the LoRa alliance, which includes the device manufacturers, end-users, and research institutions. In particular, LoRaWAN provides upper layers of the protocol stack above the radio, defining the data link layer, which leverages on the pure ALOHA medium access strategy (despite clear channel assessment is somehow permitted). Regarding the network level, the network topology is hybrid wireless and wired star-of-stars, consisting of multiple BSs (gateways) tunneling into/from a wired backhaul/backbone uplink and downlink messages. The main design goals are reduced complexity, thus lowering implementation and maintenance costs, on the wireless side. Target applications are based on uplink only, whereas downlink, despite being permitted, should be limited to increase efficient bandwidth utilization [11]. Concerning the gateways, it is interesting to note that each one executes a software (the so-called packet forwarder) to forward messages using an implementation-specific protocol. In particular, the gateway only tunnels LoRa frames, i.e., the data are opaque.

According to the previous description, each LoRaWAN comprises two tiers: one includes wireless connectivity to end devices, and the other is the backend. All the network management procedures are carried out in a centralized way in the backend. In more detail, the network reference model described in the LoRaWAN specifications consists of two or three different kinds of servers (depending on the specification release): network server (NS), application server (AS), and join server (JS), as shown in Fig. 2. Note that the implementation details are out of the scope of the specifications, while only the operations to be carried out are described.

NS is the logical entity, implementing the center of the star topology. It is in charge of checking the frame format and authentication, and providing acknowledgments if required. Additionally, it manages the LoRaWAN data link protocol features, e.g., as data rate adaptation strategy. Once authenticated, NS forwards

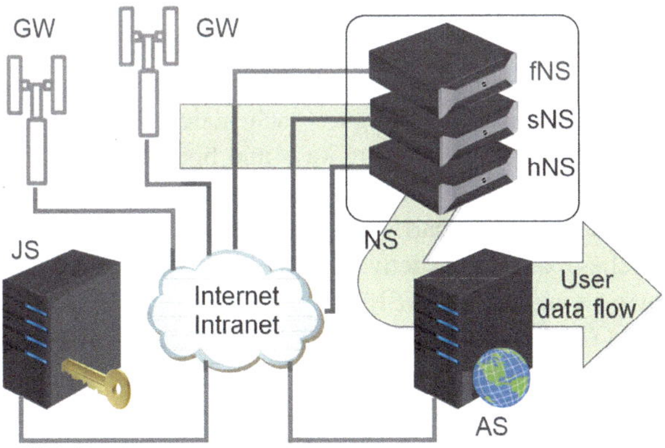

Fig. 2 LoRaWAN network reference model

an incoming uplink to the appropriate AS and queues downlink from any AS to deliver the useful payload to the proper end-device. If a JS is present, Join-request and Join-accept messages (by which the affiliation procedure is carried out) are forwarded to the end-devices and to the JS by the NS, respectively. JS, introduced in the LoRaWAN Release 1.1, is in charge of managing the affiliation of end-devices in a secure (i.e., encrypted) way. Finally, since the application-level protocol is not described in the specifications, AS has to implement it for delivering data to the final user. A comprehensive evaluation of delays along the whole route from the end-device to the end-user is provided in [12]; the capability to access the same end-node data in multiple end-user locations is described in [13].

Roaming is supported as well; accordingly, three roles exist for an NS: serving NS (sNS), home NS (hNS), and forwarding NS (fNS). Only sNS controls the data link layer behavior of the end-device, while hNS is connected to the AS, and several additional fNS can be connected to other gateways. If sNS does not change, passive roaming is carried out. On the contrary, when handover roaming is enabled, the end-device is managed by the visited network, though user data are still forwarded to the original hNS. Specifications do not provide any details of the protocols to implement interfaces between fNS and gateway, and between hNS and AS. Only the communications among NSs and JS-NS interfaces are described. In the latter case, the HTTP protocol must be used, encoding the payloads using JSON objects. As a matter of fact, several different proprietary implementations may exist. Such an architecture makes it easy to decouple the owner of the infrastructure from the owner of the data, enabling new business scenarios.

3.2 Security

Both IoT and IIoT rely on the fundamental principles of "connectivity to all" and "connectivity with all"; as a consequence, security issues are the main concern. LoRaWAN specifications already include mechanisms for ensuring confidentiality, integrity, and availability (CIA) since the very first release [14]. A preliminary step performed by the manufacturer of the end-device, or by the operator in charge of the commissioning, requires to configure each node with the root keys that will be used for deriving actual enciphering keys and both the device and the network identifiers. This information is needed to proceed with the personalization and activation procedures, usually carried out before the affiliation. In particular, two different approaches exist for the activation, namely, over-the-air activation (OTAA) or activation by personalization (ABP).

The preferred OTAA procedure is a flexible and secure way of obtaining session keys from the backend servers. The node first transmits a *join_request* message, received by the NS and further processed with information coming from the JS (if any). The JS has been introduced in LoRaWAN v1.1 to allow a complete separation between network and application domains, each one leveraging a different key. Once the message is validated, a *join_accept* message is sent back as a confirmation.

The enciphered payload of the *join_accept* is used by the end-device to derive the actual session keys from the root keys. The specifications require these keys to be device-related; thus, possible disclosure should affect only the compromised device without affecting the rest of the network. In the ABP procedure, the session keys can be preventively stored in the device by the manufacturer, which should pay attention to track and match a unique device identifier, as the DevEUI, with the keys themselves. On the other hand, if the device has not been preventively configured by the hardware manufacturer, it is responsibility of the end user application to manage and deploy these keys on the devices in the field. Nevertheless, in both cases, the session keys stored in the node must match the corresponding keys stored in the backend servers. Thus, the ABP operates with a reduced security level and should be considered only for preliminary/debugging activities.

In particular, according to LoRaWAN specifications, several session keys are defined and used. At the network level, they are used by NS for generating and checking the message integrity code (MIC) or to encipher the MAC commands. User data, on the contrary, are protected at the session-level using a session key used for de/en-ciphering the application payload. Resuming, confidentiality and integrity of MAC commands are carried at the network level, whereas user data confidentiality is implemented at the application level, as shown in Fig. 3. Obviously, security on the backend is mandatory as well, but implementation is left out to the implementer.

4 LPWAN Exploitation for Industrial Applications

As already stated in Sect. 1, the Industry 4.0 paradigm requires the interaction of devices and human operators in order to improve the overall process efficiency and automation processes. As a consequence, wireless communication systems are of

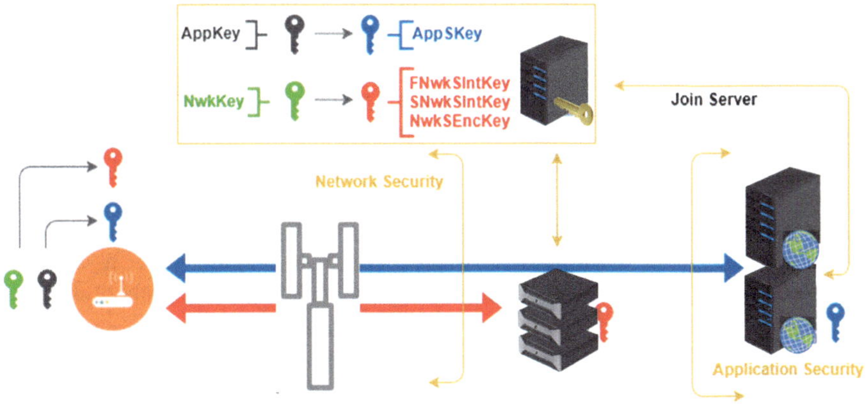

Fig. 3 LoRaWAN security model

main importance, thanks to their inherent scalability and non-invasiveness. In this respect their introduction, LPWANs have been seen as promising technologies for those non-critical industrial applications requiring broad coverage but tolerating low bandwidth and delays. As a matter of fact, the LPWAN simplicity complemented by good energy efficiency is truly interesting for cost-effective IIoT deployments. Similarly, the robustness with respect to multipath and fading is appreciated in industrial environments. Not surprisingly, manufacturers started selling rubberized versions of LPWAN devices purposely designed for the harsh industrial domain.

Comparing the LPWAN architectures with the wireless fieldbuses, purposely designed for control process applications, as the IEC62591 (WirelessHART) or the IEC62734 (ISA100.11a), several similarities can be highlighted. WirelessHART defines a Network Manager logical entity that is in charge of managing the node behavior; the ISA100.11a specifications define a System Manager, performing the same functionality. It is evident that LPWAN backend servers, such as the LoRaWAN Network Server, can provide equivalent capabilities. However, LPWANs generally lack important features, notably a synchronization mechanism [15]. As a consequence, any time information provided at the backend level results in relatively poor accuracy.

4.1 From Pure ALOHA to Scheduled MAC Protocols

Wireless industrial communication systems generally leverage on guaranteed access strategies, in which nodes are granted access to the medium according to a predefined resource assignment to easily ensure network timeliness. This strategy is adopted in many popular polling- or time division multiple access (TDMA)-based protocols. For instance, the aforementioned WirelessHART and ISA100.11a both use time-slotted channel hopping (TSCH) strategy, which adds frequency diversity (i.e., the adoption of different communication channel on a slot basis), to improve interference immunity and perform simultaneous transmissions on multiple (orthogonal) channels. Indeed, industrial applications are generally based on a limited, fixed number of devices deployed at the beginning of the plant lifetime, so that scheduling can be effectively carried out. On the contrary, LPWANs have been designed having scalability in mind, with applications possibly including thousands, if not millions, of devices. For this reason, as previously highlighted, the most common approach is pure ALOHA, which is prone to message collisions but does not require any a priori knowledge, thus minimizing implementation complexity.

Unfortunately, the performance of pure ALOHA is quite poor in crowded environments; it is well-known that the maximum throughput is about 18% of a perfectly synchronized solution. For mitigating the efficiency losses, slotted ALOHA and TDMA mechanisms have been overlayed to the original LPWAN. In the slotted access, the time is divided into slots (having a fixed length according to the maximum message length), and the nodes can access the channel only at the beginning of a slot. In the TDMA scheme, a scheduler has to be implemented in the backend to assign transmission intervals (and transmission frequency/data rates) to end-devices, thus virtually avoiding collisions. In particular, the openness of the LoRaWAN standard makes it attractive for researchers, as confirmed by the several publications aimed at demonstrating the superior performance obtainable using slotted ALOHA or scheduled mechanisms (e.g., consider [16–22]). Since the transmission policy is changed only, full compatibility with the original specifications is generally guaranteed.

The primary hypothesis for both slotted ALOHA and TDMA approaches is a common sense of time shared by all the network nodes; for this reason, some synchronization mechanisms must be implemented in order to minimize the impact of local clock drift. The MAC layer in LoRaWAN (and generally speaking in all LPWANs) supports neither synchronization nor low-level timestamping and must be purposely added. Instead, the synchronized downlink is permitted only based on reference time dissemination using downlink beacon messages to identify a repeating superframe arrangement. In [22], the LoRaWAN end-device takes a local timestamp right before sending a synchronization request. Once the gateway receives this request, the corresponding timestamp in the GW (global) time reference is taken as well and retransmitted to the end-device in the following synchronization acknowledge message. Given such a pair of timestamps, the offset between the node (local) and the GW (global) clocks is estimated. By collecting several offsets estimation, it is possible to evaluate the clock skew to correctly identify borders of slots in the local time domain.

4.2 Extending the LPWAN Coverage

As stressed in previous sections, most LPWAN solutions adopt a single-hop architecture, which is a star topology where nodes are directly connected with one (or more) BSs, thus greatly simplifying the overall system and permitting centralized management. However, in crowded environments, this single-hop massive channel access may pose additional challenges in terms of offered quality of service (QoS), limiting the overall network reliability, scalability, and flexibility. Most of the issues come from the very channel access mechanism, which, as stated in the previous Sect. 4.1, is pure ALOHA.

However, once more complex but powerful schemes as TDMA are implemented (providing time synchronization), multi-hop topologies can be easily implemented [23–25]. The advantage is that lower transmission power is permitted, still ensuring large-area coverage, at the expanse of increased latency (that is not a strict requirement in most of LPWAN target applications). Many researchers worked on this approach, as confirmed by the available literature. It must be stressed that protocol stack complexity is greatly increased as well, so that many solutions try to balance the network performance preferring the star topology and moving into the multi-hop only if QoS is degrading. Indeed, in order to preserve the node lifetime (i.e., reducing the overall power consumption, largely using deep-sleep states), the following technologies are generally required:

- network-wide time synchronization,
- TDMA-like channel access, for avoiding collisions among relaying nodes,
- adaptive transmission power levels,
- simple, flexible, and scalable joining procedure,
- energy-aware, adaptive, and resilient routing protocol.

It is interesting to note that the availability of two receive windows for downlink communications provided by the LoRaWAN solution paves the way for a relayer-based approach for transparently doubling the coverage in both uplink and downlink directions. In particular, in [26], it is suggested the exploitation of a so-called e-Node, which transparently forwards copies of any incoming uplink, leveraging on the deduplication mechanism natively implemented by NS. On the contrary, on the downlink, the message is forwarded in the second receiving window so that if the end-node is able to immediately receive the message in the first window, overhearing is avoided; otherwise, if direct link is not permitted due to the actual link budget, a second opportunity is naturally provided.

4.3 Limitations and Future Directions

Since LPWANs have been originally designed for applications tolerating low-data rate and sporadic communications, their usage can be easily extended to condition monitoring of industrial equipment, facilities, and environments. The recent research activity is mainly focused on overcoming limitations imposed by the simple but not-so-efficient MAC protocols, showing that performance comparable with standard solutions currently used in process control (e.g., WirelessHART and ISA100.11a [20]) can be obtained. On the contrary, there are no real advantages in managing cyclic communications with refresh time shorter than 1 second. LPWANs are the quick-and-dirty solution for filling the current gap in mMTC communications waiting for large deployments of 5G networks. It has to be considered that Rel-16 of 3GPP specifications is focused on IIoT-related enhancements, and several activities of the 3GPP are ongoing for supporting time-sensitive communication integration and deterministic applications [27]. However, the two approaches can coexist, and possible adoption of 4G/5G technologies as LPWANs backhaul/backbone has already been proposed [28].

5 LPWAN Simulations

Performance evaluation of any wireless technology is important to understand its suitability against key performance indicators laid by applications. It also plays a vital role in the design and optimization of the wireless links and the overall network performance. Usually, evaluation is required both from the link-level and the system-level perspective, often carried out by theoretical modeling, simulations, or empirical measurements. All these tools are employed, although independently, but often empirical measurements in simple scenarios are used as a benchmark to perform full-scale simulations and fine-tune analytical models. In LPWAN systems, designed to collect data from a massive number of devices, system-level performance of the protocols is imperative to metric their performance in terms of coverage, scalability, and reliability. Also, these technologies often provide various degrees-of-freedom in the selection of MAC/PHY layer parameters. Therefore, the design and optimization of the network performance require fine-tuning these parameters under dynamic interaction among the devices at the protocol level, which is only possible by extensive simulations.

In contrast to simulations, test-beds cannot scale due to cost and experimental repeatability issues. Additionally, mostly measurements (to find path loss, packet loss rate) are carried out with a limited number of devices communicating with the network at a time. However, the network behavior is different when many devices use the network simultaneously, and therein the interference becomes the primary source of packet loss and coverage reduction (outage) [29, 30]. In this case, the deployments are though helpful to predict the coverage boundaries, e.g., for SFs in LoRa, but they cannot help to analyze the network scalability. Meanwhile, the theoretical models usually study a limited aspect of the physical and medium access layer using assumptions, limitations, and abstractions that still need extensive validation through simulations. Moreover, MAC layer features and enhancements, such as new MAC design, adaptive data rate (ADR) scheme, power-control algorithms, SF allocation scheme, the mutual interaction between downlink and uplink traffic, and retransmission scheme, are difficult to model analytically and usually lag behind the simulation-based analysis.

Industrial applications demand timeliness and reliability, whereas the LPWAN technologies are not feasible for real-time monitoring unless low-scale deployment is of interest. Instead, it is suitable for smart-city applications, metering, and logistics (tracking). Despite the fact, there are many studies that show that the design of a MAC is feasible to support lazy-control and soft-realtime applications as discussed in Sect. 4.1. Although there is no extensive simulation framework for evaluating the industrial use cases and all these studies are concluded based on the limited test-beds, we summarize the simulation environments available in the literature that could be exploited to evaluate the full potential of these proposals.

In what follows, the discussion is focused on the available design and validation tools for LoRaWAN, which, as previously stated, is currently the most diffused example of LPWANs. Considered tools range from radio planning applications to link- and system-level simulation environments. In particular, their designs, abstractions, and details of implementation are briefly sketched out. In the end, we highlight the potential research directions, especially for designing a unified solution to evaluate the LoRa network performance. For details on test-beds- and theoretical-based studies, the reader can refer to [31].

5.1 Radio Planning Tools

Radio network planning, concerned with the topographic maps and a precise propagation modeling, is important for network planning and deployment to provide adequate coverage in the area of interest. A measurement-based analysis of coverage is performed in [32, 33], and a generalization on propagation models is achieved. However, it is vital to have a radio planning tool that takes into account the topographic information of the deployment area and, together with propagation models, provides coverage information rapidly. Two prominent radio planning tools are from CloudRF [34] and ATDI [35].

CloudRF is an online service for modeling radio propagation, with its core in open source tool *SPLAT* – a terrestrial RF path and terrain analysis tool. CloudRF supports LoRa with a low receiver threshold of -140 dBm with a dedicated sub-noise floor color schema. Also, it uses various propagation models (including but not limited to Irregular Terrain Model, Okumura-Hata, Cost231-Hata, Ericsson 999, ECC33, ITU-R P.525) suitable for the UHF spectrum. Incorporating high-resolution LIDAR data enables to plan in urban areas with a high degree of precision. For wide-area plots, there is 30 m terrain data worldwide, which includes clutter like buildings and vegetation.

ATDI provides a commercial radio coverage prediction and planning tool for LoRa that takes into account the urban buildup and building impact (shadowing, absorption) predictions for indoor device deployments.

5.2 Link-Level Simulations

The link-level simulations are crucial to analyze the performance of LoRa modulation, especially under the impact of self-interference. In [36], a link-layer simulator, known as PhySimulator, is designed to analyze the impact of self-interference in LoRa. PhySimulator investigates the performance of a reference device receiving a useful LoRa signal in the presence of interfering signals using the same or different SFs. Especially, it quantifies the signal-to-interference ratio (SIR) thresholds for which interference rejection of other LoRa signals does not work for all combinations of SFs. The simulator exposes the non-orthogonality of SFs; that is, it shows that the collisions between packets using different SFs can indeed cause packet loss [36]. On the other hand, a BER model for LoRa modulation in AWGN channels is proposed in [37], which is then further used in system-level simulations.

5.3 System-Level Simulations

System-level simulations (SLS) are usually performed to determine the overall network performance. SLS allows the system designers to evaluate the impact of various parameters such as node density or traffic loads, interaction between uplink and downlink traffic, macrodiversity, and medium access protocols on the system performance. Meanwhile, it allows the abstraction of many link-level details that would otherwise add time complexity. In LPWAN systems, it is infeasible to conduct a testbed-based scalability analysis, which is also reproducible, with a high density of devices. In this respect, many system-level results, covering different aspects of LoRaWAN system, are presented in the literature using custom-built simulators such as ns-3 compatible LoRaWAN module [37, 38], python-based [39], and OMNeT++-compatible [40]. In these simulators, various components and algorithms for link and medium access layers are implemented at different scales, while the generic components of a LoRaWAN-specific simulator can be outlined as follows:

1. *Propagation model:* takes into account the path loss model, Fading, and shadowing phenomenon.
2. *Reception model:* defines the packet success based on bit error rate (BER) curves or signal-to-noise ratio (SNR) and SIR thresholds.
3. *Interference model:* considers the effect of concurrent transmissions on a reference transmission. Specific to LoRa, capture effect plays an important role where both the time and power capture are relevant. In addition, both the co-SF and inter-SF spreading interference need to be considered.
4. *Medium access layer:* although ALOHA is a de facto medium access choice, slotted-ALOHA and LBT/CSMA are also appealing to enhance scalability as discussed in Sect. 4.1.
5. *Uplink/downlink traffic:* to transfer sensory information from the field devices to the gateway (i.e., uplink communication) is of primary interest in LoRaWAN, it also provides support for downlink for various functions, e.g., network management (handshaking, network joining, exchange of security keys), and adapting communication parameters.

In the following, we outline the three main flavors of system-level simulators for LoRaWAN.

LoRaSim

LoRaSim [39] is a python-based discrete-event simulator, which is designed to analyze the scalability of a LoRa network under periodic uplink only traffic. It supports the simulation of scenarios with multiple gateways and directional antenna; however, it does not support the downlink traffic. For uplink traffic, the devices can

Algorithm 1 Pseudocode for LoRaSim packet collision model

Require: Received signal with spreading factor s, on channel h,
Set of overlapping signal $\mathcal{I}(s, h)$,
Co-SF SIR threshold $\delta(s)$.

```
SIR_success = True;
for i ∈ I(s, h) do
   if Preamble condition satisfied then
      if Signal Power / Interferer Power(i) < δ(s) then
         SIR_success = False;
      end if
   else
      SIR_success = False;
   end if
end for
return SIR_success
```

be configured to use any possible value of transmission parameters (SF, frequency channel, bandwidth, coding rate, and transmission power).

In LoRaSim, the successful reception of the uplink packets depends on the selected transmission parameters and multiple other factors, including distance-dependent path loss, fading, packet collisions, and receiver sensitivity of the devices. LoRaSim uses the long-distance path loss model and log-normal shadowing. The packet collision model in LoRaSim is based on two main assumptions: (a) two transmissions in orthogonal channels (i.e., transmissions at different frequency channels or different spreading factors) do not collide; (b) in non-orthogonal channels (i.e., using same channel and SF), a collision is marked when two packets overlap in time; however, the stronger of the two packets is still decoded successfully given that the SIR difference between the packets is more than 6 dB and at least 5 symbols in the preamble are detected. The collision model mainly considers the time- and power-capture effect of LoRa as experimentally characterized in [39]. A pseudocode of the packet collision model is given in Algorithm 1.

There are many simulators that are derived from LoRaSim including LoRaEnergySim [41], LoRaWANSim [42], and LoRaFREE [43]. LoRaEnergySim includes an energy consumption model missing from LoRaSim, while LoRaWANSim extends LoRaSim to extend support for ACKs and downlink reception. LoRaFREE incorporates the impact of imperfect orthogonality of spreading factors, and the duty cycle limitation at both the devices and the gateway. Moreover, LoRaFREE supports bidirectional communication by adding the downlink support and the retransmission strategy for confirmed uplink transmissions. It also provides energy consumption profiling.

FLoRa

FLoRa [40], which stands for Framework for LoRa, is a simulation framework to perform end-to-end simulations of LoRaWAN networks, which includes the accurate modeling of the backhaul link to NS. FLoRa is based on the OMNeT++ network simulator and uses components of the INET framework. It allows the creation of a LoRa network using the modules as LoRa nodes, gateway(s), and a network server. Application logic can be created as an independent module, which is connected to the network server. The other salient features include (a) dynamic management/configuration of parameters using ADR [40]; (b) node-wide collection of energy consumption statistics, where the time and power collision model is the same as in Algorithm 1; and (c) support for multiple gateways.

ns-3 LoRaWAN Module

LoRaWAN module is implemented in ns-3 by multiple independent researchers, where ns-3 is a generic open-source network simulator. The two prominent ones are reported in [37] and [38], which support a packet collision model with co-SF and inter-SF interference and a LoRa network with multiple gateways. The ns-3 module in [37] supports both the confirmed uplink and the downlink traffic via a simple network server, while the module in [38], although originally lacked the downlink traffic as well as confirmed uplink traffic, now supports both.

In LoRaSim and the other simulators (except LoRaFREE) derived from it, the collision model assumed perfect orthogonality between SFs. However, the ns-3 modules consider the same channel transmissions over a different SF as interference. The main difference between the two is how the link-layer performance, i.e., collisions, is modeled. In [37], a BER model in additive white Gaussian noise (AWGN) channel is developed based on Matlab simulations. The BER model considers the interference from different SFs by calculating instant SNR values. On the contrary, in [38], the outage condition is based on the SIR, which is calculated for each spreading factor using the cumulative interference. Each interference power is weighted by the amount of overlap with the useful signal. A transmission is successful only if the SIR calculated independently for each spreading factor is above the minimum threshold. The underlying algorithm is given in Algorithm 2.

In ns-3, other than implementing standard ALOHA-based MAC, CSMA and p-CSMA schemes with LBT are implemented and evaluated in [44] and [45], respectively. Table 1 summarizes the salient features of the simulators reported in the literature.

Algorithm 2 Pseudocode for ns-3 packet collision model

Require: Received signal with spreading factor s, on channel h
Set of overlapping signal $\mathcal{I}(s, h)$
Co-SF and Inter-SF SIR threshold $\delta(s, s')$

```
SIR_success = True;
for s' = {7, ..., 12} do
    for i ∈ I(s', h) do
        Cumulative Interference Energy(s') += Time Overlap(i) · Interferer Power(i);
    end for
    if Signal Energy / Cumulative Interference Energy(s') < δ(s, s') then
        SIR_success = False;
    end if
end for
return SIR_success
```

5.4 *Stress-Test Tools for Network Server*

The simulators discussed earlier are mainly concerned with the testing of the LoRa PHY layer and LoRaWAN protocol. Therefore, the LoRa network up to the gateways is only being evaluated by SLS simulators. There are other unique tools that are designed to test the scalability of the network server (NS). In this respect, two tools that stand out are Mbed LoRaWAN Stack Simulator [46] and LoRahammer [47].

Mbed simulator provides a virtual environment to run LoRaWAN stack without a PHY layer. It uses a fake LoRa radio driver, yet allows to measure the performance of the NS functionality. To do this, the fake radio intercepts the packet (to get the encrypted packet and select data rate and frequency) whenever the LoRaWAN stack wants to drive the radio. The packet is delivered directly to NS, which cannot differentiate a fake radio from the real one. This approach includes two-way data, acknowledgments, and OTAA joins function. This notable infrastructure helps engineers develop, test, and deploy LoRaWAN devices.

LoRahammer is designed to perform stress testing of NS. In a large IoT network, NS must handle millions of messages per second. In order to test the NS design capability to handle these messages in a timely manner, LoRahammer simulates the behavior of traffic from a large infrastructure with massive devices.

Figure 4 shows a comparison of the Mbed LoRaWAN simulator and LoRahammer with respect to a real-life LoRaWAN network.

Table 1 Salient features of the LoRaWAN system-level simulators (SLS)[a]

Simulator	Interf. model[b]	Capture effect	MAC	Ack	Traffic dir.	ADR	Energy	Backend sim.[c]	Dev. env.[d]
LoRaSim [39]	Co-SF	Alg. 1	ALOHA	✗	U[e]	✗	✗	✗	Python
LoRaEnergySim [41]	Co-SF	Alg. 1	ALOHA	✓	U	✓	✓	✗	Python
LoRaWANSim [42]	Co-SF	Alg. 1	ALOHA	✓	U/D[f]	✓	✓	✗	Python
LoRaFREE [43]	Co/inter-SF	Alg. 1	ALOHA	✓	U/D	Partial	✓	✗	Python
FLoRa [40]	Co-SF	Alg. 1	ALOHA	✓	U/D	✓	✓	✓	OMNeT++
NS-3 [37]	Co/inter-SF	BER	ALOHA	✓	U/D	✗	✗	✗	C++
NS-3 [38]	Co/inter-SF	Alg. 2	ALOHA	✓	U/D	✓	✓	✗	C++
NS-3 [44]	Co-SF	Alg. 1	CSMA	✗	U	✗	✓	✗	C++
NS-3 [45]	Co-SF	Alg. 1	p-CSMA	✗	U	✗	✗	✗	C++

[a] No LoRaWAN SLS supports mobility
[b] Interference model
[c] Backend simulation
[d] Development environment
[e] Uplink
[f] Downlink

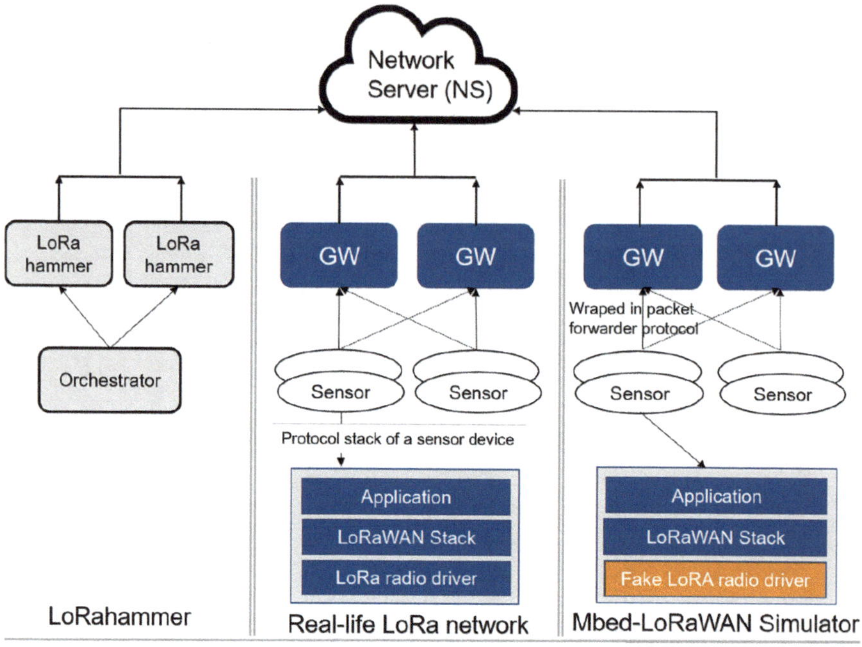

Fig. 4 Working principle of Mbed simulator and LoRahammer for testing a Network Server (NS)

5.5 *Limitations and Future Directions*

There is a lack of system-level simulation environments for LPWANs[1] except for LoRaWAN. The simulators for LoRa/LoRaWAN are growing rapidly with the adoption and maturity of the technology, but there is still a lack of a comprehensive framework that covers all important aspects to understand the potential of the technology fully; a few examples are:

- All available simulators use simplified path-loss, fading, and shadowing, while it is required to build a system-level simulator on realistic propagation models with digital terrain models, clutter, and buildings.
- Many scheduled-MAC protocols for industrial use cases and scenarios are proposed and tested in limited details, but a detailed system-level analysis is needed.
- Mostly the simulation environments consider class-A devices; however there is a need for a simulation setup with heterogeneous device roles

[1]A Sigfox simulator can be found at https://github.com/maartenweyn/lpwansimulation but not many details are available.

- There is no simulator which models the device mobility. LoRa is susceptible to mobility, which worsens the performance as it cannot cope with the network dynamics. Therefore, it is important to incorporate mobility models in the simulators.

Considering these requirements, there must be an effort to build a simulation framework, where the various inputs from the research community can be integrated (as a module) and tested in a unified fashion such that each new proposal could be compared fairly and squarely.

6 Conclusions

In this chapter, we presented an overview of emerging LPWAN technologies, and discussed their potential for industrial IoT, especially for delay-tolerant monitoring applications demanding wide-coverage and low-power operation of devices. We explained the basic properties and protocol stack functionalities of different LPWAN flavors, including NB-IoT, SigFox, and LoRa/LoRaWAN, and shifted our focus to LoRaWAN for its predominant position in the industry/research. We discussed the limitations of LoRaWAN medium access and indoor coverage and established directions for future works. In the end, we discussed the significance of performance analysis of LPWANs based on simulations and summarized available options to conduct simulations at different levels in a LoRa network, including radio planning, link-level and system-level simulations, and end-to-end stress testing of a LoRaWAN network.

References

1. Hatler M, Gurganious D, Kreegar J (2018) Industrial LPWAN: a market dynamics report, Published Nov 2018. https://www.onworld.com/iLPWAN/, cited 24 Sept 2019
2. Xu J, Yao J, Wang L, Ming Z, Wu K, Chen L (2018) Narrowband Internet of things: evolutions, technologies, and open issues. IEEE Internet Things J 5(3):1449–1462
3. Raza U, Kulkarni P, Sooriyabandara M (2017) Low power wide area networks: an overview. IEEE Commun Surv Tutorials 19(2):855–873
4. Xie Z, Xu R, Lei L (2014) A study of clear channel assessment performance for low power wide area networks. In: WiCOM, pp 311–315, Sept 2014
5. Vejlgaard B, Lauridsen M, Nguyen H, Kovacs IZ, Mogensen P, Sorensen M (2017) Coverage and capacity analysis of Sigfox, LoRa, GPRS, and NB-IoT. In: IEEE VTC Spring, Sydney, 2017, pp 1–5
6. Ratasuk R, Vejlgaard B, Mangalvedhe N, Ghosh A (2016) NB-IoT system for M2M communication. In: IEEE WCNC, Doha, pp 1–5
7. Wang YE, Lin X, Grovlen A, Sui Y, Bergman J (2016) A primer on 3GPP narrowband internet of things. IEEE Commun Mag 55(3):117–123
8. Ratasuk R, Mangalvedhe N, Xiong Z, Robert M, Bhatoolaul D (2017) Enhancements of narrowband IoT in 3GPP Rel-14 and Rel-15. In: IEEE CSCN, Helsinki, pp 60–65

9. Lavric A, Petrariu AI, Popa V (2019) Long range SigFox communication protocol scalability analysis under large-scale, high-density conditions. IEEE Access 7:35816–35825
10. Ferrari P, Flammini A, Rizzi M, Sisinni E, Gidlung M (2017) On the evaluation of LoRaWAN virtual channels orthogonality for dense distributed systems. In: IEEE international workshop on measurements and networking (M&N), Naples, 27–29 Sept 2017, pp 85–90
11. Rizzi M, Ferrari P, Sisinni E, Flammini A (2017) Evaluation of the IoT LoRaWAN solution for distributed measurement applications. IEEE Trans Instrum Meas 66(12):3340–3349
12. Fernandes Carvalho D, Ferrari P, Sisinni E, Depari A, Rinaldi S, Pasetti M, Silva D (2019) A test methodology for evaluating architectural delays of LoRaWAN implementations. Pervasive Mob Comput 56:1–17
13. Fernandes Carvalho D, Depari A, Ferrari P, Flammini A, Rinaldi S, Sisinni E (2019) On the evaluation of application level delays in public LoRaWAN networks. In: IEEE International Workshop on Measurements and Networking (M&N), Catania, 8–10 July 2019, pp 1–6
14. Butun I, Pereira N, Gidlund M (2019) Security risk analysis of LoRaWAN and future directions. Future Internet 11:3
15. Rizzi M, Depari A, Ferrari P, Flammini A, Rinaldi S, Sisinni E (2019) Synchronization uncertainty versus power efficiency in LoRaWAN networks. IEEE Trans Instrum Meas 68(4):1101–1111
16. Beltramelli L, Mahmood A, Österberg P, Gidlund M (2020) LoRa beyond ALOHA: an investigation of alternative random access protocols. IEEE Trans Ind Informat https://doi.org/10.1109/TII.2020.2977046
17. Haxhibeqiri J, Moerman I, Hoebeke J (2019) Low overhead scheduling of LoRa transmissions for improved scalability. IEEE Internet Things J 6(2):3097–3109, April 2019 https://doi.org/10.1109/JIOT.2018.2878942
18. Reynders B, Wang Q, Tuset-Peiro P, Vilajosana X, Pollin S (2018) Improving reliability and scalability of LoRaWANs through lightweight scheduling. IEEE Internet Things J 5(3):1830–1842
19. Bonafini F, Depari A, Ferrari P, Flammini A, Pasetti M, Rinaldi S, Sisinni E, Gidlund M (2019) Exploiting localization systems for LoRaWAN transmission scheduling in industrial applications. In: IEEE WFCS, Sundsvall, 27–29 May 2019, pp 1–8
20. Piyare R, Murphy AL, Magno M, Benini L (2018) On-demand LoRa: asynchronous TDMA for energy efficient and low Latency communication in IoT. Sensors 18(11):3718
21. Rizzi M, Ferrari P, Flammini A, Sisinni E, Gidlund M (2017) Using LoRa for industrial wireless networks. In: IEEE WFCS, Trondheim, 31 May–2 June 2017, pp 1–4
22. Polonelli T, Brunelli D, Marzocchi A, Benini L (2019) Slotted ALOHA on LoRaWAN-design, analysis, and deployment. Sensors 19:838
23. Ebi C, Schaltegger F, Rüst A, Blumensaat F (2019) Synchronous LoRa mesh network to monitor processes in underground infrastructure. IEEE Access 7:57663–57677
24. Lee H, Ke K (2018) Monitoring of large-area IoT sensors using a LoRa wireless mesh network system: design and evaluation. IEEE Trans Instrum Meas 67(9):2177–2187
25. Aslam MS, Khan A, Atif A, Hassan SA, Mahmood A, Qureshi HK, Gidlund M (2019) Exploring multi-hop LoRa for green smart cities. IEEE Netw https://doi.org/10.1109/MNET.001.1900269
26. Sisinni E, Fernandes Carvalho D, Ferrari P, Flammini A, Cabral Silva DR, Da Silva I (2018) Enhanced flexible LoRaWAN node for industrial IoT. In: IEEE WFCS, Imperia, 13–15 June 2018
27. Ghosh A, Maeder A, Baker M, Chandramouli D (2019) 5G evolution: a view on 5G cellular technology beyond 3GPP release 15. IEEE Access 7:127639–127651
28. Navarro-Ortiz J, Sendra S, Ameigeiras P, Lopez-Soler JM (2018) Integration of LoRaWAN and 4G/5G for the Industrial Internet of things. IEEE Commun Mag 56(2):60–67
29. Mahmood A, Sisinni E, Guntupalli L, Rondón R, Hassan SA, Gidlund M (2019) Scalability analysis of a LoRa network under imperfect orthogonality. IEEE Trans Ind Inf 15(3):1425–1436

30. Beltramelli L, Mahmood A, Gidlund M, Österberg P, Jennehag U (2018) Interference modelling in a multi-cell LoRa system. In: WiMob, Limassol, pp 1–8
31. Haxhibeqiri J, Poorter ED, Moerman I, Hoebeke J (2018) A survey of LoRaWAN for IoT: from technology to application. Sensors 18(11):3995
32. Petajajarvi J, Mikhaylov K, Roivainen A, Hanninen T, Pettissalo M (2015) On the coverage of LPWANs: range evaluation and channel attenuation model for LoRa technology. In: ITST, Copenhagen, pp 55–59
33. Anjum M, Khan MA, Ali Hassan S, Mahmood A, Gidlund M (2019) Analysis of RSSI fingerprinting in LoRa Networks. In: IWCMC, Tangier, pp 1178–1183
34. CloudRF (2019) Available online: https://cloudrf.com/LoRa_planning, cited on 1 Nov 2019
35. ATDI (2019) Available online: https://atdi-group.com/, cited on 1 Nov 2019
36. Croce D, Gucciardo M, Mangione S, Santaromita G, Tinnirello I (2019) Impact of LoRa imperfect orthogonality: analysis of link-Level performance. IEEE Commun Lett 22:796–799. Simulator available: http://lora.tti.unipa.it/, cited on 1 Nov 2019
37. Abeele FVD, Haxhibeqiri J, Moerman I, Hoebeke J (2017) Scalability analysis of large-scale LoRaWAN networks in ns-3. IEEE Internet Things J 4:2186–2198. Source code: https://github.com/timolex/ns-3-dev-with-imec-idlab-lorawan-module, cited on 1 Nov 2019
38. Magrin D, Centenaro M, Vangelista L (2019) Performance evaluation of LoRa networks in a smart city scenario. In: IEEE ICC, Paris, 21–25 May 2017, pp 1–7. Source code: https://github.com/DvdMgr/lorawan, cited on 1 Nov 2019
39. Bor M, Roedig U, Voigt T, Alonso J (2019) Do LoRa low-power wide-area networks scale? In: ACM International Conference on Modeling, Analysis and Simulation of Wireless and Mobile Systems, New York, pp 59–67. Simulator available: https://www.lancaster.ac.uk/scc/sites/lora/lorasim.html, cited on 1 Nov 2019
40. Slabicki M, Premsankar G, Francesco MD (2019) Adaptive configuration of lora networks for dense IoT deployments. In: IEEE/IFIP Network Operations and Management Symposium, Taipei, 2018, pp 1–9. Simulator available: https://flora.aalto.fi/, cited on 1 Nov 2019
41. LoRaEnergySim (2019) Available online: https://github.com/GillesC/LoRaEnergySim, cited on 1 Nov 2019
42. Pop A, Raza U, Kulkarni P, Sooriyabandara M (2017) Does bidirectional traffic do more harm than good in LoRaWAN based LPWA networks? In: IEEE GLOBECOM, Singapore, 2017, pp 1–6
43. Abdelfadeel KQ, Zorbas D, Cionca V, Pesch D (2019) FREE-Fine-grained scheduling for reliable and energy-efficient data collection in LoRaWAN. IEEE Internet Things J 7(1):669–683, Jan 2020. Simulator Available online: https://github.com/kqorany/FREE, cited on 1 Nov 2019
44. Duda A, To TH (2018) Simulation of LoRa in NS-3: improving LoRa performance with CSMA. In: IEEE ICC, Kansas City, 20–24 May 2018
45. Kouvelas N, Rao V, Prasad R (2018) Employing p-CSMA on a LoRa network simulator. arXiv 2018, arXiv:1805.12263
46. Mbed LoRaWAN stack simulator (2019) Available online: https://github.com/janjongboom/mbed-simulator, cited on 1 Nov 2019
47. LoRahammer (2019) Available online: https://github.com/itkSource/lorhammer, cited on 1 Nov 2019

Power Measurement Framework for LPWAN IoT

Hua Wang, André Sørensen, Maxime Remy, Nicolaj Kjettrup, Jimmy Jessen Nielsen, and Germán Corrales Madueño

1 Introduction

The concept of Internet of Things (IoT) is that every device has access to the Internet and is able to collect or distribute information over the network. It is expected that the number of connected IoT devices will experience massive yearly growth in the coming years [1]. Examples of IoT applications include smart metering, smart cities, smart factories, asset tracking, logistics, etc. The IoT devices may be deployed in those hard-to-access locations, for example, basements where the water meters are placed. Since changing the battery of those IoT devices in hard-to-access locations can be costly and/or impractical, it often ends up that the battery lifetime determines the lifetime of the device. Therefore, the main requirements for IoT device are wide-area coverage, low power consumption, and low cost.

Low-power wide-area network (LPWAN) has gained increasingly popularity in recent years both from industries and academia to address the requirements imposed by IoT applications. Many LPWAN technologies have been proposed including both proprietary and standardized approaches operating in the licensed as well as unlicensed bands. Among them, Sigfox, LoRa, and Narrowband IoT (NB-IoT) are the solutions that attracted highest interest. The 3rd Generation Partnership Project (3GPP) has standardized two LPWAN technologies targeted for IoT applications: Long-Term Evolution for Machines (LTE-M) and NB-IoT. Both LTE-M and NB-IoT are developed based on Long-Term Evolution (LTE), but are targeted for

H. Wang (✉) · A. Sørensen · M. Remy · N. Kjettrup
Keysight Technologies, Aalborg, Denmark
e-mail: hua.wang@keysight.com; andre.soerensen@keysight.com; maxime.remy@keysight.com; nicolaj.kjettrup@keysight.com

J. Jessen Nielsen · G. Corrales Madueño
Aalborg University, Aalborg, Denmark
e-mail: jjn@es.aau.dk; gco@es.aau.dk

N. H. Mahmood et al. (eds.), *Wireless Networks and Industrial IoT*,
https://doi.org/10.1007/978-3-030-51473-0_6

different use cases. NB-IoT is focused on lowering the device complexity, increasing the battery life, and improving the coverage at the cost of the latency, bandwidth, and mobility [2]. On the other hand, LTE-M provides a hybrid solution between LTE and NB-IoT with larger bandwidth, lower latency, and higher mobility at the cost of other areas, such as complexity and coverage. In this chapter, we only focus on NB-IoT.

The performance of an IoT device can be measured from various domains, such as throughput, coverage, power consumption, etc. This chapter is focused on the energy domain. A 10-year battery lifetime for a predefined traffic profile is required by 3GPP [3]. To validate this claim, it is important to provide a power consumption model which can be used to estimate the battery lifetime of an IoT device. The model should be simple and easy to use so that users can configure the parameters according to his application's specific requirements, such as transmission parameters and traffic profile. This is critical for the further evolution and market penetration of IoT, as developers, researchers, and mobile network operators need to know what each IoT technology can provide in terms of battery lifetime for a given use case.

The power consumption model for regular broadband LTE network has been addressed by many researchers. A review of those models can be referred to [4]. However, only a few studies have been published in recent years focusing on the power consumption modelling for IoT devices [5–7]. In this chapter, a more comprehensive and flexible power consumption model is presented based on measurements with the aim to accurately estimate the battery lifetime for any given coverage scenarios and traffic profiles. This model, in order to be as accurate as possible, based itself on the tested device power consumption for each state (e.g. TX) and an estimated time spent in said states based on the 3GPP standards. Detailed modelling of each user equipment (UE) state and procedure will be described. By combining the components of different states and procedures, the power consumption of any UE behaviour can be modelled, and the battery lifetime can be estimated with predefined traffic pattern.

The rest of this chapter is structured as follows. Section 2 introduces the states and procedures of an IoT device. It is followed by a presentation of the testbed setup used to estimate the different states' power consumption of the UE and the modelling of those states and procedures. The considered coverage scenario and traffic profile are presented in Sect. 3, together with the battery lifetime estimation model. The measurement results are presented in Sect. 4. Conclusions and future work are drawn in Sect. 5.

2 Power Consumption Model

This section describes the proposed power consumption model for LTE-M and NB-IoT. The UE procedures and states will be presented first, followed by the detailed modelling of each state and the main procedures.

2.1 Procedures

The UE procedures are certain actions performed by the UE with specific purposes and functionalities. The main procedures used in NB-IoT are described as follows [8]:

- **Synchronization:** It is used for the UE to synchronize its clock and frequency to the network when the UE powers up or wakes up from sleep, by using Primary Synchronization Signal (PSS) and Secondary Synchronization Signal (SSS). After the UE is synchronized to the network, it can perform the random access procedure.
- **Random Access (RA):** The RA procedure is used when the UE has the intention to communicate with the network but has no allocated resources. It initiates the communication with the network and is followed by either an attach, a resume, or a service request procedure.
- **Attach:** The attach procedure follows right after the random access procedure. It is used when the UE connects to the network for the first time.
- **Service Request:** The service request procedure also follows right after the random access procedure. It is similar to the attach procedure but with fewer steps. It is used when the UE has already been registered in the network, requests for data transmission, but has no allocated resources.
- **Connection Resume:** The connection resume procedure has been introduced as a part of the User Plane Cellular IoT (CIoT) Evolved Packet System (EPS) Optimization. It can be used as a replacement for the service request procedure, if the connection has been suspended instead of released.
- **Detach:** The detach procedure is used to tell the network that the UE no longer wants to access it and so the connection should be terminated. Hereafter, the RRC release procedure occurs, after which the UE is no longer connected to the network.
- **Release:** The release procedure is used when the UE doesn't have any activity in the network but still would like to be registered in the network. After the release procedure, The UE can still be contacted by paging. But if the UE wants to transmit data, it has to do a service request procedure.
- **Connection Suspend:** The connection suspend procedure has been introduced as a part of the User Plane CIoT EPS Optimization. It can be used as a replacement for release. The resume procedure can be used for re-establishing the connection if the connection has been suspended.
- **Tracking Area Update (TAU):** The TAU procedure is used to indicate to the network that the UE is still alive. It occurs at the beginning of each power saving mode (PSM) cycle. If there is no TAU from the UE, the network assumes the UE is shut down and deregister it.

2.2 States

The states of the UE can be defined from a network perspective or from a device perspective.

From a network perspective, the UE can be in different network status which defines what available resources the UE has and how reachable the UE is. The network status of the UE is determined by EPS Mobility Management (EMM) and EPS Connection Management (ECM) protocols.

The EMM protocol indicates if the UE is registered in the network or not. When the UE is powered on, it starts in EMM-DEREGISTERED. It can then perform an attach procedure to get into EMM-REGISTERED. The UE moves from EMM-REGISTERED to EMM-DEREGISTERED when either a detach procedure is performed or the TAU expires.

The ECM protocol indicates whether the UE has established signalling to the EPC or not. The UE changes from ECM-IDLE to ECM-CONNECTED by performing either an attach, a service request, or a resume procedure. To change back from ECM-CONNECTED to ECM-IDLE, the UE performs either a release, a connection suspend, or a detach procedure. Figure 1 illustrates an example of different procedures performed by the UE and the corresponding network status.

From a device perspective, the UE can be in different states such as uplink transmission (TX), downlink reception (RX), idle, etc. A state transition diagram and the associated procedures is shown in Fig. 2. It is assumed that the UE uses either Power Saving Mode (PSM) or Extended Discontinuous Reception (eDRX) for the power saving. The states in grey indicate the UE is in EMM-DEREGISTERED, and the states in white indicate the UE is in EMM-REGISTERED. For UE in EMM-REGISTERED, it is ECM-CONNECTED if the UE is in state Connected Mode Discontinuous Reception (cDRX), TX, or RX. It is ECM-IDLE if the UE is in state DRX, eDRX, or PSM.

Once the power consumption of each state and the associated procedure is known, the power consumption of any UE behaviour can be modelled by combining the corresponding states and procedures illustrated in Fig. 2. The following next two sections detail the modelling of different states and procedures.

<table>
<tr><td>UE Procedure</td><td>Off</td><td>On/Sync</td><td>Attach</td><td>Release</td><td>Service Request</td><td>Detach</td></tr>
<tr><td>EMM Status</td><td colspan="2">DEREGISTERED</td><td colspan="3">REGISTERED</td><td>DEREGISTERED</td></tr>
<tr><td>ECM Status</td><td colspan="2">IDLE</td><td>CONNECTED</td><td>IDLE</td><td>CONNECTED</td><td>IDLE</td></tr>
</table>

Fig. 1 Example of UE procedures and the corresponding network status

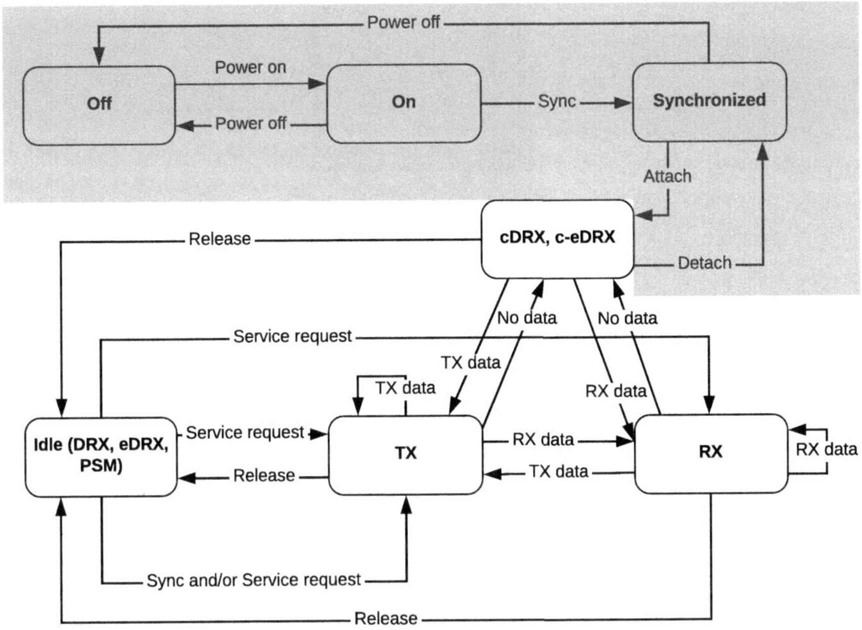

Fig. 2 The state transition diagram for LTE-M and NB-IoT

2.3 *Testbed Setup*

A testbed is developed for to both characterize the power consumption of the UE state presented in Fig. 2 and to validate the model proposed in this chapter.

The device under test (DUT) is connected to a NB-IoT or LTE-M base station to measure the power consumption in terms of the voltage and current level. Figure 3 shows the measurement setup. The DUT's antenna port is connected, via cables, to a Keysight E7515A UXM Wireless, which is a standard-compliant base station emulator supporting both NB-IoT and LTE-M protocols with debugging capabilities. The DUT is also connected to a Keysight N6705B DC Power Analyzer which acts as both a power supply and a sensor for battery drain measurements. The measurement setup is controlled using Keysight's Test Automation Platform (TAP), which provides interfaces to both the measurement equipments and the DUT and orchestrates the behaviour of different components. Besides, a power measurement tool is developed to synchronize the protocol logs from the UXM and the measurement logs from the power analyzer with 0.2 ms resolution.

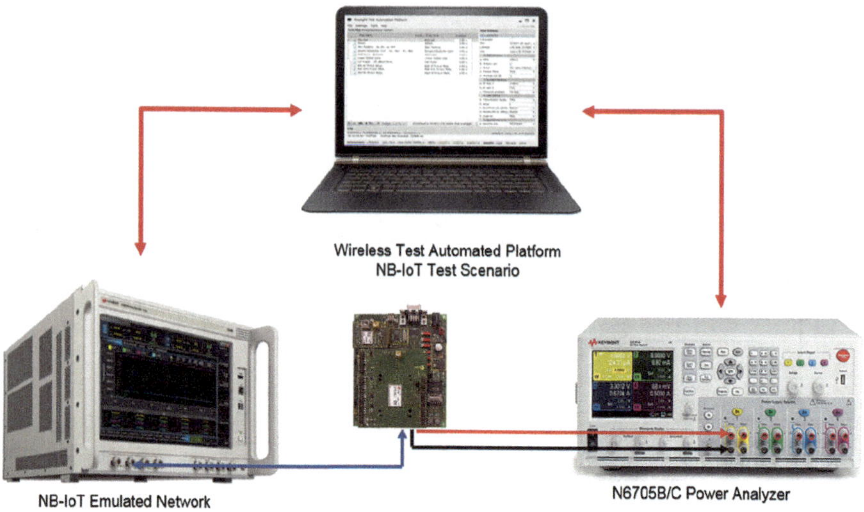

Fig. 3 Illustration of measurement setup

2.4 Modelling of States

The states and the corresponding state transition diagram have been introduced in Sect. 2.2. In this section, the power consumption model for each state is given.

Modelling of the TX State

In TX state, the device is transmitting in the uplink. Its energy consumption is characterized by its power consumption and the length of the transmission. The power consumption is affected by the transmission power; thus a characterization of the DUT's power consumption across the different transmission power is required for the model. The transmission power from the device is controlled by the uplink power control specified in [9].

Figure 4 shows the measured power consumption as a function of the device uplink transmit power for two different NB-IoT and LTE-M devices. Similar trend is observed for both NB-IoT and LTE-M devices. It can be seen that the power consumption curve can be split into two parts. The first part is when the power amplifier is not required, resulting in almost linear increase of the power consumption. The second part is when the power amplifier is used. In this case, the power consumption rises exponentially. The reason for exponential increase in the power consumption is because the efficiency of the power amplifier decreases with the increase of the output power.

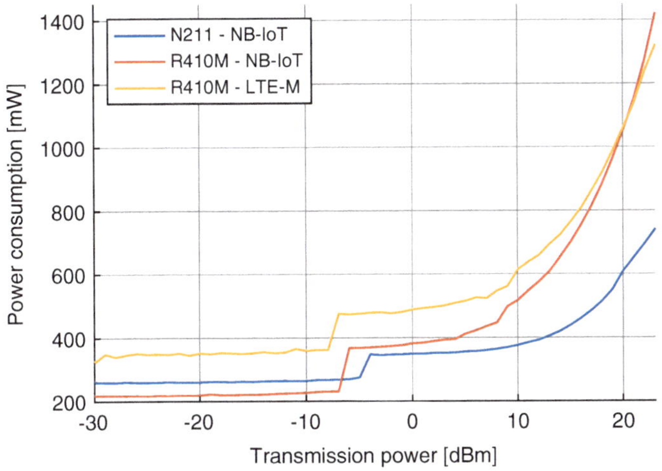

Fig. 4 Measured power consumption as a function of uplink transmit power

The energy consumption for the TX state can be modelled as:

$$E_{\mathrm{TX}} = t_{\mathrm{TX}} P_{\mathrm{TX}} + t_{\mathrm{TX_{gaps}}} P_{\mathrm{TX_{gaps}}} \tag{1}$$

where E_{TX} is the energy spent in data transmission, t_{TX} is time spent in data transmission, P_{TX} is the power consumption during data transmission, $t_{\mathrm{TX_{gaps}}}$ is time spent in transmission gaps, and $P_{\mathrm{TX_{gaps}}}$ is the power consumption during transmission gaps.

In NB-IoT, there is a mandatory gap of 0.04 s after 0.256 s of continuous transmission [10]. The power consumption during TX gaps is much lower than the power consumption of data transmission. These gaps are introduced during the transmission to allow the low-quality oscillators to resynchronize with the network. While in LTE-M, these gaps are optional.

From the measurements, it is observed that the device transmission power is independent of the selection of modulation and coding scheme (MCS), the number of allocated physical resource block (PRB)s or Resource Unit (RU)s, and the number of repetitions. But these parameters do determine the transmission time t_{TX} and can be calculated as:

$$t_{\mathrm{TX}}^{\mathrm{NB-IoT}} = \left\lceil \frac{\mathrm{Payload}}{\mathrm{TBS}} \right\rceil \cdot \mathrm{RU_{length}} \cdot \mathrm{RU_{allocated}} \cdot \mathrm{Rep} \tag{2}$$

where Payload is data size in bits, $\mathrm{RU_{length}}$ is the length of a RU, $\mathrm{RU_{allocated}}$ is the number of allocated RUs, Rep is the number of repetitions, and $\lceil\rceil$ is the ceil function. The Transportation Block Size (TBS) is in bits and is determined by the MCS and the number of allocated RUs.

The time spent in transmission gaps depends on t_{TX} and can be calculated as:

$$t_{\mathrm{TX_{gaps}}}^{\mathrm{NB-IoT}}(t_{\mathrm{TX}}) = \left\lfloor \frac{t_{\mathrm{TX}}}{t_{\mathrm{TX_{max}}}} \right\rfloor \cdot t_{\mathrm{TX_{gap}}} \tag{3}$$

where $t_{\mathrm{TX_{max}}}$ is the maximum continuous transmission time allowed, $t_{\mathrm{TX_{gaps}}}$ is the duration of a gap, and $\lfloor\rfloor$ is the floor function. According to [10], $t_{\mathrm{TX_{max}}}$ and $t_{\mathrm{TX_{gaps}}}$ for NB-IoT are 256 and 40 ms, respectively.

Modelling of the RX State

In RX state, the device is receiving data in the downlink. Its energy consumption is characterized by its power consumption and the length of the reception. The energy consumption for the RX state can be modelled as:

$$E_{\mathrm{RX}} = t_{\mathrm{RX}} P_{\mathrm{RX}} + t_{\mathrm{RX_{gaps}}} P_{\mathrm{RX_{gaps}}} \tag{4}$$

where t_{RX} and P_{RX} indicate the time spent and the power consumption in data reception, respectively, and $t_{\mathrm{RX_{gaps}}}$ and $P_{\mathrm{RX_{gaps}}}$ indicate the time spent and the power consumption in reception gaps respectively. Reception gaps in the downlink channel occurs due to the reception of System Information Block (SIB) and other control signalling [9, 10]. Similar to the TX state, the power consumption in reception gaps is lower than the power consumption of data reception, and the power consumption in RX state is independent of the MCS, number of allocated subframes, and number of repetitions.

Unlike the TX model where the gaps form an obvious pattern, the reception gaps are harder to include in the model due to the fact that they are dependent on both the length of the reception and its start. Thus a general estimate of the number of gaps occurring in a reception is done. The reception time t_{RX} and the reception gap time $t_{\mathrm{RX_{gaps}}}$ can be calculated as:

$$t_{\mathrm{RX}} = \mathrm{SF_{RX}} \cdot \mathrm{SF_{length}}$$

$$\mathrm{SF_{RX}} = \left\lceil \frac{\mathrm{Payload}}{\mathrm{TBS}} \right\rceil \cdot \mathrm{SF_{allocated}} \cdot \mathrm{Rep}$$

$$t_{\mathrm{RX_{gaps}}}(\mathrm{SF_{RX}}) = \left\lceil \mathrm{SF_{RX}} \cdot \left(\frac{1}{\mathrm{SF_{av}}} - 1 \right) \right\rceil \cdot \mathrm{SF_{length}} \tag{5}$$

where $\mathrm{SF_{allocated}}$ is the number of allocated subframes per TBS, $\mathrm{SF_{RX}}$ is the required number of subframes for the payload, and $\mathrm{SF_{av}}$ is the fraction of subframes available for data reception. To keep the model simple, only the most recurring gaps are taken into account, namely, Master Information Block (MIB), SIB1, Narrowband Primary

Synchronization Signal (NPSS), and Narrowband Secondary Synchronization Signal (NSSS). When using this simplification, roughly 14 subframes are available for the Narrowband Physical Downlink Shared Channel (NPDSCH) out of 20 subframes [10, 11]. The rest six subframes are reserved by other channels, such as the physical broadcast channel, SIBs, primary and secondary synchronization signals, and others.

Modelling of the DRX State

DRX was introduced in Release 8 to prolong the battery lifetime of the device. This is achieved by introducing the DRX cycle, during which the device alternates between active checking of the paging (with their lengths configured by the onDurationTimer) and inactivity [12].

The DRX cycle can be divided into two periods: active and idle. For the active period, it was measured that some devices have a short synchronization before entering in the active period. This synchronization is added to lower the device power consumption during the sleep period. From the measurement, it is noted that the energy consumption for the synchronization is device specific and is not affected by the number of paging repetitions and the DRX cycle length. Thus, it can be modelled as a fixed energy cost based on the measurements. After the device has been synchronized with the network, the device monitors the paging in Physical Downlink Control Channel (PDCCH). The number of paging the device has to monitor is indicated by the paging repetitions [11, 13]. This will of course impact the battery lifetime and should be included in the model. When the device has finished listening to the paging, it goes into sleep. Figure 5 shows an example of measured power consumption levels in the DRX cycle. The duration of a DRX cycle is given by the network parameter defaultPagingCycle which is given in number of radio frames.

The energy consumption of the DRX can then be modelled as:

$$E_{\mathrm{DRX}_{\mathrm{cycle}}} = E_{\mathrm{DRX}_{\mathrm{sync}}} + E_{\mathrm{paging}} \cdot \mathrm{Rep}_{\mathrm{paging}} + P_{\mathrm{DRX}_{\mathrm{Sleep}}} \cdot t_{\mathrm{DRX}_{\mathrm{Sleep}}}$$

$$t_{\mathrm{DRX}_{\mathrm{Sleep}}} = t_{\mathrm{DRX}} - (t_{\mathrm{DRX}_{\mathrm{onDuration}}} + t_{\mathrm{DRX}_{\mathrm{sync}}}) \quad (6)$$

where $t_{\mathrm{DRX}_{\mathrm{sync}}}$ and $E_{\mathrm{DRX}_{\mathrm{sync}}}$ are the measured time and energy consumption for the synchronization in a DRX cycle, $t_{\mathrm{DRX}_{\mathrm{Sleep}}}$ and $P_{\mathrm{DRX}_{\mathrm{Sleep}}}$ are the time and power consumption in DRX sleep state, E_{paging} is the energy spent on a single paging occasion, $\mathrm{Rep}_{\mathrm{paging}}$ is the number of paging repetitions in a DRX cycle, $t_{\mathrm{DRX}_{\mathrm{onDuration}}}$ is the time spent on monitoring the paging in a DRX cycle, and t_{DRX} is the length of a DRX cycle.

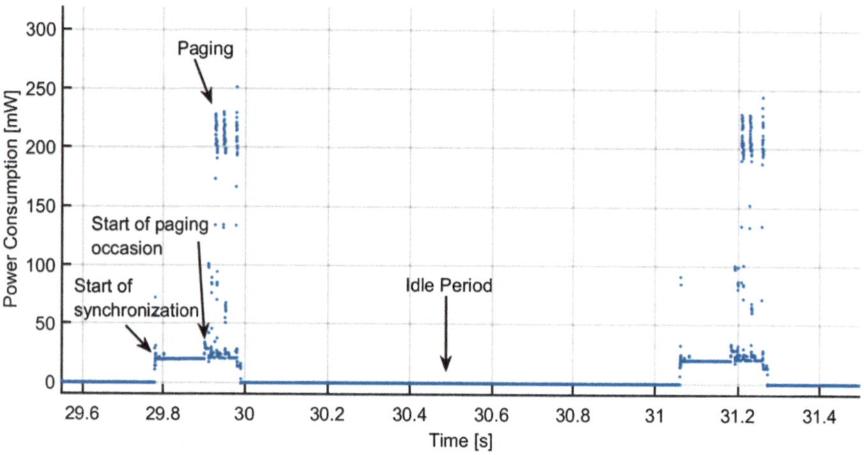

Fig. 5 Example of power consumption in the DRX cycle

Modelling of the cDRX State

cDRX is the equivalent of DRX in ECM-CONNECTED (DRX is in ECM-IDLE). However, as the device is in connected status, it does not perform paging, but rather it monitors the PDCCH. After the device has monitored the PDCCH for a number of subframes, which are specified by the OnDurationTimer and the UE-specific search space (USS) repetitions, the device goes into cDRX sleep state [11, 14]. Assuming that the OnDuration stays constant across the device life, the energy consumption of the cDRX can be modelled as:

$$E_{\text{cDRX}_{\text{cycle}}} = E_{\text{cDRX}_{\text{onDuration}}} \cdot \text{Rep}_{\text{USS}} + t_{\text{cDRX}_{\text{sleep}}} \cdot P_{\text{cDRX}_{\text{sleep}}}$$

$$t_{\text{cDRX}_{\text{sleep}}} = t_{\text{cDRX}_{\text{cycle}}} - t_{\text{cDRX}_{\text{onDuration}}} \cdot \text{Rep}_{\text{USS}} \tag{7}$$

where $t_{\text{cDRX}_{\text{onDuration}}}$ and $E_{\text{cDRX}_{\text{onDuration}}}$ are the time and energy spent in the OnDuration of cDRX, while Rep_{USS} is the USS repetitions. $t_{\text{cDRX}_{\text{cycle}}}$ and $t_{\text{cDRX}_{\text{sleep}}}$ are the length of a cDRX cycle and the time spent in sleep state, respectively, and $P_{\text{cDRX}_{\text{sleep}}}$ is the power consumption when the device is in cDRX sleep state. It should be noted that this model neglects the inactivity timer which triggers when entering into cDRX is neglected in this model. The main reason for it is that in most traffic profiles, the device is not expected to quickly alternate between RX and cDRX as it is taxing for the battery.

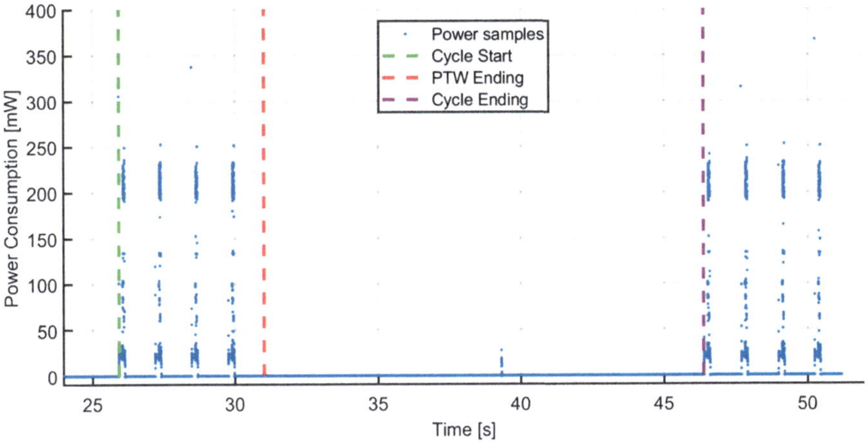

Fig. 6 Example of eDRX cycle with DRX

Modelling of the eDRX State

eDRX is a mechanism that can extend the cycle length of the two DRX (i.e. DRX and cDRX) to allow increased sleeping duration, thus further reducing the energy consumption. While eDRX does not have any important impact on the cDRX model, a new model needs to be developed for DRX [11]. Figure 6 shows an example of an eDRX cycle for DRX, which is mainly composed of two parts: the paging time window (PTW) and the sleep. In the PTW, the device behaves as being in DRX, and the number of DRX cycles depends on the DRX cycle length and the PTW length [13]. For example in Fig. 6, there are four DRX cycles within the PTW. The last part of eDRX is the sleep, and the duration is dependent on the PTW length and eDRX cycle length.

The energy consumption of the eDRX can be modelled as:

$$E_{\mathrm{eDRX_{cycle}}} = E_{\mathrm{DRX_{cycle}}} \cdot \left\lceil \frac{T_{\mathrm{PTW}}}{t_{\mathrm{DRX_{cycle}}}} \right\rceil + P_{\mathrm{eDRX_{sleep}}} \cdot t_{\mathrm{eDRX_{sleep}}}$$

$$t_{\mathrm{eDRX_{sleep}}} = t_{\mathrm{eDRX_{cycle}}} - T_{\mathrm{PTW}} \tag{8}$$

where T_{PTW} is the PTW length; $t_{\mathrm{eDRX_{sleep}}}$ and $P_{\mathrm{eDRX_{sleep}}}$ are the time and power consumption in the eDRX idle period, respectively; $t_{\mathrm{eDRX_{cycle}}}$ is the length of a eDRX cycle; and $E_{\mathrm{DRX_{cycle}}}$ is the energy spent in a DRX cycle. The ceil of the ratio between the PTW and the DRX cycle length is used to take into account that most of the DRX cost is upfront with the monitoring of the paging.

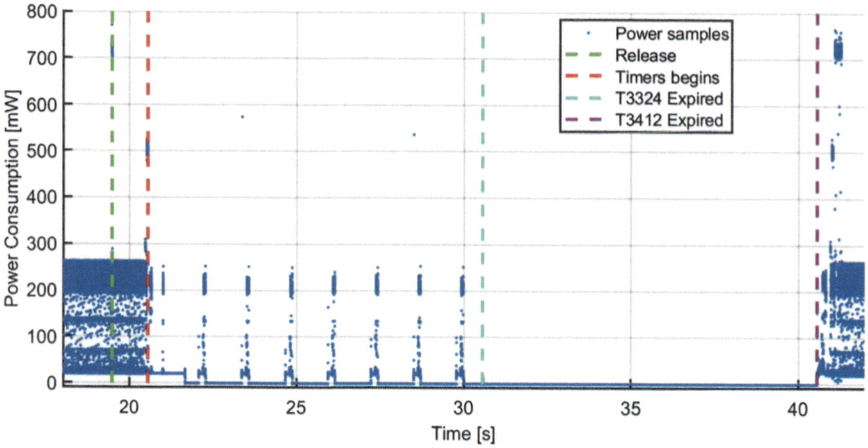

Fig. 7 Example of PSM cycle

Modelling of the PSM State

An example of a PSM cycle is shown in Fig. 7. There are two important timers associated with PSM, T_{3412} and T_{3324}, which determines the period when the device is reachable and the period when the device is sleeping. In the reachable period, the UE can either be operating with DRX or eDRX, and it has been verified empirically that the power consumption equals DRX or eDRX [15]. Therefore, the energy consumption model for DRX or eDRX can be reused within the period of T_{3324}. After the expiration of T_{3324}, the device shuts down all Access Stratum (AS) functions and goes into sleep. The device stays in deep sleep until the T_{3412} timer expires, after which the device will perform a TAU [16]. The TAU has a big impact on the battery lifetime, and the model for it will be described in section "Modelling of the Service Request, Attach, Release, and TAU Procedures". From an energy consumption point of view, ideally the configuration of TAU periodicity (i.e. T_{3412}) should be that the UE will wake up from PSM rather due to uplink data transmission than due to the periodicity of TAU.

The energy consumption of PSM can be modelled as:

$$E_{\mathrm{PSM_{cycle}}} = \left\lceil \frac{T_{3324}}{t_{\mathrm{DRX_{cycle}}}} \right\rceil \cdot E_{\mathrm{DRX_{cycle}}} + t_{\mathrm{PSM_{sleep}}} \cdot P_{\mathrm{PSM_{sleep}}}$$

$$t_{\mathrm{PSM_{sleep}}} = T_{3412} - T_{3324} \tag{9}$$

where $P_{\mathrm{PSM_{sleep}}}$ is the power consumption during sleep period.

Table 1 Key parameters for the power consumption model in each state

States	Key parameters
TX	Payload, TX power, TBS, allocated RUs, Repetitions
RX	Payload, TBS, allocated subframes, Repetitions
DRX	DRX Cycle, OnDurationTimer, Paging Repetitions
cDRX	cDRX Cycle, OnDurationTimer
eDRX	eDRX Cycle, Paging Time Window
PSM	T_{3324}, T_{3412}

Key Parameters of States

The power consumption model for each state was introduced from sections "Modelling of the TX State" to "Modelling of the PSM State". The key parameters that will affect the power consumption in each state are summarized in Table 1. The proper configuration of these parameters is important not only to ensure good radio performance (such as connectivity and throughput) but also to save energy, i.e. extend batter lifetime. The guidelines on how to configure the parameters related to TX, which is the most energy consuming state, will be detailed in Sect. 3.1.

2.5 *Modelling of Procedures*

In Sect. 2.1, the main procedures used in LTE-M and NB-IoT have been introduced. Each procedure is composed of a sequence of uplink and downlink messages. By combining the message size and the associated TX state model (uplink transmission) or RX state model (downlink reception), the energy consumption in each procedure can be calculated.

In this section, five of the most commonly used procedures are considered: the synchronization, the attach, the control plane service request, the radio resource control (RRC) release, and the TAU.

Modelling of the Synchronization Procedure

For the synchronization, it is difficult to provide a general model that fits for all devices as the synchronization procedure varies from device to device and is heavily dependent on the implementation. Instead, the synchronization procedure is modelled from measurements. Table 2 lists the measured energy and time consumption for the synchronization procedure of a specific NB-IoT device a u-blox SARA-N211. The measurement starts with the start of synchronization and ends before the occurrence of the first Physical Random Access Channel (PRACH).

Table 2 Measured energy and time consumption for the synchronization procedure, from NB-IoT device SARA-N211

Procedure	Energy consumption for the synchronization	Duration for the synchronization
Attach	325 mJ	3500 ms
Service request/PSM	160 mJ	2200 ms

Modelling of the Service Request, Attach, Release, and TAU Procedures

The uplink and downlink messages associated with the service request, attach, release, and TAU procedures are listed in Table 3. The message size is obtained from a NB-IoT device SARA-N211. The energy consumption for each procedure can be calculated by using the TX/RX state model described in Sect. 2.4 in combination with the associated messages listed in Table 3. It is worth mentioning that by using Control Plane Cellular IoT EPS Optimization procedure, the user data can be transmitted within the services request, thereby reducing the overhead by skipping the EPS bearers establishment [17].

To simplify the models of the procedures, delays and signalling for the individual messages have been excluded. Based on that assumption, the procedures can be modelled as a summation of the energy consumption by each message given as:

$$E_{\text{Procedure}} = \sum_{i=1}^{I} E_{\text{TX}}(d_i) + \sum_{j=1}^{J} E_{\text{RX}}(d_j) \tag{10}$$

$$t_{\text{Procedure}} = \sum_{i=1}^{I} t_{\text{TX}}(d_i) + \sum_{j=1}^{J} t_{\text{RX}}(d_j) \tag{11}$$

where I is the number of uplink messages, d_i is the data size of message i, J is the number of downlink messages, and d_j is the data size of message j.

3 Battery Lifetime Modelling

A key requirement for an IoT device is 10 years battery lifetime for a predefined traffic profile, as specified by 3GPP. Therefore it is important to propose a model which can estimate the battery life of an IoT device. The following three prerequisites are required to estimate the energy consumption of an IoT device:

- **Power Consumption Model of the Modem**: As described in Sect. 2.2, the UE transits between different states based on the state machine and will go through different procedures depending on the type of actions the UE has to

Table 3 Messages transmitted in different procedures, the message size is measured from NB-IoT device SARA-N211

Procedures	Messages	UL/DL	Message size (bits)
Service request	Random access response	DL	72
	RRC connection request	UL	72
	RRC connection setup	DL	144
	RRC connection complete	UL	424
	Service request & data	UL	data + 128
	Service accept	DL	176
	ACK	UL	32
	Total downlink	DL	424
	Total uplink	UL	data + 672
Attach	Random access response	DL	104
	RRC connection request	UL	88
	RRC connection setup	DL	304
	RRC connection complete	UL	424
	Attach request	UL	256
	Identity request	DL	96
	Identity response	UL	176
	Authentication request	DL	432
	Authentication response	UL	264
	Security command	DL	328
	Security complete	UL	240
	Attach accept	DL	1080
	UE enquiry	DL	208
	UE capability	UL	128
	Attach complete	UL	240
	EMM info	DL	488
	Total downlink	DL	2848
	Total uplink	UL	1848
Release	RRC release	DL	72
	ACK	UL	32
TAU	Random access response	DL	72
	RRC connection request	UL	72
	RRC connection setup	DL	144
	RRC connection complete	UL	424
	TAU request	UL	144
	TAU accept	DL	448
	TAU complete	UL	80
	RRC Release	DL	72
	ACK	UL	32
	Total downlink	DL	736
	Total uplink	UL	752

perform. The power consumption model is responsible for estimating the power consumption in each state and procedure that the UE has gone through.

- **Configuration of the Modem**: The actual calculation in the model depends on the settings of the PHY transmission parameters such as TBS, which further depends on MCS, coding rate, the number of allocated RUs/SFs, and the number of repetitions. The configuration of these parameters in the modem depends on the location of the UE, i.e. whether the UE is in good, bad, or extreme coverage conditions.
- **Traffic Profile**: The traffic profile determines how often the UE transmits/receives and how large the transmitted/received data is. It has a big impact on the lifetime of a device.

Once those three prerequisites have been obtained, the battery lifetime can be estimated. The power consumption model has been described in detail in Sect. 2. This section presents how to configure the device to reflect the different coverage scenarios and what is the typical traffic profile of an IoT device, together with the battery lifetime estimation model.

3.1 NB-IoT Device Configurations

The configuration of PHY transmission parameters in terms of MCS and transmission format depends on the received signal-to-noise ratio (SNR) and the target Block Error Rate (BLER). For Long-Term Evolution (LTE) systems, the target BLER is set to be 10% for data channels. The calculation of SNR requires the estimation of the coupling loss. To compare the performance of a NB-IoT device with different coverage levels, three coupling loss values of 144, 154, and 164 dB are selected in this study, representing three coverage scenarios of good, bad, and extreme. These three coverage levels are based on the Maximum Coupling Loss of LTE, LTE-M, and NB-IoT, respectively [18, 19].

It is assumed that NB-IoT device uses a single subcarrier of 15 KHz in the uplink. The UE is assumed to use the maximum transmission power of 23 dBm for uplink transmissions. The noise figure at the receiver is assumed to be 3 dB. There is only one antenna element for transmit and receive in the UE. With these assumptions, the link budget can be calculated for each coverage scenario, and the results are summarized in Table 4.

Once the received SNR is calculated for each coverage scenario, the optimal configuration of MCS targeting for a 10% BLER can be found either from link level performance curves (e.g. BLER vs. SNR for different MCS) or from analytical approximations.

The combination of MCS and resource assignment in terms of number of subframes (for downlink) or resource units (for uplink) determines the TBS. The standard specifies the available transmission formats in terms of number of

Table 4 Received SNR for NB-IoT in three different coverage scenarios

Parameter	Coverage scenario		
	Good	Bad	Extreme
Maximum TX power	23 dBm	23 dBm	23 dBm
Targeted coupling loss	144 dB	154 dB	164 dB
Occupied channel bandwidth	15 KHz	15 KHz	15 KHz
Receiver noise figure	3 dB	3 dB	3 dB
Thermal noise density	−174 dBm/Hz	−174 dBm/Hz	−174 dBm/Hz
Effective noise power	−129.24 dBm	−129.24 dBm	−129.24 dBm
Received SNR	8.24 dB	−1.76 dB	−11.76 dB

Table 5 TBS table for NPUSCH [9]

TBS index	RUs index (I_{RU})/number of RUs assigned							
	0/1	1/2	2/3	3/4	4/5	5/6	6/8	7/10
0	16	32	56	88	120	152	208	256
1	24	56	88	144	176	208	256	344
2	32	72	144	176	208	256	328	424
3	40	104	176	208	256	328	440	568
4	56	120	208	256	328	408	552	680
5	72	144	224	328	424	504	680	872
6	88	176	256	392	504	600	808	1000
7	104	224	328	472	584	712	1000	
8	120	256	392	536	680	808		
9	136	296	456	616	776	936		
10	144	328	504	680	872	1000		
11	176	376	584	776	1000			
12	208	440	680	1000				

subframes or resource units vs. TBS index for NB-IoT [9]. Table 5 shows the TBS table for NPUSCH.

The amount of allocated resources in terms of number of subframes or resource units can be calculated once the payload size and the MCS (i.e. TBS index) have been determined, by looking up Table 5. An example of PHY transmission configuration with different scenarios for NB-IoT device is listed in Table 6, assuming a payload size of 800 bits.

3.2 *Traffic Profile*

The traffic model defines how often the UE transmits/receives and how big the transmitted/received data is. It plays a critical role in estimating the battery life of a device. The traffic patterns for different verticals have been defined in [20]. The IoT

Table 6 Example PHY transmission configuration for NB-IoT assuming payload size of 800 bits

Parameters	Coverage scenario		
	Good	Bad	Extreme
TBS Index for NPUSCH and NPDSCH	10	2	0
TBS Index for signalling messages	2	2	0
Number of RUs	5	20	30
Repetitions for NPUSCH and NPDSCH	1	4	32
Repetitions for signalling messages	2	8	64

Table 7 IoT use cases and traffic profiles for some of the verticals [21–24]

Verticals	Use cases	Traffic pattern	Periodicity	Latency	UL/DL
Smart city	Traffic management	UE-initiated Periodic	10 min	5 s	Mostly UL
	Smart parking	UE-initiated Event-driven	Irregular Infrequent	10 s	Mostly UL
	Light automation	UE-initiated Event-driven	Irregular Infrequent	15 s	Mostly UL
	Urban condition monitoring	UE-initiated Periodic	15 min	5 s	Mostly UL
	Waste management	UE-initiated Event-driven	Irregular Infrequent	30 s	Mostly UL
	Structural health monitoring	UE-initiated Periodic	15 min	5 s	Mostly UL
Smart energy	Electrical grid	Network-initiated Periodic	Hourly/daily	10 ms	Mostly UL
	Smart metering	UE-initiated Periodic	Hourly/daily	15 s	Mostly UL
	Distribution grid	Network-initiated Periodic	Hourly/daily	100 ms	Mostly UL
Smart transport	Vehicle tracking	UE-initiated Periodic	30 s	10 s	Mostly UL
	Fleet management	UE-initiated Periodic	1 s	10 ms	Mostly UL
	Shipment monitoring	UE-initiated Periodic	15 min	5 s	Mostly UL

use cases and traffic profiles associated with some of the verticals are presented in Table 7.

It can be seen that most of the traffic is uplink dominated with periodic traffic pattern. Therefore in this study, a deterministic uplink traffic model is used, resembling the behaviour of sensor devices, where data is transmitted towards the eNB periodically with a predefined interval.

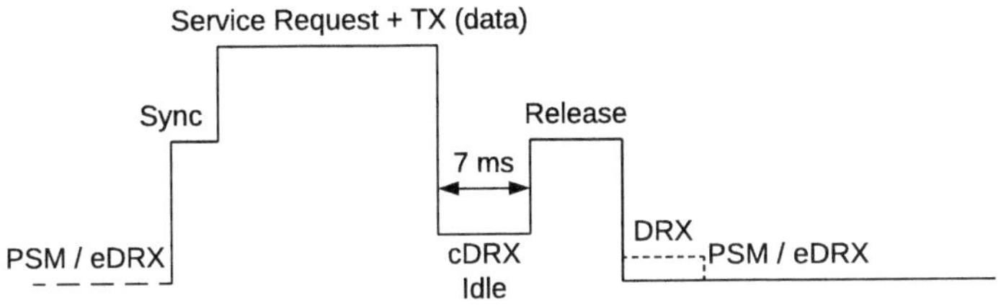

Fig. 8 An example of uplink UE transmission cycle for NB-IoT

Table 8 Different types of delay tolerated for NB-IoT

Delay type	Max delay [ms]
TX → DCI	3
DCI → TX	8
RX → DCI	12
DCI → RX	4
TX → ACK	3
RX → ACK	12

A transmit cycle is defined as the time interval from the start of a transmission to the time instance just before the start of the next transmission. The UE will go through certain procedures in a cycle, where multiple messages between the UE and eNB have to be exchanged to establish and release a connection. An example of uplink UE transmission cycle for NB-IoT is illustrated in Fig. 8, assuming that the UE is in EMM-REGISTERED state. If the UE is in EMM-DEREGISTERED state, e.g. the UE is powered up for the first time, the attach procedure shall be used instead of the service request.

3.3 *Battery Lifetime Estimation*

Once the power consumption model for each UE state and procedure has been obtained, the traffic profile has been defined, and the PHY transmission parameters have been determined according to different coverage scenarios, the average power consumption of the device can be calculated.

The total energy consumption during a cycle where only one transmission occurs, denoted as t_{cycle}, can be divided into the energy consumption in the active period and in the idle period. The energy consumption in the active period consists of synchronization, connection setup, data transmission, connection release, and the delays occurring between each message. The delays between different message transmissions [1] are summarized in Table 8. It is assumed that for NB-IoT the UE is using a power consumption close to the cDRX sleep ($P_{\text{cDRX}_{\text{sleep}}}$) during these delays.

The energy consumption in the active period of a UE cycle as exampled in Fig. 8 can be calculated as:

$$E_{\text{active}} = E_{\text{conn}} + E_{\text{TX}} + E_{\text{RX}} + E_{\text{release}} + E_{\text{delay}} \tag{12}$$

$$E_{\text{delay}} = t_{\text{delay}} \cdot P_{\text{cDRX}_{\text{sleep}}}$$

$$E_{\text{conn}} = \begin{cases} E_{\text{Sync}} + E_{\text{ServiceRequest}} & \text{if PSM is used} \\ E_{\text{ServiceRequest}} & \text{if DRX or eDRX is used} \end{cases}$$

The values of E_{Sync}, $E_{\text{ServiceRequest}}$, E_{TX}, E_{RX}, and E_{release} can be obtained following the power consumption models described in Sects. 2.4 and 2.5.

The energy consumption in the idle period of a UE cycle can be calculated as:

$$E_{\text{idle}} = \begin{cases} E_{\text{DRX}_{\text{cycle}}} \cdot \lceil \frac{T_{3324}}{t_{\text{DRX}_{\text{cycle}}}} \rceil + t_{\text{PSM}_{\text{sleep}}} \cdot P_{\text{PSM}_{\text{sleep}}} & \text{if PSM is used} \\ E_{\text{DRX}_{\text{cycle}}} \cdot \lceil \frac{T_{\text{PTW}}}{t_{\text{DRX}_{\text{cycle}}}} \rceil + t_{\text{eDRX}_{\text{sleep}}} \cdot P_{\text{eDRX}_{\text{sleep}}} & \text{if eDRX is used} \end{cases} \tag{13}$$

$$t_{\text{PSM}_{\text{sleep}}} = t_{\text{cycle}} - \left(t_{\text{conn}} + t_{\text{TX}} + t_{\text{RX}} + t_{\text{release}} + t_{\text{delay}} + T_{3324}\right)$$

$$t_{\text{eDRX}_{\text{sleep}}} = t_{\text{cycle}} - \left(t_{\text{conn}} + t_{\text{TX}} + t_{\text{RX}} + t_{\text{release}} + t_{\text{delay}} + T_{\text{PTW}}\right)$$

where $E_{\text{DRX}_{\text{cycle}}}$ is the energy consumption of one DRX cycle.

By combining Equations (12) and (13), the average power consumption of the modem in a transmit cycle can be calculated as:

$$P_{\text{modem}} = \frac{E_{\text{active}} + E_{\text{idle}}}{t_{\text{cycle}}} \tag{14}$$

and the battery lifetime of the device can be estimated by:

$$L = \frac{C_{\text{bat}} \cdot SF_{\text{bat}}}{P_{\text{modem}} + P_{\text{device}}} \tag{15}$$

where C_{bat} is the battery capacity, SF_{bat} is the battery safety factor accounting for self-discharge, and P_{device} is the sensor circuitry average power consumption, i.e. all but without the modem.

4 Measurement Results

This section presents the validation of the power consumption model, as well as the battery lifetime estimation of an NB-IoT device Sara-N211.

Table 9 Measured power or energy consumption for NB-IoT device Sara-N211 in different states

State	Sub-state	Measurements for N211
TX	TX @ 23 dBm (P_{TX}) [mW]	742.858
	TX gaps ($P_{\mathrm{TX}_{\mathrm{gaps}}}$) [mW]	153.6
RX	RX (P_{RX}) [mW]	222.134
	RX gaps ($P_{\mathrm{RX}_{\mathrm{gaps}}}$) [mW]	177.422
cDRX	Idle ($P_{\mathrm{cDRX}_{\mathrm{Sleep}}}$) [mW]	21.337
	On energy ($E_{\mathrm{cDRX}_{\mathrm{onDuration}}}$) [mJ]	0.885
	On duration ($t_{\mathrm{cDRX}_{\mathrm{onDuration}}}$) [ms]	7.926
DRX	Idle ($P_{\mathrm{DRX}_{\mathrm{Sleep}}}$) [$\mu$W]	4.07
	DRX sync energy ($E_{\mathrm{DRX}_{\mathrm{sync}}}$) [mJ]	0.01
	DRX sync time ($t_{\mathrm{DRX}_{\mathrm{sync}}}$) [ms]	247.5
PSM	Idle ($P_{\mathrm{PSM}_{\mathrm{sleep}}}$) [$\mu$W]	9.5

4.1 Characterization of the Modem

A number of test cases have been executed to characterize the power consumption of the modem in different states. The measured power or energy consumption of a NB-IoT device Ublox Sara-N211 is summarized in Table 9, which serves as inputs to the analytical power consumption models described in Sects. 2 and 3.

4.2 Model Validation

To validate the analytical model, the energy consumption of an NB-IoT device has been measured for different configurations. The device transmits every hour with a fixed payload size of 576 bits. The battery capacity is set to be 5 Wh, assuming an ideal case without the self-discharge effect. Since the focus of this work is on the modem power consumption, the power consumption of the sensor circuitry is assumed to be zero, which means that all of the available battery capacity is allocated to the modem. The selected MCS (i.e. TBS index) ranges from 0 to 10, and the number of repetitions ranges from 1 to 8, with different combinations between the two parameters. PSM is used to conserve energy due to the relatively low power consumption during the sleep period as compared to eDRX. All measurements are performed in LTE band 20 ($\sim$800 MHz) using a single, in-band 15 KHz subcarrier. The measurement settings used for model validation and battery lifetime estimation are summarized in Table 10.

Figure 9 shows the measured and estimated energy consumption per transmit cycle (1 h) for a NB-IoT device Sara-N211 with different configurations. It can be seen that the energy consumption decreases as the MCS (i.e. TBS index)

Table 10 Measurement settings used for the model validation and battery lifetime estimation

Parameter	Setting
UE transmit cycle (t_{cycle})	1 h
Payload size	576 bits
Battery capacity (C_{bat})	5 Wh
Battery safety factor (SF_{bat})	1
Sensor average power consumption (P_{device})	0 W
Uplink Tx bandwidth	1 subcarrier of 15 KHz
TBS Index for NPUSCH and NPDSCH	{0, 2, 10}
TBS Index for signalling messages	{0, 2}
Repetitions for NPUSCH and NPDSCH	{1, 2, 4, 8}
Repetitions for signalling messages	{Repetitions for data} × 2
Power saving technique	PSM
T_{3324} timer	60 s
T_{3412} timer	2 h

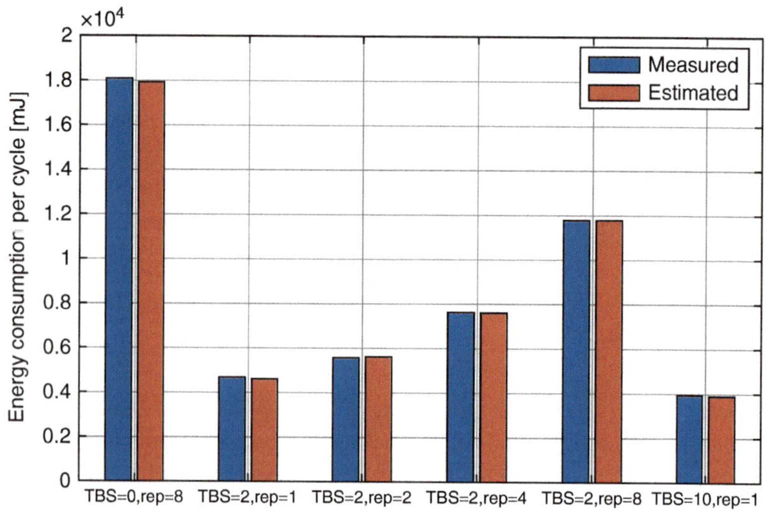

Fig. 9 The measured and estimated energy consumption per transmit cycle (1 h) for NB-IoT device Sara-N211 with different configurations

increases. Though not shown here, it is found from the measurements that the power consumption is independent on the MCS for both uplink and downlink. However, the selection of MCS determines how long the device has to stay in transmit state, which impacts the overall power consumption significantly. Also it is shown clearly in the figure that increasing the number of repetitions would increase the total energy consumption as expected. Figure 9 demonstrates that the proposed power consumption model matches very well with the measurement results.

Table 11 Estimated battery lifetime for NB-IoT device Sara-N211 with different transmit intervals and coverage scenarios

Transmit cycle	Good	Bad	Extreme
1 h ($T_{3412} = 2$ h)	1.35 year	1.12 year	0.51 year
6 h ($T_{3412} = 2$ h)	5.47 year	4.15 year	1.53 year
24 h ($T_{3412} = 2$ h)	9.00 year	6.18 year	1.89 year
24 h ($T_{3412} = 6$ h)	10.83 year	8.13 year	3.58 year

4.3 *Battery Lifetime Estimation*

Next we apply Eq. (15) and settings in Table 10 to estimate the battery lifetime with different transmit cycles and coverage scenarios, e.g. good, bad, and extreme, as defined in Table 6. The estimated battery lifetime is listed in Table 11. With 1 h transmit cycle, the lifetime for NB-IoT device Sara-N211 can last for 1.35 year and 0.51 year in the good and extreme scenarios, respectively. Increasing the transmit cycle to 24 h, the lifetime increases up to 9 years in the good scenario, getting close to the 10-year battery lifetime requirement specified by 3GPP [3]. Note that in Table 10, the TAU procedure with periodicity (i.e. T_{3412} timer) of 2 h is considered in the battery lifetime estimation. The TAU procedure is energy expensive and will dominate the total energy consumption when the transmit cycle length is much longer than the TAU periodicity, especially in unfavourable radio conditions. That explains why in the extreme scenario, the battery lifetime doesn't increase very much when the transmit cycle increases from 6 to 24 h. Increasing the periodicity of TAU will decrease the total energy consumption within one transmit cycle, as the occurrence of TAU decreases. In that way, the overall battery lifetime estimate can exceed 10 years. The last row in Table 10 lists the estimated battery lifetime when T_{3412} is increased to 6 h, which shows that the 10-year battery lifetime requirement can be satisfied.

5 Conclusions

It is expected that the number of IoT devices will experience massive growth in the coming years. Since many of the IoT devices will be deployed in hard-to-access locations, ensuring a long battery lifetime is of great importance. 3GPP specifies a 10-year battery lifetime requirement for a predefined traffic profile. To validate this claim, a proper power consumption model is needed. This chapter presented an empirical power consumption model for IoT device battery lifetime estimation. Specifically, the focus is on 3GPP standardized LPWAN technology NB-IoT.

It starts with the introduction of the states and the procedures, followed by the detailed modelling of each state and the main procedures. The power consumption of any UE behaviour can be modelled by combining the corresponding states and procedures from the state transition diagram.

Besides the power consumption model, the next two prerequisites required for the estimation of the battery lifetime are the configuration of the modem and the applied traffic profile. The configuration of PHY transmission parameters depends on the received SNR and the target BLER. An example of three coverage scenarios, namely, good, bad, and extreme, is given with specific configurations of PHY transmission parameters. For the traffic profile, uplink periodic traffic pattern is assumed which resembles the traffic profile of most IoT devices. Once the three prerequisites have been determined, the energy consumption of an IoT device within a transmit cycle can be calculated, and the battery lifetime can be estimated.

The measurement setup has been described, and validation measurements have been performed with different configurations. The results show that the proposed empirical power consumption model matches very well with the measurement results. The battery lifetime is estimated with different transmit intervals and coverage scenarios. It is shown that with proper configuration of the traffic profile, coverage scenario, as well as network configuration parameters (e.g. TAU periodicity), the battery lifetime can last for 10 years as required by 3GPP.

This chapter only considers the power consumption for NB-IoT. Other LPWAN technologies such as Sigfox and LoRa could also be interesting to model and compare with NB-IoT. In addition, only the power consumption model for the modem is considered. Other hardware such as the sensors, the actuators, and the processor also need to be taken into account when estimating the battery lifetime. Furthermore, for accurate estimation of the battery lifetime, the capacity and leakage of the battery should also be taken into account. Those can be the future work of power measurement for LPWAN IoT devices.

References

1. Liberg O, Sundberg M, Wang E et al (2017) Cellular internet of things. Elsevier. ISBN: 978-0-128-12458-1
2. Mangalvedhe N, Ratasuk R, Ghosh A (2016) NB-IoT deployment study for low power wide area cellular IoT. In: Proceedings of IEEE 27th annual international symposium on personal, indoor, and mobile radio communications (PIMRC), Sept 2016
3. 3GPP, Cellular system support for ultra low complexity and low throughput internet of things; (Release 13), 3rd generation partnership project (3GPP), TR 45.820, Version 2.0.0
4. Lauridsen M (2015) Studies on mobile terminal energy consumption for LTE and future 5G. PhD thesis, Aalborg University, Jan 2015
5. Lauridsen M, Krigslund R, Rohr M, Madueno G (2018) An empirical NB-IoT power consumption model for battery lifetime estimation. In: IEEE 87th vehicular technology conference (VTC Spring), July 2018
6. Duhovnikov S, Baltaci A, Gera D et al (2019) Power consumption analysis of NB-IoT technology for low-power aircraft applications. In: IEEE 5th world forum on internet of things (WF-IoT), Apr 2019
7. Casals L, Mir B, Vidal R, Gomez C (2017) Modeling the energy performance of LoRaWAN. Sensors 17(10):1–30

8. 3GPP, General packet radio service (GPRS) enhancements for evolved universal terrestrial radio access network (E-UTRAN) access (Release 14), 3rd generation partnership project (3GPP), TS 23.401, Version 14.11.0
9. 3GPP, Evolved universal terrestrial radio access (E-UTRA); Physical layer procedures (Release 14), 3rd generation partnership project (3GPP), TS 36.213, Version 14.2.0
10. 3GPP, Evolved universal terrestrial radio access (E-UTRA); Physical channels and modulation (Release 14), 3rd generation partnership project (3GPP), TS 36.211, Version 14.13.1
11. 3GPP, Evolved universal terrestrial radio access (E-UTRA); Radio resource control (RRC) (Release 14), 3rd generation partnership project (3GPP), TS 36.331, Version 14.12.0
12. Holma H, Toskala A (2012) LTE for UMTS: evolution to LTE-advanced. Wiley, ISBN: 978-0-470-66000-3
13. 3GPP, Evolved universal terrestrial radio access (E-UTRA); User equipment (UE) procedures in idle mode (Release 14), 3rd generation partnership project (3GPP), TS 36.304, Version 14.7.0
14. 3GPP, Evolved universal terrestrial radio access (E-UTRA); Medium access control (MAC) protocol specification (Release 14), 3rd generation partnership project (3GPP), TS 36.321, Version 14.12.0
15. 3GPP, Mobile radio interface Layer 3 specification; Core network protocols; Stage 3 (Release 14), 3rd generation partnership project (3GPP), TS 24.008, Version 14.9.0
16. 3GPP, Non-access-stratum (NAS) protocol for evolved packet system (EPS); Stage 3 (Release 14), 3rd generation partnership project (3GPP), TS 24.301, Version 14.10.0
17. Andres-Maldonado P, Ameigeiras P, Prados-Garzon J et al (2017) Narrowband IoT data transmission procedures for massive machine-type communications. IEEE Netw 31(6):8–15
18. Vos G, Bergman J, Bitran Y et al (2017) Coverage analysis for LTE-M CAT-M1 devices white paper
19. Ericsson (2017) Ericsson Mobility Report [Online]. Available: https://www.ericsson.com/assets/local/mobility-report/documents/2017/ericsson-mobility-report-june-2017-north-america.pdf
20. Mocnej J, Pekar A, Seah WK, Zolotova I (2018) Network traffic characteristics of the IoT application use cases. Technical Report Series
21. Lorca J, Solana B, Barco R et al. Scenarios, KPIs, use cases and baseline system evaluation. One5G Deliverable D2.1 [Online]. Available: https://one5g.eu/wp-content/uploads/2017/12/ONE5G_D2.1_finalversion.pdf
22. Penza M, Suriano D, Villani MG et al (2014) Towards air quality indices in smart cities by calibrated low-cost sensors applied to networks. In: Sensors
23. Sivanathan A, Sherratt D, Gharakheili H et al (2017) Characterizing and classifying IoT traffic in smart cities and campuses. In: IEEE conference on computer communications workshops (INFOCOM WKSHPS)
24. Lee S, Tewolde G, Kwon J (2014) Design and implementation of vehicle tracking system using GPS/GSM/GPRS technology and smartphone application. In: IEEE world forum on internet of things (WF-IoT)

Dynamic Resource Management in Real-Time Wireless Networks

Tianyu Zhang, Tao Gong, Xiaobo Sharon Hu, Qingxu Deng, and Song Han

1 Introduction

In recent years, we have been witnessing the Internet of Things (IoT) paradigm making its way into industry with purposely designed solutions. A number of industrial IoT (IIoT) technologies have been deployed in areas such as industrial automation, process control, environmental monitoring, security surveillance, and others 7 [1]. Depending on the application domains, many tasks performed by IIoT systems are safety- and mission-critical and thus have stringent requirements on the communication fabric to provide hard real-time performance and reliable information delivery [2]. Taking the process automation industry as an example, control-related applications are sensitive to packet loss and jitter and require transmission reliability of 99.99% [3] and bounded delay at millisecond level (10–100 ms).

In contrast to traditional wired industrial networks, wireless communication technologies are gaining rapid adoption in different industrial sectors, thanks to their easier deployment, reduced maintenance cost, and enhanced mobility of devices [4–7]. This paradigm shift makes real-time wireless networks (RTWNs) become the foundation of many current IIoT applications [8–10].

T. Zhang · X. Sharon Hu
University of Notre Dame, Notre Dame, IN, USA
e-mail: tzhang4@nd.edu; shu@nd.edu

T. Gong · S. Han (✉)
University of Connecticut, Storrs, CT, USA
e-mail: tao.gong@uconn.edu; song.han@uconn.edu

Q. Deng
Northeastern University, Shenyang, China
e-mail: dengqx@mail.neu.edu.cn

N. H. Mahmood et al. (eds.), *Wireless Networks and Industrial IoT*,
https://doi.org/10.1007/978-3-030-51473-0_7

However, RTWNs face unique challenges that distinguish them from traditional industrial control systems [5]. First of all, it is challenging to meet the stringent timing requirements of control tasks running in RTWNs. Traditional carrier-sense multiple access with collision avoidance (CSMA/CA) mechanism, which is a probabilistic scheme for channel access, can cause unexpected transmission collision and delayed message delivery. In contrast, RTWNs typically adopt time-division multiple access (TDMA)-based media access control (MAC) mechanisms to achieve deterministic end-to-end message delivery. Packet scheduling in RTWNs thus plays a critical role in achieving the desired performance but is a challenging problem especially when RTWNs start to be deployed over large geographic areas.

Secondly, almost all RTWNs need to deal with unexpected *disturbances* since industrial systems are usually open to environment forces. This further exacerbates the challenge of developing effective techniques for resource management in RTWNs. Unexpected disturbances in general can be classified into external disturbances of the physical plants (e.g., sudden pressure change in an oil pipeline) and internal disturbances within the network infrastructure (e.g., link failure due to multiuser interference or weather-related changes in channel signal-to-noise ratio (SNR)). To assure stable and safe operations in the presence of external disturbances, corresponding control tasks must increase their demands to the network resources (e.g., providing higher sampling and control rates). On the other hand, internal disturbances may impact the network fabric and trigger faults in the network which can also reduce the network's capacity used to respond to external disturbances. Therefore, disturbances not only impact the RTWN's demand for real-time network resources but also the supply of those resources.

To handle unexpected disturbances without reserving overly pessimistic amount of resources, carefully designed dynamic decision-making must be incorporated. However, finding the right level of dynamic decision-making is not trivial as it is a trade-off between efficient usage of network resources (e.g., no wasted bandwidth) and achievable quality of service (QoS) (e.g., the number of messages missing end-to-end timing constraints).

In this chapter, we present in detail a suite of dynamic resource management techniques in RTWNs to tackle the above challenges. We first discuss a hybrid dynamic packet scheduling framework, referred to as HD-PaS, to handle external disturbances which cause abruptly increased network traffic [11, 12]. HD-PaS overcomes the disadvantage of traditional centralized resource management methods under which a centralized control node undertakes all the work to handle external disturbances. By offloading the computation from the centralized controller node to local nodes, HD-PaS only executes a lightweight algorithm in the controller node to determine the corresponding response to the external disturbance. In this way, better QoS can be achieved. To further handle the internal disturbances, in the second part of this chapter, we introduce a reliable dynamic packet scheduling framework, called RD-PaS [13]. RD-PaS can not only dynamically react to online network traffic changes caused by external disturbances but also construct reliable static and dynamic schedules to deal with packet loss caused by internal disturbances. Both HD-PaS and RD-PaS rely on a centralized controller node to dynamically make online decisions. Such centralized resource management approaches will cause scalability issue when the network size grows. To address this issue, at the end of the chapter, a fully distributed dynamic packet scheduling framework, referred to as FD-PaS, is introduced [14, 15]. The key challenge in the design of a dynamic and distributed RTWN resource management approach lies in how to generate timely responses to the disturbances without leveraging any centralized coordination. FD-PaS incorporates several key advances in both algorithm design and data link layer protocol design to enable individual nodes to make online decisions locally and achieve guaranteed response time to unexpected disturbances.

The remainder of this chapter is organized as follows. A typical RTWN system model is introduced in Sect. 3. We present the designs of HD-PaS, RD-PaS, and FD-PaS resource management frameworks in Sects. 4, 5, and 6, respectively. In Sect. 7, we introduce the implementation of FD-PaS on a RTWN testbed to show its applicability in real-world RTWNs. Section 8 concludes the chapter.

2 Resource Management Researches in RTWNs

Network resource management in RTWNs in the presence of unexpected disturbances has drawn a lot of attention in recent years. Traditional static packet scheduling approaches (e.g., [16–18]), where decisions are made offline or only get updated infrequently can support deterministic real-time communication, but either cannot properly handle unexpected disturbances or must make rather pessimistic assumptions. Many centralized dynamic scheduling approaches for handling internal disturbances have been proposed (e.g., [19–21]). Studies on addressing external disturbances are relatively few and mostly rely on centralized decision-making. The approach in [22] stores a predetermined number of link layer schedules in the system and chooses the appropriate one when disturbances are detected. However, this approach is either incapable of handling arbitrary disturbances or needs to make some approximation. Both [23] and [24] support admission control in response to adding/removing tasks for handling disturbances in the network. They however do not consider scenarios when not all tasks can meet their deadlines. The protocol in [25] proposes to allocate reserved slots for occasionally occurring emergencies (i.e., disturbances) and allow regular tasks to steal slots from the emergency schedule when no emergency exists. However, how to satisfy the deadlines of regular tasks in the presence of emergencies is not considered.

The IEEE 802.15.4e Time Slotted Channel Hopping (TSCH) network, due to the combination of time-division multiplexing (TDM), time synchronization, and the time formatted into slotframes, results in a deterministic wireless MAC standard. In recent years, a number of algorithms have been designed for packet scheduling in TSCH networks, in both centralized (e.g., [26–28]) and distributed manner (e.g., [29–31]). Most of those approaches, however, assume static network topologies and fixed network traffic which limit their applications in dynamic networks. To overcome this drawback, [32] proposes Orchestra, a distributed scheduling solution that schedules packet transmissions in TSCH networks to support real-time applications. However, Orchestra does not consider real-time constraint, i.e., ignores the hard deadlines associated with tasks running in the network. It only provides best effort but no guarantee on the end-to-end latency of each task.

Recently, an IETF working group, named 6TiSCH, has been formed to investigate IPv6 connectivity over the TSCH mode of IEEE 802.15.4e protocol [33]. 6TiSCH architecture offers a suite of features for achieving industrial-grade deterministic performance for end-to-end communication. Among these features, Track mechanism is one of the most promising one which is established over the network for flows to create guarantees of minimum and maximum end-to-end latency. Track is essentially the result of a network resource reservation for certain multi-hop paths which are subject to workload changes caused by disturbances. And 6TiSCH allows other packets to reuse the reserved slots within the Tracks when they are not currently used. However, at the time of writing, there is no specification about how to create and manage Tracks to guarantee end-to-end packet deadlines in the presence of disturbances.

3 RTWN System Model

In a typical RTWN, multiple sensors and actuators are wirelessly connected to a controller node directly or through relay nodes. We refer to non-controller nodes as device node in this chapter. All device nodes have routing capability and are equipped with a single omnidirectional antenna to operate on a single channel in half-duplex mode.[1] The network is modeled as a directed graph $G = (V, E)$, where the node set $V = \{\{V_0, V_1, \dots\}, V_c\}$ and V_c represents the controller node. V_c connects to all the nodes via some routes and is responsible for executing relevant control algorithms. V_c also contains a network manager which is responsible for network configuration and resource allocation.

The system runs a fixed set of tasks $\mathcal{T} = \{\tau_0, \tau_1, \dots, \tau_n\ \tau_{n+1}\}$. τ_i $(0 \le i \le n)$ is a unicast task following a designated routing path with H_i hops. It periodically generates a packet which originates at a sensor node, passes through the controller node, and delivers a control message to an actuator. The k-th instance of task τ_i, referred to as packet $\chi_{i,k}$, is associated with release time $r_{i,k}$ and deadline $d_{i,k}$. τ_{n+1} is a broadcast task running on the controller node and disseminates necessary network configuration to all nodes in the network by following a predetermined broadcast graph [16]. Following industrial practice, RTWNs adopt time-division multiple access (TDMA)-based data link layer. Every node follows a given schedule to transmit or receive packets, and the transmission of the packet on each hop must

[1] In practice, RTWNs usually apply multichannel communication. For simplicity, in this chapter, we illustrate the resource management frameworks under a single channel assumption.

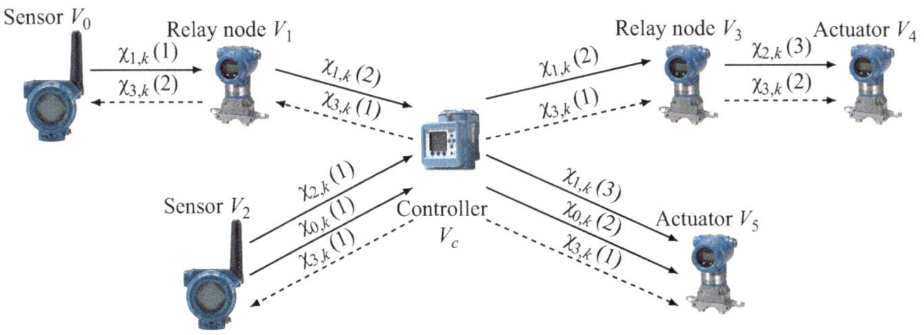

Fig. 1 An example RTWN with three unicast tasks and one broadcast task

be completed in a single time slot. Figure 1 gives an example RTWN with three unicast tasks and one broadcast task running on seven nodes.

When external disturbances (e.g., sudden change in temperature or pressure) occur, many industrial applications would require more frequent sampling and control actions, which in turn increase network resource demands. To capture such abrupt increase in network resource demands, many task models can be applied. In this chapter, we adopt the rhythmic task model [34] which has been shown to be effective for handling disturbances in event-triggered control systems [35]. In the rhythmic task model, each unicast task τ_i has two states: *nominal state* and *rhythmic state*. In the nominal state, τ_i follows a nominal period P_i and a nominal relative deadline $D_i (\leq P_i)$, which are all constants. When an external disturbance occurs, τ_i enters the rhythmic state in which its period and relative deadline are first reduced in order to respond to the disturbance and then gradually return to their nominal values by following some nondecreasing pattern. We use vectors $\overrightarrow{P_i} = [P_{i,x}, x = 1, \ldots, R]^T$ and $\overrightarrow{D_i} = [D_{i,x}, x = 1, \ldots, R]^T$ to represent the periods and relative deadlines of τ_i when it is in the rhythmic state. As soon as τ_i enters the rhythmic state, its period and relative deadline adopt sequentially the values specified by $\overrightarrow{P_i}$ and $\overrightarrow{D_i}$, respectively. τ_i returns to the nominal state when it starts using P_i and D_i again.

To simplify the notation, we refer to any task currently in the rhythmic state as *rhythmic task* and denote it as τ_0, while task τ_i $(1 \leq i \leq n)$ is a *periodic task* which is currently not in the rhythmic state. As shown in Fig. 2, when τ_0 enters the rhythmic state, we also say that the system switches to the *rhythmic mode*. The system returns to the *nominal mode* when the external disturbance has been handled, typically some time after τ_0 returns to the nominal state. Since disturbances may cause catastrophe to the system, the rhythmic task has a hard deadline when the system is in the rhythmic mode, while periodic tasks can tolerate occasional deadline misses.

Without loss of generality, one can assume that τ_0 enters the rhythmic state at a certain release time (denoted as $t_{n \to r}$) and returns to the nominal state at another one (denoted as $t_{r \to n}$) after a certain number of rhythmic periods specified by $\overrightarrow{P_i}$.

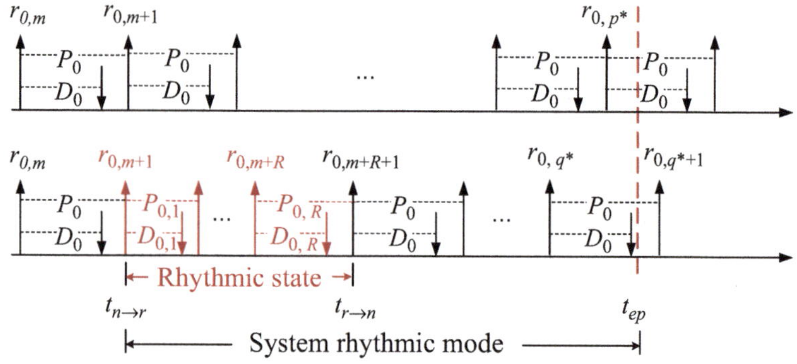

Fig. 2 Timing parameters of the rhythmic task τ_0 in the system rhythmic mode. Top and bottom subfigures denote the nominal and actual release times and deadlines of τ_0, respectively

Any packet of τ_0 released in the system rhythmic mode is referred to as a *rhythmic packet*, while the packets of task τ_i $(1 \leq i \leq n)$ are *periodic packets*. The delivery of packet $\chi_{i,k}$ at the h-th hop is referred to as a transmission.

Traditionally, RTWNs employ link-based scheduling (LBS) to allocate time slots for individual tasks where each slot is allocated to a link by specifying the sender and receiver [36]. If packets from different tasks share a common link and are both buffered at the same sender, their transmission order is decided by a node-specified policy (e.g., FIFO). This approach introduces uncertainty in packet scheduling and may violate the end-to-end (e2e) timing constraints on packet delivery. To tackle this problem, *transmission-based scheduling (TBS)* and *packet-based scheduling (PBS)* are proposed in [11] and [37], respectively, to construct deterministic schedules. Each of these two scheduling models has its own advantages and disadvantages and is preferred in different usage scenarios as discussed in [37].

In the TBS model, each time slot is allocated to the transmission of a specific packet at a particular hop or kept idle. Once the network schedule is constructed, packet transmission in each time slot is unique and fixed. In the PBS model, each time slot is allocated to a specific packet or kept idle. Within each time slot assigned, every node along the packet's routing path decides the action to take (e.g., transmit, receive, or idle), depending on whether the node has received the packet or not. Table 1 gives an example of the time slot allocation for task τ_0 ($V_2 \rightarrow V_c \rightarrow V_5$) in Fig. 1. In the TBS model, each time slot is allocated to a dedicated hop. Particularly, both slot 0 and slot 1 are allocated to the first hop in case the first transmission fails in the first slot. In the PBS model, slots are allocated to each packet of τ_0. Specifically, sensor node V_2 uses the first slot to transmit its first hop transmission, while the second slot (slot 1) can be used to transmit both hops depending on whether the first transmission succeeds in slot 0.

When an internal disturbances occurs, packet transmissions may fail, which can significantly affect the timely delivery of real-time packets. To capture such packet loss, packet delivery ratio (PDR) is used to represent the probabilistic transmis-

Table 1 An example of time slot allocation in TBS model and PBS model

	Slot 0	Slot 1	Slot 2
TBS model	$V_2 \rightarrow V_c$	$V_2 \rightarrow V_c$	$V_c \rightarrow V_5$
PBS model	$V_2 \rightarrow V_c$	$V_2 \rightarrow V_c$ $V_c \rightarrow V_5$	$V_c \rightarrow V_5$

Table 2 Task parameters for the motivational example

Task	Routing path	P_i	D_i	$\overrightarrow{P_i}$	$\overrightarrow{D_i}$
τ_0	$V_2 \rightarrow V_c \rightarrow V_5$	10	9	$[4, 6]^T$	$[3, 5]^T$
τ_1	$V_1 \rightarrow V_c \rightarrow V_5$	10	8	N/A	N/A
τ_2	$V_2 \rightarrow V_c \rightarrow V_3 \rightarrow V_4$	10	7	N/A	N/A
τ_3	$V_c \rightarrow *$[a]	10	10	N/A	N/A

[a] Task τ_3 is a broadcast task, which has two hops. The first hop is from V_c to V_1, V_2, V_3, V_5, and the second hop is from V_1 (V_3) to V_0 (V_4)

sion success rate on each link. To handle internal disturbances, a retransmission mechanism is commonly employed in RTWNs [38, 39]. Specifically, if a sender node does not receive any ACK from the receiver node within the current slot, it automatically retransmits the packet in the next possible time slot. Table 1 gives an example schedule with the retransmission mechanism.

To quantify the reliability requirement of the e2e packet delivery for each task, a *required* e2e PDR for τ_i is introduced. The transmission of any packet of τ_i is reliable if and only if the achieved e2e PDR of τ_i is larger than or equal to the required value.

4 HD-PaS Framework

In this section, we first give a motivational example to show the drawbacks of traditional centralized scheduling approaches in handling external disturbances. We then describe the hybrid dynamic packet scheduling framework HD-PaS.

4.1 Motivational Example

Consider an example RTWN shown in Fig. 1. It consists of four tasks (τ_0, τ_1, τ_2, and τ_3) running on seven nodes ($V_0, \ldots, V_5$ and V_c) where V_0 and V_2 are sensor nodes, V_4 and V_5 are actuator nodes, V_1 and V_3 are relay nodes, and V_c is the controller node. Task τ_0 is the rhythmic task. Task τ_1 and τ_2 are periodic tasks and task τ_3 is the broadcast task. Their routing paths, periods and relative deadlines, as well as $\overrightarrow{P_0}$ and $\overrightarrow{D_0}$ for τ_0 are given in Table 2.

The periodic tasks and rhythmic task are synchronous, and all of their first packets $\chi_{0,1}$, $\chi_{1,1}$, and $\chi_{2,1}$ are released at time slot 0. When the system starts, it uses the predetermined static schedule which is of length 10 repeatedly as shown in Fig. 3a. Here, (i, h) within a slot indicates that this slot is allocated to the h-th hop of task τ_i. Suppose that at time slot 10, τ_0 enters the rhythmic state (i.e., $t_{n\to r} = 10$). Based on $\overrightarrow{P_0}$ and $\overrightarrow{D_0}$ in Table 2, τ_0 returns to the nominal state at time slot $t_{r\to n} = t_{n\to r} + P_{0,1} + P_{0,2} = 20$. If we continue to use the static schedule after $t_{n\to r}$, the rhythmic packet $\chi_{0,2}$ released at time slot 10 would miss its deadline at time slot 13. Therefore, the network cannot properly respond to the period and deadline changes of τ_0 using the static schedule.

In a centralized dynamic scheduling approach, the controller node constructs a temporary schedule for the network in the rhythmic mode and broadcasts the differences between the dynamic and static schedule to each device node. The system resumes the static schedule when it returns to the nominal mode. Figure 3b shows one possible dynamic schedule, which can accommodate all rhythmic and periodic packets, but introduces six updated slots (11, 12, and 15–18) with respect to the static schedule in Fig. 3a. These updated slots must be piggybacked to a broadcast packet and propagated to all nodes in the RTWN. Since the payload size of a broadcast packet is always bounded, the maximum number of allowed updated slots (NUT) is limited. Suppose, in this example, the limit equals to 4. Then the dynamic schedule in Fig. 3b cannot be piggybacked to one broadcast packet. The online scheduling framework (OLS) proposed in [35] considers the constraint on NUT and drops some periodic packets to satisfy such constraint. Figure 3c shows

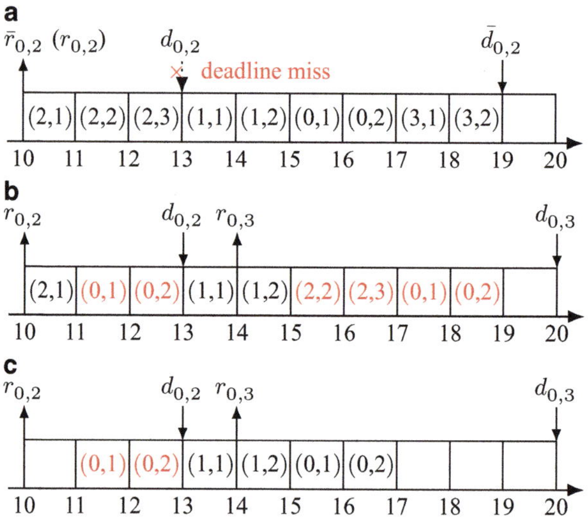

Fig. 3 (**a**) An example static schedule; (**b**) A dynamic schedule with six updated slots; (**c**) The dynamic schedule created by OLS. $\bar{r}_{0,2}$, and $\bar{d}_{0,2}$ denote the release time and deadline of $\chi_{0,2}$ when τ_0 is in the nominal state

the dynamic schedule constructed by OLS, which updates only two slots, with the cost of dropping one periodic packet ($\chi_{2,2}$) to satisfy both the timing constraint of rhythmic task τ_0 and the NUT constraint.

4.2 *Overview of HD-PaS*

It can be observed that in a centralized approach, the restriction on NUT is mainly due to the choice that the controller undertakes all the work to handle disturbances, while other device nodes only need to update its own schedule according to the slot update information received from the controller. Such an approach implicitly assumes that device nodes have no local computing capability, which however is not true for RTWNs nowadays. Therefore, a hybrid dynamic packet scheduling framework, referred to as HD-PaS, is proposed to leverage local computing capability at individual nodes to achieve better performance in terms of fewer dropped packets and more feasible task sets than centralized approaches.

Figure 4 gives an overview of how HD-PaS works. After the system initialization, when a broadcast packet containing the task and routing information is received

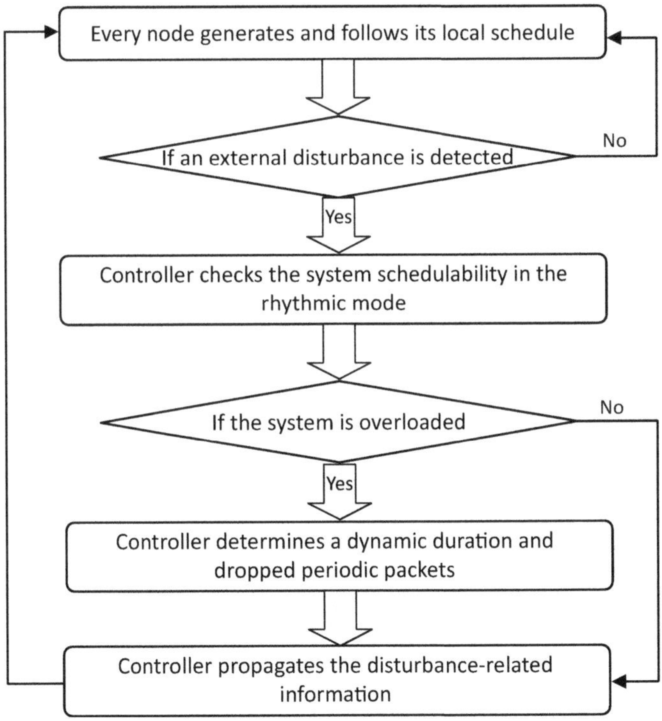

Fig. 4 An overview of the HD-PaS framework

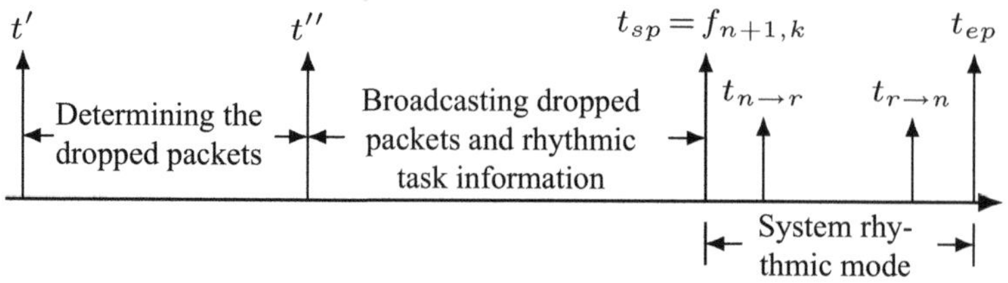

Fig. 5 Network operations after a disturbance is reported to the controller V_c. t' denotes the time when V_c receives the rhythmic event request. t'' denotes the time when V_c sends the first hop of the broadcast packet

from the controller, each node generates the local schedule according to a designated scheduling policy and then follows the schedule to operate (e.g., EDF). To generate such a schedule, each node uses the task and routing information to determine for each time slot whether it should send/receive a particular packet or stay idle. This can be done by simply following the EDF simulation process. Since every node maintains the same task and routing information, the schedules generated at different nodes are consistent. All packets can meet their nominal deadlines as long as the nominal task set utilization is less than or equal to 1 according to EDF [40].

When external disturbance is detected, the sensor node sends a rhythmic event request via τ_0 (following the current schedule) to the controller. Upon receiving the request at time t' (see Fig. 5)[2], V_c first checks the schedulability of the system assuming τ_0 will be in the rhythmic state. If the system is overloaded, the controller determines the time duration of the system rhythmic mode and the dropped periodic packets to guarantee the deadlines of all rhythmic packets. V_c then piggybacks the start time of the duration (t_{sp} in Fig. 5) and the task information about dropped packets and the rhythmic task (task ID with corresponding $\overrightarrow{P_i}$ and $\overrightarrow{D_i}$) to a broadcast packet and disseminates it to all nodes in the network at time t''. Otherwise, only t_{sp} and the rhythmic task information needs to be broadcast. Thus, instead of broadcasting the entire updated schedule, HD-PaS only piggybacks the indices of the packets to be dropped to the broadcast packet when the system is overloaded. Upon receiving such broadcast information at or before t_{sp}, each node generates its local schedule accordingly using the updated information, and the system enters the rhythmic mode at the start point t_{sp}.

In order to ensure HD-PaS works properly, several challenges need to be tackled. First, ideally, each node could construct and store the entire schedule for the hyperperiod in the nominal mode. This is, however, not practical due to limited computing capability and memory on device nodes. Second, since constructing a schedule takes time, such computation should not occur when the node is supposed

[2] A system does not go to the rhythmic mode immediately after a disturbance is detected. It only enters the rhythmic mode (and τ_0 enters the rhythmic state) after each device receives the broadcast packet at the start point t_{sp}.

to send or receive packets. Third, to allow fast response to disturbances, an efficient method is needed at the controller node to determine which packets to drop.

HD-PaS contains two key functions, local schedule generation and dynamic schedule generation, to address the above challenges. We shall describe their details in the following subsections.

4.3 Local Schedule Generation

The basic idea of local schedule generation is to incrementally construct the global schedule of the whole network and store the portion of the schedule that it involves in as its local schedule during run time. Specifically, HD-PaS follows two design principles when constructing the schedule at each node: (i) construct one segment of the entire EDF schedule at a time to avoid generating the whole schedule, and (ii) perform local schedule generation in the idle slots of each node. The questions to be answered include (1) what these segments should be, (2) what is the upper bound on the lengths of these segments, and (3) what need to be maintained from one segment to the next to support the incremental computation. Below we first introduce a few definitions and then present how HD-PaS answers the above questions.

Definition (Local Busy Slot) A time slot is a local busy slot for node V_j if V_j either sends or receives a transmission belonging to any unicast task in the time slot. □

Definition (Local Idle Slot) A time slot is a local idle slot for node V_j if it is not a local busy slot for V_j. □

Definition (Schedule Segment) A schedule segment, denoted as SS_j, is a segment of the local schedule constructed by node V_j at a time. SS_j starts either at each time slot when V_j receives a broadcast packet or at the first local idle slot after V_j completes a sequence of consecutive local busy slots. □

In HD-PaS, each node V_j generates its local schedule incrementally according to the definition of SS_j. A local schedule is constructed by simply following the EDF policy to determine that each time slot in SS_j should send/receive which packet or be idle. HD-PaS uses the broadcast slot to generate a schedule segment. Generating a schedule segment must be completed within a local idle slot or a broadcast slot to guarantee that HD-PaS works properly. This guarantees that each schedule segment is computed according to the up-to-date system status information since the current segment ends before *any broadcast slot* in which a disturbance propagation may happen. The next segment starts from the broadcast slot, and the computation is performed within this broadcast slot after extracting the up-to-date system information. This approach automatically satisfies the schedule design principle (ii) as it uses local idle slots to derive the schedule. (Note that a broadcast slot at V_j is also a local idle slot for V_j based on the definition of local idle slot.)

Considering the example RTWN in Sect. 4.1, Fig. 6 shows the first three schedule segments of device node V_3. The top is the global schedule of the whole network,

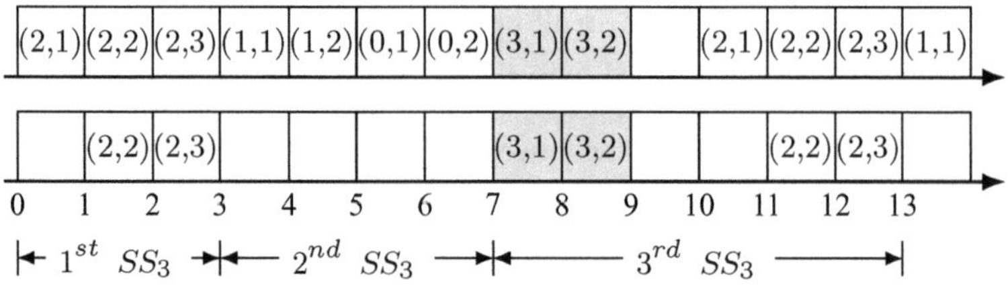

Fig. 6 Schedule segments of node V_3. The top (bottom) is the global (local) schedule of the system (V_3). Blank slots in the top (bottom) schedule are global (local) idle slots, respectively. Gray slots are assigned for transmissions of broadcast packets

while the bottom represents the local schedule generated at V_3. Gray slots in the schedules are assigned for transmissions of broadcast packets and blank slots are idle slots. In the global schedule, slot 9 is an idle slot according to EDF policy. In V_3's local schedule, however, there exist totally 8 idle slots since V_3 does not transmit or receive any packet within these time slots.

To facilitate local schedule computation, each local node needs to maintain certain data. First, the generated schedule must be stored. To save memory, only the information associated with the slots (e.g., slot 1, 2, 11, and 12 in Fig. 6) is stored. Second, to generate the local schedule, V_j must have the full knowledge of the task parameters, including P_i, D_i, and H_i in the nominal state.[3] Please note that V_j only needs to generate the local schedule involving itself. It does not need to store the route information of all tasks but only thosc portions that traverse through itself.

4.4 Dynamic Schedule Generation

The local schedule generation function can work properly when the system is in the nominal mode. However, when the system enters the rhythmic mode, the rhythmic task adopts new periods and deadlines. Some periodic packets may have to be dropped to ensure all rhythmic packets delivered by their deadlines. The information in the schedule table would not be sufficient for each node to generate a consistent local schedule. Though information such as all possible rhythmic task parameters can be stored locally, the device nodes need to know which task is entering the rhythmic state and what packets should be dropped if any. In this case, HD-PaS utilizes its dynamic schedule generation function to handle the rhythmic event when an external disturbance is detected.

[3]Upon detection of the external disturbance(s), specifications of the rhythmic task(s) are received from the controller node.

As shown in Fig. 5, when an external disturbance is detected and reported to the controller node at t', the dynamic schedule generation function is about to find an updated schedule for the system to use in the rhythmic mode such that (i) all rhythmic packets meet their deadlines and (ii) the minimum number of periodic packets are dropped. This updated schedule should be used in the time interval defined by the start point and end point of the rhythmic mode, i.e., $[t_{sp}, t_{ep})$ (see Fig. 5). Suppose the broadcast packet $\chi_{n+1,k}$ reaches all nodes at time slot t^*. The start point t_{sp} can be directly set as t^* from which all nodes in the network can switch to follow an updated dynamic schedule consistently. However, the selection of t_{ep} is not trivial, and it must guarantee that all packets released after it meet their nominal deadlines if no task enters the rhythmic state again. Thus, the main problem to solve by the controller is to determine (i) the end point of the rhythmic mode, t_{ep}, and (ii) which periodic packets to drop in the updated schedule such that the QoS degradation is minimized (i.e., minimum dropped packets).

When an external disturbance occurs, the duration of the system rhythmic mode should be as short as possible so that the system can promptly return to its nominal mode. However, a shorter system rhythmic mode can lead to worse performance since more periodic packets may have to be dropped to properly handle the external disturbance. That is, the choice of end point t_{ep} impacts not only the length of the system rhythmic mode but also the number of dropped periodic packets. Thus, a well-designed end-point selection method is required.

Theoretically, any time slot within $[t^*, t^u_{ep}]$[4] can be an end point. However, selecting the time slot in this set that leads to the minimum number of dropped packets can be time-consuming. To tackle this challenge, HD-PaS observes that using any time slot between two successive release times as t_{ep} leads to more dropped packets than just using the later release time as t_{ep}. Thus, the search space can be significantly reduced by only examining the release times of all packets in $[t^*, t^u_{ep}]$.

With an end point (candidate) being selected, the controller needs to determine the minimum number of dropped packets while guaranteeing the real-time deliveries of critical rhythmic packets. To achieve this, HD-PaS maps this problem to the problem of minimizing the number of late jobs, which has been well studied [41–43] in real-time scheduling literature and can be solved in polynomial time using Lawler's algorithm [41]. Readers are referred to [11, 12] for the details of the solution.

---

[4] t^u_{ep} is a user-specified parameter to bound the maximum allowed latency for handling the current rhythmic event.

5 RD-PaS Framework

By relying on a well-designed hybrid dynamic packet scheduling mechanism, HD-PaS is able to construct a temporary dynamic schedule to handle external disturbances in RTWNs. However, HD-PaS makes an assumption that all wireless links are reliable without any internal disturbances occurred in the network. Such assumption simplifies the algorithm design and analysis, but it is not realistic in real-life settings especially in noisy and harsh industrial environments.

In this section, we introduce a reliable dynamic packet scheduling framework, called RD-PaS, for handling both external and internal disturbances in RTWNs. Essentially, RD-PaS is an enhanced version of HD-PaS as they both consist of local and dynamic schedule generation functions.

RD-PaS advances HD-PaS by considering the lossy nature of wireless links caused by internal disturbances in RTWNs. As stated in the system model section, packet transmission may fail when internal disturbance occurs. Any packet loss can not only block the delivery of the current real-time message but also cause network bandwidth waste since all subsequent allocated slots of the packet are unused. On the other hand, RTWNs always require packets to satisfy stringent requirements in terms of reliability. That is, the achieved e2e packet delivery ratio must be larger than a desired value after which the packet is said to be reliable. Therefore, RD-PaS deploy retransmission mechanisms to deal with such conflict between task demand and network supply. Specifically, if a sender node does not receive any ACK from the receiver node within the current slot, it automatically retransmits the packet in the next possible time slot. Then, the key questions need to be answered by RD-PaS are (i) how many (extra) retransmission slots should be assigned to each packet of tasks to guarantee both of their timing and reliability requirements without any redundant network resource request, and (ii) how to adjust the schedule when external disturbance occurs to still guarantee the reliable and timely transmissions of the critical rhythmic packets while achieving the minimum reliability degradation on other packets?

Note that the second question bears similarity to the dynamic schedule generation problem in HD-PaS as they both adjust the schedule dynamically to accommodate the increased network resource demand caused by external disturbance. The main difference is that RD-PaS has more flexibility in schedule adjustment. Specifically, any relatively unimportant periodic packet can release network bandwidth by dropping a portion of retransmission slots instead of dropping the whole packet. Therefore, RD-PaS formulates a similar reliable dynamic scheduling problem and presents a heuristic to solve it. As such, in the following, we mainly discuss the answer to the first question, i.e., the retransmission slots allocation of RD-PaS. Further, we describe the solution by assuming that the RTWN applies transmission-based scheduling (TBS) model. Readers are referred to [13] for the solution based on the packet-based scheduling (PBS) model.

5.1 Retransmission Slot Allocation

In the TBS model, each specific time slot is assigned to an individual packet transmission. Considering the lossy nature of wireless links, when a transmission is not successful, retransmissions are needed, which require extra time slots. To reduce the demand on network resources, it is required to minimize the number of extra slots for each packet while satisfying its reliability requirement. Since the reliability requirements for all packets released from a certain task are commonly the same, RD-PaS determines the slot allocation for each task, and this solution, once determined, can be applied to all packets of the task. Therefore, at the highest level, the problem needs to be solve is as follows.

Retransmission Slot Allocation Problem: Given a RTWN where each link has an associated PDR, and a task set in which each task has a single routing path, determine the minimum number of extra slots needed by each task τ_i for satisfying its reliability requirement.

To solve this problem, RD-PaS proposes to first determine whether a given number of extra time slots for each task can satisfy the reliability requirement and then search for the optimal number of extra time slots for every task.

Definition (Retry Vector) A retry vector of task τ_i, denoted by $\overrightarrow{R}_i$, represents the specific slot allocation on each transmission of τ_i. □

For example, consider task τ_2 with three hops $V_2 \rightarrow V_c \rightarrow V_3 \rightarrow V_4$ in Table 2. A possible retry vector $\overrightarrow{R}_2 = [2, 3, 4]$ denotes that two slots are assigned to τ_2's first hop, three slots are assigned to the second hop, and four slots are assigned to the third hop, respectively. The total number of slots allocated to each packet of τ_2 equals to 9. Given the packet delivery ratios of all the links along the routing path of task τ_i and its retry vector, the e2e PDR of τ_i, denoted as λ_i, can be derived as:

$$\lambda_i = \prod_{h=0}^{H_i-1} 1 - (1 - \lambda^L_{[h]})^{R_i[h]}, \tag{1}$$

where $\lambda^L_{[h]}$ represents the PDR value of link for transmitting the h-hop of τ_i and $R_i[h]$ is the number of slots assigned to the h-hop according to the retry vector.

Recall that the objective of RD-PaS is to find the minimum total number of slots allocated to task that satisfies its reliability requirement. And, for a given number of slots, say w, assigned to τ_i, the number of possible slot allocations, i.e., retry vectors, equals to $\binom{w-1}{H_i-1}$. That is, it is nontrivial to determine the optimal retry vector which leads to the largest packet delivery ratio of τ_i. To tackle this, RD-PaS proposes an optimal algorithm to incrementally generate a set of optimal retry vectors, and the final obtained $\overrightarrow{R}_i$ is set as the retransmission slot allocation for τ_i. The basic idea of the algorithm is to add one extra retransmission slot to the hop which yields the maximum reliability gain. Such a greedy approach indeed leads to an exact solution which is proved in [13].

6 FD-PaS Framework

In the above sections, we describe HD-PaS and RD-PaS frameworks for handling both external and internal disturbances in RTWNs. However, they essentially take a centralized approach to handle disturbances which suffers the following limitations. First, they are subject to single-point failure since they both rely on a centralized controller node to make online decisions. If the controller fails during network operation, any occurred disturbances that are not properly handled may cause catastrophe to the system. Second, as observed from a motivating example to be described in the following subsection, centralized approaches require a feedback process which is slow to handle unexpected disturbances especially in large RTWNs.

To overcome these drawbacks, this section introduces a fully distributed dynamic packet scheduling framework called FD-PaS to handle both internal and external disturbances in RTWNs. FD-PaS makes online decisions locally without any centralized control point when disturbances occur. In the following, we first give a motivational example to show that centralized approaches incur long latency for handling disturbances.

6.1 Motivational Example

Consider the RTWN (shown in Fig. 1) with three tasks (τ_0, τ_1 and τ_2) running on seven nodes ($V_0, \ldots, V_5$ and V_c) with V_0 and V_2 being sensors, V_4 and V_5 being actuators, V_1 and V_3 being relay nodes, and V_c being the controller node and functioning as the gateway in centralized approaches. Note that since centralized approaches rely on the controller node to disseminate the dynamic schedule, a broadcast task τ_3 is needed. The tasks' routing paths, periods, and relative deadlines are given in Table 3.

Assume all tasks are synchronized and first released at time slot 0, and each node employs an EDF scheduler to construct its local schedule (see Fig. 7). Suppose

Table 3 Task parameters for the motivational example

Task	Routing path	P_i $(= D_i)$
τ_0	$V_2 \rightarrow V_c \rightarrow V_5$	9
τ_1	$V_0 \rightarrow V_1 \rightarrow V_c \rightarrow V_5$	9
τ_2	$V_2 \rightarrow V_c \rightarrow V_3 \rightarrow V_4$	10
τ_3	$V_c \rightarrow *$	18

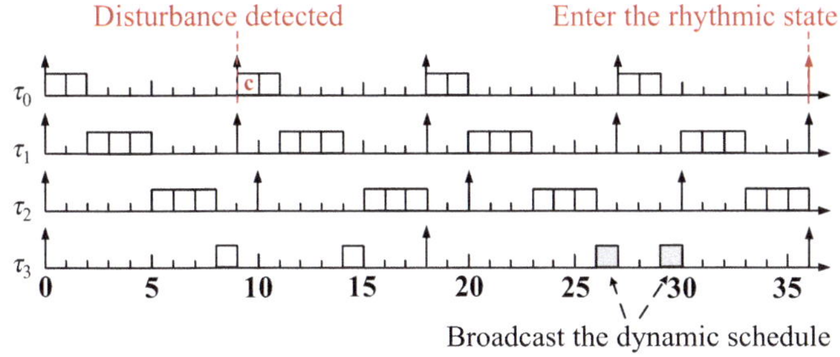

Fig. 7 Local EDF schedules of the tasks in the example. The block with symbol c denotes the transmission of the rhythmic event request. The shaded blocks denote the transmissions of the broadcast task to propagate the dynamic schedule generated at the controller node to the network

at time slot 9, an external disturbance is detected and sensor V_2 sends a rhythmic event request via τ_0 to the controller node. V_c then determines the time slot $t_{n\rightarrow r}$ when τ_0 is going to enter its rhythmic state. In order to achieve fast response to the disturbance, $t_{n\rightarrow r}$ should be set to be as early as possible, but later than the time slot when all nodes in the network receive the dynamic schedule. In this example, V_c has to wait till time slot 26 to broadcast the constructed dynamic schedule. Only after the broadcast packet reaches all nodes at 30, τ_0 can enter its rhythmic state at the nearest release time slot 36. Therefore, for this example, although the disturbance is detected by the sensor at time slot 9, the system cannot enter the rhythmic mode starting to handle the disturbance until slot 36, which is three nominal periods later.

From the above example, one can readily see that the centralized approaches suffer from a considerably long response time to the disturbances especially for large RTWNs. Moreover, centralized approaches rely on a single point (the controller node) in the network to make online packet scheduling decisions. These are the two main roadblocks in scaling up the packet scheduling framework to handle disturbances in large-scale RTWNs.

6.2 Overview of FD-PaS

In order to achieve fast response to external disturbances in RTWNs, the key idea of FD-PaS is to make dynamic, local schedule adaptation at each node along the path of the rhythmic task while avoiding transmission collisions from other nodes that still follow their static schedules in the system rhythmic mode.

Figure 8 gives an overview of the execution model of FD-PaS. After network initialization, each node locally generates a static schedule, S, using the local schedule generation mechanism in HD-PaS and follows S to transmit packets. When an external disturbance is detected by a sensing task, say τ_0 at t', a notification is propagated to all the nodes responsible for handling the disturbance, denoted

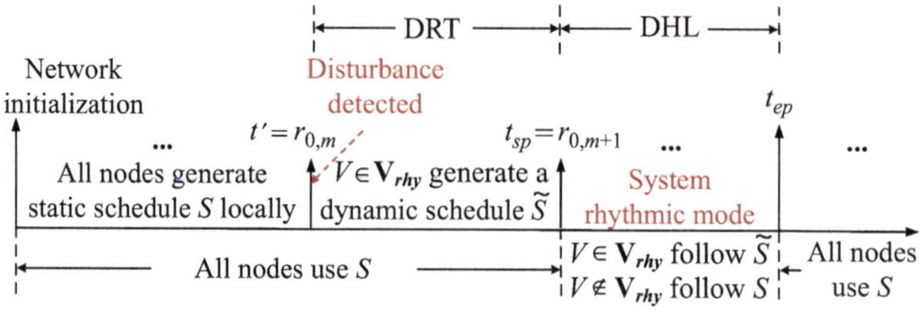

Fig. 8 Overview of the execution model of FD-PaS

as $\mathbf{V}_{rhy}$. Upon receiving the notification, each node $V_j \in \mathbf{V}_{rhy}$ determines the time duration of the network being in the rhythmic mode and generates a dynamic schedule $\tilde{S}$ for handling the disturbance. Starting from the next release time, one nominal period of τ_0 after detecting the disturbance, the nodes in $\mathbf{V}_{rhy}$ follow $\tilde{S}$, while all other nodes keep using static schedule S to transmit periodic packets. Thus, by not relying on a broadcast packet to disseminate the dynamic schedule generated by a centralized point in the network, FD-PaS is able to significantly reduce the response time of reacting to disturbances. For ease of discussion, in the rest of the chapter, we refer to *disturbance response time* (DRT) as the time duration from t' to the start time of the system rhythmic mode and *disturbance handling latency* (DHL) as the time duration of the system rhythmic mode (see Fig. 8).

To ensure that FD-PaS works properly as stated above, two fundamental questions need to be answered: (i) which nodes belong to $\mathbf{V}_{rhy}$? and (ii) how do these nodes receive the external disturbance information from the sensing node? Recall that when an external disturbance occurs, the rhythmic task will enter its rhythmic state following reduced periods and deadlines. An updated schedule is needed to accommodate the increased workload. To ensure that each (re)transmission of the rhythmic task can be successful, both the sender and the receiver of its packet must follow the same schedule. Thus, all nodes along the routing path of the rhythmic task must know the disturbance information to generate a consistent dynamic schedule and should be included in $\mathbf{V}_{rhy}$. For example, $\mathbf{V}_{rhy} = \{V_2, V_c, V_5\}$ for the example in Fig. 1 if τ_0 enters the rhythmic state. When a disturbance is detected at a release time of τ_0, its information can be piggybacked onto the current packet and transmitted to all nodes in $\mathbf{V}_{rhy}$. Propagating disturbance information in this manner guarantees that all nodes in $\mathbf{V}_{rhy}$ receive the disturbance information within one nominal period of τ_0, i.e., P_0, since the static schedule ensures that each task is assigned with the required number of transmission and retransmission slots along its routing path within P_0 in order to meet the e2e timing and reliability requirements.

According to such disturbance propagation mechanism, only the nodes on the path of the rhythmic task are included in $\mathbf{V}_{rhy}$. Nodes in $\mathbf{V}_{rhy}$ construct their local schedules individually and employ them in the dynamic duration to handle the disturbance. All other nodes in the network follow the original static schedule. With this distributed execution model, inconsistencies between the dynamic and static schedules in the system rhythmic mode may arise, which would result in transmission collisions. To ensure that the disturbances are handled appropriately, in FD-PaS, the transmissions of critical rhythmic packets need to be always successful even in the presence of collision with other periodic packets. To achieve this, FD-PaS proposes an enhancement to the IEEE 802.15.4e standard [36], called Multi-Priority MAC (MP-MAC), to avoid transmission collisions.

6.3 Avoiding Transmission Collisions

In conventional RTWNs such as WirelessHART [38] and 6TiSCH [39], TDMA-based data link layer is widely adopted to provide synchronized and collision-free channel access. In addition, most of those protocols employ the clear channel assessment (CCA) operation at the beginning of each transmission for collision avoidance. CCA, however, cannot prioritize packet transmissions. When multiple transmissions happen in the same time slot sharing the same destination, it cannot guarantee the more important packets (e.g., rhythmic packets) are granted the access to the channel.

To tackle this challenge, FD-PaS proposes an enhancement to the IEEE 802.15.4e standard [36], called Multi-Priority MAC (MP-MAC), to support prioritization of packet transmissions in RTWNs. Figure 9 gives a comparison of the slot timing of 802.15.4e (top) and MP-MAC (bottom). In a 802.15.4e time slot, the sender transmits a packet and the receiver responds with an acknowledgment (ACK) if the packet is successfully received.[5] The packet transmission starts at *TxOffset* after the start of the time slot, while the ACK starts at *TxAckDelay* after the completion of the packet transmission. A long Guard Time (LGT) and a short Guard Time (SGT) are used by the receiver and sender, respectively, to tolerate clock drift and radio/CPU operation delays. With this standard design of 802.15.4e, if multiple senders transmit packets in the same time slot, they are not aware of the other transmissions and thus will cause interference. The slot timing of MP-MAC is presented at the bottom of Fig. 9. In MP-MAC, instead of being set as a constant, *TxOffset* is varied to implicitly indicate the priority of the packet (shown as red dashed lines). A packet with a higher priority is associated with a shorter *TxOffset* to start the transmission earlier. In addition, a CCA operation will be performed before each transmission to ensure that there is no higher priority packet transmission present in the channel. This enhancement ensures that only the highest priority packet (with the shortest *TxOffset*) is transmitted, and all lower priority transmissions yield to it.

Similar to the guard times, the *TxOffset* values for different priorities need to be set sufficiently apart so that different senders and receivers have consensus on the priorities. In MP-MAC, *PriorityTick* is defined as the difference between two

[5]No acknowledgment is provided for broadcast and multicast packets.

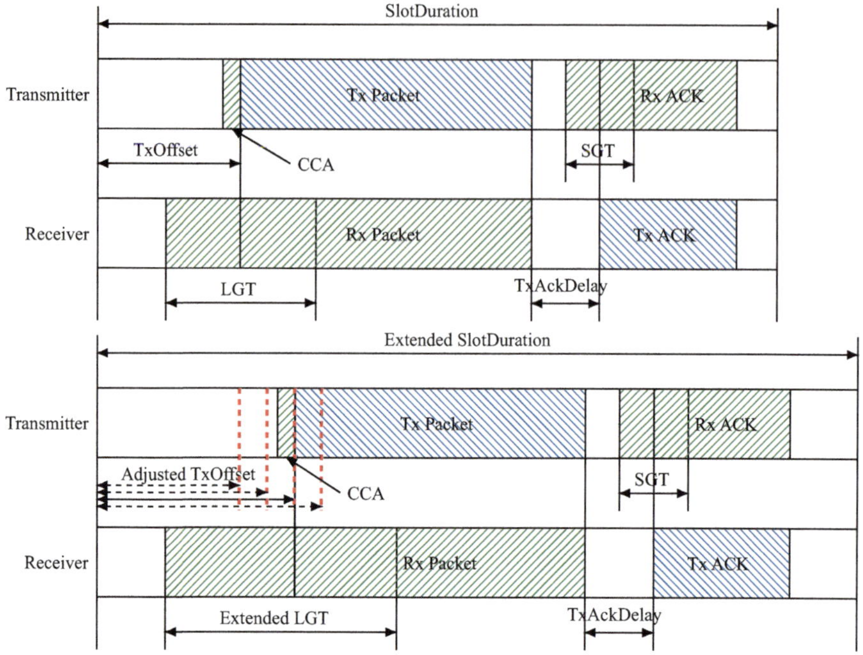

Fig. 9 Slot timing of 802.15.4e (top) and MP-MAC (bottom)

consecutive *TxOffset*s. To support k different priorities in MP-MAC, the length of the time slot, compared to the standard design, needs be extended by $(k - 1) \times PriorityTick$. A longer *PriorityTick* can ensure successful packet prioritization, but either leads to longer SlotDuration and reduced network throughput or smaller number of supported priorities if the size of the time slot is fixed.

MP-MAC ensures that once the dynamic schedules are generated locally, the nodes in $\mathbf{V}_{rhy}$ can follow those schedules to handle the external disturbance without transmission collisions with other nodes in the network. Since all the nodes in $\mathbf{V}_{rhy}$ receive the same disturbance information, the dynamic schedules generated locally at these nodes are all consistent. However, since all nodes not belonging to $\mathbf{V}_{rhy}$ still follow the original static schedule to transmit periodic packets, periodic transmissions cannot be adjusted in the dynamic schedule. If any periodic transmission is replaced by a rhythmic one supported by MP-MAC, the reliability of this packet is degraded. Therefore, the objective of the dynamic schedule generation in FD-PaS is to minimize the reliability degradation for all periodic packets while satisfying the timing and reliability requirements of rhythmic ones. It is proved that the dynamic schedule generation is an NP-hard problem, and an optimal ILP solution and an efficient heuristic are both introduced to solve it. Details of the problem formulation and ILP solution can be found in [15].

7 Implementation

In this section, we introduce an implementation of FD-PaS on a real-time wireless network testbed to show its applicability in real-world RTWNs. The testbed is based on OpenWSN [44] with required enhancements to support FD-PaS. OpenWSN is an open-source implementation of the 6TiSCH protocol suite [45] which aims to enable IPv6 over the TSCH (Time Synchronized Channel Hopping) mode of IEEE 802.15.4e. An OpenWSN network typically consists of multiple OpenWSN devices, an OpenWSN Root, and an OpenLBR (Open Low-Power Border Router). The Root and OpenLBR communicate through a wired connection (e.g., UART), using OpenBridge protocol. They together form the controller node.

The OpenWSN network adopts a TDMA-based data link layer. Adaptive synchronization mechanisms [46] are incorporated to ensure network-wide time synchronization among all device nodes. A time slot can be one of the following five types: OFF, TX, RX, SerialRX, and SerialTX. When an IPv6 packet is generated by a device node, it is compressed to a 6LoWPAN packet and then transmitted in a dedicated TX slot to its neighbor on the path to the Root. This process repeats on the neighbor node until the packet reaches the Root. The Root forwards the packet to OpenLBR in an SerialTX slot, where the 6LoWPAN packet is decompressed and sent to Linux kernel for forwarding. If the destination of the packet is within the same network, the packet is forwarded back to OpenLBR. OpenLBR compresses it again to an 6LoWPAN packet and adds the 6LoWPAN source routing header, by examining the network topology stored in the RPL routing module. The 6LoWPAN packet is then sent to the Root in the next SerialRX slot. The Root transmits the packet over the air in the next available TX slot to its neighbor as specified in the source routing header. This process repeats until the packet reaches the final destination.

Our testbed consists of seven wireless devices (TI CC2538 SoC + SmartRF evaluation board). One of them is configured as the root node (controller node) and the rest are device nodes to form a multi-hop RTWN. A CC2531 sniffer is used to capture the packet. A 8-Channel Logic Analyzer is used to record device activities by physical pins, in order to accurately measure the timing information among different devices.

7.1 Software Stack Enhancement to Support FD-PaS

To implement FD-PaS on the testbed, the following four software modules are added. One of them (i.e., Manager) is implemented on OpenLBR. The other three modules are implemented in the application layer of the OpenWSN stack.

EDF Local Scheduler (EDF-LS) implements the distributed local schedule generation algorithm on the nodes. In the first idle slot of each schedule segment, it cleans up the existing link schedule, constructs the local schedule in the new schedule segment, and installs them to the slotframe.

Packet Generator (P-Gen) constructs periodic packets according to its task information. It is invoked before the end of a time slot when MAC layer finishes all activities. If a packet needs to be transmitted in the next slot, P-Gen samples the sensor and prepares the packet. A broadcast packet, however, is initiated by the controller node in a broadcast slot instead of by P-Gen and forwarded by the nodes according to a calculated broadcast graph.

Receiver binds to a UDP port to receive packets from the Manager. It handles three types of messages: (1) the link info message to install broadcast slots, (2) the task info message for the P-Gen and EDF-LS modules to construct periodic packets and schedule segments, respectively, in the nominal mode, and (3) the rhythmic event response message for the P-Gen and EDF-LS modules to construct rhythmic packets and schedule segments, respectively, in the rhythmic mode.

Manager is responsible for installing the broadcast graph in the network and initializing the P-Gen and EDF-LS modules in the device nodes. It also runs the end-point selection and packet dropping algorithms to handle rhythmic events. The results along with the rhythmic task information are broadcast to the device nodes for constructing local schedules.

7.2 Functionality of FD-PaS in a Multi-task Multi-hop RTWN

FD-PaS is deployed on a seven-node multi-hop network as shown in Fig. 1. The system running in the network consists of three tasks, $\tau_0 = \{\{V_0, V_1, V_c, V_3, V_4\}, 15, 8\}$, $\tau_1 = \{\{V_2, V_c, V_3\}, 30, 6\}$ and $\tau_2 = \{\{V_1, V_c, V_5\}, 20, 4\}$. For each task, the first element denotes the routing path and the second one denotes its period (relative deadline). The third element represents the total number of slots assigned to τ_i, including both transmission and retransmission slots, in the static schedule. Assume that τ_0 is the rhythmic task when an external disturbance occurs and the rhythmic period (deadline) $\overrightarrow{P_0}(\overrightarrow{D_0}) = [12, 12, 12, 12, 12]$. The system starts running in the nominal mode at slot 1 and then switches to the rhythmic mode from slot 61. A Logic Analyzer is used to capture the radio activities from a pin of each device during slot 1–120.

The captured results on the testbed are illustrated in Fig. 10. Specifically, Fig. 10a summarizes the legends. Figure 10b shows the system nominal mode during time slot 1–60. Figure 10c demonstrates the system rhythmic mode using FD-PaS during time slot 61–120. In Fig. 10b and c, seven waveforms represent the radio activities (transmitting, receiving, or listening) for all the seven nodes, as labeled on the left side of the figures. Each falling or rising edge of the waveform in the *Slot* row (lower part of the figures) marks the start of a new slot. In the bottom *Schedule* row, slot assignments are indicated using different colors and patterns. Each colored small block indicates the release time of the corresponding task at a certain node. Each

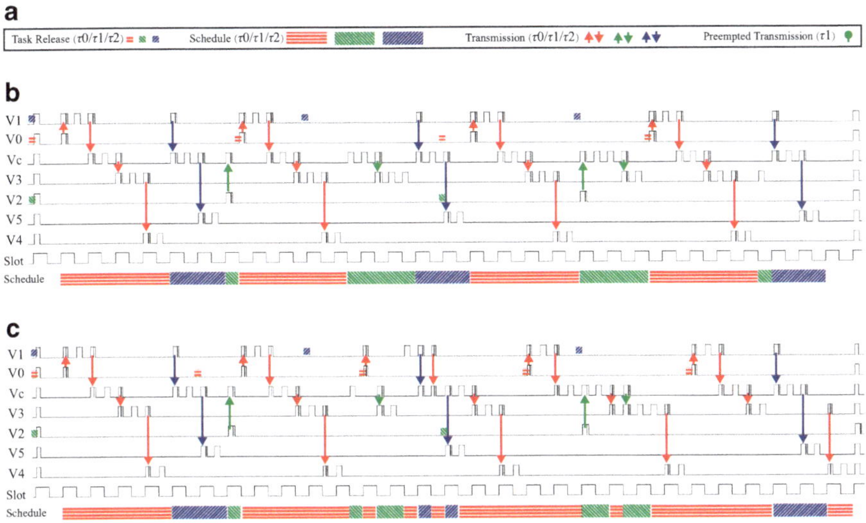

Fig. 10 Slot information and radio activities in the test case captured by Logic Analyzer. (**a**) Legends used in the figures. (**b**) Nominal mode (time slot 1–60). (**c**) Rhythmic mode using FD-PaS (time slot 61–120)

transmission is denoted by a colored arrow of which the starting and ending points represent the sending and receiving nodes, respectively. In the rhythmic mode, a colored circle denotes a dropped periodic transmission preempted by a rhythmic one. For example, in Fig. 10b, τ_1 releases its first packet at slot 1 and is transmitted from V_2 to V_c at slot 15. External disturbance is emulated by triggering a button during network operation.

Figure 10b illustrates radio activities of the system in the nominal mode (1–60 slots), after which the system switches to the rhythmic mode upon receiving an external disturbance. Given by the static schedule, each packet $\chi_{i,k}$ is allocated with extra slots for retransmission in the system nominal mode. But according to the testbed result shown in Fig. 10b, each transmission successes in its first assigned time slot without using any retransmission slot. During the disturbed rhythmic mode (slot 61–120), task τ_0 releases five packets as indicated in Fig. 10c. To accommodate the increased workload in the system rhythmic mode, FD-PaS determines to reduce the number of slots assigned to τ_1's packets both from 6 to 4, even though both packets of τ_1 still have chances to be successfully transmitted to the destination as illustrated in Fig. 10c.

8 Conclusion

In this chapter, we presented a set of dynamic resource management frameworks, from centralized approach to fully distributed one, to handle internal and external disturbances in RTWNs. In general, centralized approaches are proven to be sufficient for small-scale installations and can achieve high performance (in terms of lower packet degradation) in handling disturbances. This is due to the fact that centralized management relies on a controller node, which has a global view on the network, to make online decisions. On the other hand, scalability becomes a significant limitation for a centralized approach as RTWNs start to be deployed over large geographic areas. Therefore, we introduce a fully distributed scheduling framework called FD-PaS. Unlike centralized approaches where dynamic schedules are generated in the controller node and disseminated to the entire network, FD-PaS makes online decisions to handle disturbances locally without any centralized control. Such a fully distributed framework not only significantly improves the scalability but also provides guaranteed fast response to external disturbances. To demonstrate the applicability, we also introduce an implementation of FD-PaS on a real-world 6TiSCH network testbed.

Acknowledgments The work reported herein is supported by the National Science Foundation under NSF Award IIP-1919229.

References

1. Da Xu L, He W, Li S (2014) Internet of things in industries: a survey. IEEE Trans Ind Inform
2. Tramarin F, Mok AK, Han S (2019) Real-time and reliable industrial control over wireless lans: algorithms, protocols, and future directions. Proc IEEE
3. Åkerberg J, Gidlund M, Björkman M (2011) Future research challenges in wireless sensor and actuator networks targeting industrial automation. In: 9th IEEE International Conference on Industrial Informatics
4. Sisinni E, Saifullah A, Han S, Jennehag U, Gidlund M (2018) Industrial internet of things: challenges, opportunities, and directions. IEEE Trans Ind Inform
5. Lu C, Saifullah A, Li B, Sha M, Gonzalez H, Gunatilaka D, Wu C, Nie L, Chen Y (2015) Real-time wireless sensor-actuator networks for industrial cyber-physical systems. Proc IEEE
6. Willig A (2008) Recent and emerging topics in wireless industrial communications: a selection. IEEE Trans Ind Inform
7. Willig A, Matheus K, Wolisz A (2005) Wireless technology in industrial networks. Proc IEEE
8. Hei X, Du X, Lin S, Lee I (2013) Pipac: patient infusion pattern based access control scheme for wireless insulin pump system. In: INFOCOM
9. Gatsis K, Ribeiro A, Pappas GJ (2014) Optimal power management in wireless control systems. IEEE Trans Autom Control
10. Karbhari VM, Ansari F (2009) Structural health monitoring of civil infrastructure systems. Elsevier
11. Zhang T, Gong T, Gu C, Ji H, Han S, Deng Q, Hu XS (2017) Distributed dynamic packet scheduling for handling disturbances in real-time wireless networks. In: RTAS

12. Zhang T, Gong T, Han S, Deng Q, Hu XS (2018) Distributed dynamic packet scheduling framework for handling disturbances in real-time wireless networks. IEEE Trans Mobile Comput
13. Gong T, Zhang T, Hu XS, Deng Q, Lemmon M, Han S (2019) Reliable dynamic packet scheduling over lossy real-time wireless networks. In: ECRTS
14. Zhang T, Gong T, Yun Z, Han S, Deng Q, Hu XS (2018) Fd-pas: a fully distributed packet scheduling framework for handling disturbances in real-time wireless networks. In: RTAS
15. Zhang T, Gong T, Han S, Deng Q, Hu XS (2019) Fully distributed packet scheduling framework for handling disturbances in lossy real-time wireless networks. IEEE Trans Mobile Comput
16. Han S, Zhu X, Mok AK, Chen D, Nixon M (2011) Reliable and real-time communication in industrial wireless mesh networks. In: RTAS
17. Leng Q, Wei Y-H, Han S, Mok AK, Zhang W, Tomizuka M (2014) Improving control performance by minimizing jitter in RT-WiFi networks. In: RTSS
18. Saifulah A, Lu C, Xu Y, Chen Y (2010) Real-time scheduling for WirelessHART networks. In: RTSS
19. Crenshaw TL, Hoke S, Tirumala A, Caccamo M (2007) Robust implicit EDF: a wireless mac protocol for collaborative real-time systems. ACM Trans Embed Comput Syst
20. Shen W, Zhang T, Gidlund M, Dobslaw F (2013) SAS-TDMA: a source aware scheduling algorithm for real-time communication in industrial wireless sensor networks. Wirel Netw
21. Ferrari F, Zimmerling M, Mottola L, Thiele L (2012) Low-power wireless bus. In: SenSys
22. Sha M, Dor R, Hackmann G, Lu C, Kim T-S, Park T, Self-adapting mac layer for wireless sensor networks. In: RTSS (2013)
23. Chipara O, Wu C, Lu C, Griswold WG (2011) Interference-aware real-time flow scheduling for wireless sensor networks. In: ECRTS
24. Zimmerling M, Mottola L, Kumar P, Ferrari F, Thiele L (2017) Adaptive real-time communication for wireless cyber-physical systems. ACM Trans Cyber-Phys Syst
25. Li B, Nie L, Wu C, Gonzalez H, Lu C, Incorporating emergency alarms in reliable wireless process control. In: ICCPS (2015)
26. Palattella MR, Accettura N, Grieco LA, Boggia G, Dohler M, Engel T (2013) On optimal scheduling in duty-cycled industrial iot applications using ieee802. 15.4 e tsch. IEEE Sensors J
27. Soua R, Minet P, Livolant E (2012) Modesa: an optimized multichannel slot assignment for raw data convergecast in wireless sensor networks. In: IPCCC
28. Soua R, Livolant E, Minet P (2013) Musika: a multichannel multi-sink data gathering algorithm in wireless sensor networks. In: IWCMC
29. Tinka A, Watteyne T, Pister K (2010) A decentralized scheduling algorithm for time synchronized channel hopping. In: ADHOCNETS
30. Morell A, Vilajosana X, Vicario JL, Watteyne T (2013) Label switching over ieee802. 15.4 e networks. Trans Emerging Telecommun Technol
31. Soua R, Minet P, Livolant E (2016) Wave: a distributed scheduling algorithm for convergecast in ieee 802.15. 4e tsch networks. Trans Emerging Telecommun Technol
32. Duquennoy S, Al Nahas B, Landsiedel O, Watteyne T (2015) Orchestra: robust mesh networks through autonomously scheduled TSCH. In: SenSys
33. Thubert P, Watteyne T, Struik R, Richardson M (2015) An architecture for ipv6 over the TSCH mode of ieee 802.15. 4. Working Draft, IETF Secretariat, Internet-Draft draft-ietf-6tisch-architecture-08
34. Kim J, Lakshmanan K, Rajkumar R (2012) Rhythmic tasks: a new task model with continually varying periods for cyber-physical systems. In: ICCPS
35. Hong S, Hu XS, Gong T, and Han S (2015) On-line data link layer scheduling in wireless networked control systems. In: ECRTS
36. De Guglielmo D, Anastasi G, Seghetti A (2014) From IEEE 802.15. 4 to IEEE 802.15. 4e: A step towards the internet of things. In: Advances onto the Internet of Things
37. Brummet R, Gunatilaka D, Vyas D, Chipara O, Lu C (2018) A flexible retransmission policy for industrial wireless sensor actuator networks. In: ICII

38. Song J, Han S, Mok A, Chen D, Lucas M, Nixon M, Pratt W (2008) Wirelesshart: applying wireless technology in real-time industrial process control. In: RTAS
39. Dujovne D, Watteyne T, Vilajosana X, Thubert P, 6tisch: deterministic ip-enabled industrial internet (of things). IEEE Commun Mag (2014)
40. Liu CL, Layland JW (1973) Scheduling algorithms for multiprogramming in a hard-real-time environment. J ACM (JACM)
41. Lawler E, New and improved algorithms for scheduling a single machine to minimize the weighted number of late jobs. Preprint, Computer Science Division, University of California
42. Moore JM (1968) An n job, one machine sequencing algorithm for minimizing the number of late jobs. Manag Sci
43. Baptiste P (1999) An $O(n^4)$ algorithm for preemptive scheduling of a single machine to minimize the number of late jobs. Oper Res Lett
44. Watteyne T, Vilajosana X, Kerkez B, Chraim F, Weekly K, Wang Q, Glaser S, Pister K (2012) OpenWSN: a standards-based low-power wireless development environment. Trans Emerging Telecommun Technol
45. Watteyne T, Palattella M, Grieco L (2015) Using IEEE 802.15.4e time-slotted channel hopping (TSCH) in the internet of things (IoT): Problem statement, RFC 7554, May 2015
46. Stanislowski D, Vilajosana X, Wang Q, Watteyne T, Pister KS (2014) Adaptive synchronization in ieee802. 15.4e networks. IEEE Trans Ind Inform

Pervasive Listening: A Disruptive Network Design for Massive Low-Power IoT Connectivity

Benoît Ponsard and Christophe Fourtet

1 Introduction

Industrial Internet of Things (IIoT) covers a variety of use cases and, hence, a variety of communication requirements. For example, process automation needs radio communications with high reliability; latency as low as a few milliseconds and high throughput are mandatory for real-time systems. The cellular industry designates these requirements as ultra-reliable and low latency communication (URLLC). Partially addressed in 3GPP release 15, URLLC enhancements are in the 3GPP roadmap of release 16 [1].

This chapter focuses on another type of IIoT uses cases, where nation-wide mobility and low power in devices are key characteristics. These use cases, that have been addressed by Sigfox [2] since 2012, can be seen as predecessors of the massive machine-type communication (mMTC), which is now considered by the cellular industry as one of the three generic services of 5G. Hereafter, we give three examples of mMTC use cases, and show that their communication constraints are different from URLLC.

Returnable Industrial Packaging
Industries that have mass production are interested in IoT sensors for improving the logistics of parts towards assembly lines. Just-in-time delivery of parts is often critical. Small parts are delivered in bulk, but major parts with significant size or value are delivered in returnable industrial packaging (RIP). In the car industry, RIP is used between final assembly lines and tier-one subcontractors of car manufacturers, or between manufacturer factories. Examples of parts that are

B. Ponsard (✉) · C. Fourtet
SIGFOX, Labege, France
e-mail: benoit.ponsard@sigfox.com; christophe.fourtet@sigfox.com

N. H. Mahmood et al. (eds.), *Wireless Networks and Industrial IoT*,
https://doi.org/10.1007/978-3-030-51473-0_8

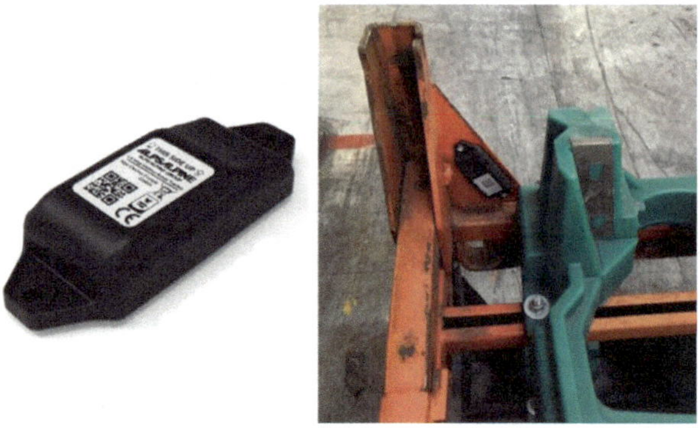

Fig. 1 Tracker and RIP equipped with tracker. (Photos courtesy of respective copyright owners)

delivered in critical returnable industrial packaging are windscreens, car seats, car doors, dashboards, etc.

Returnable industrial packaging is assigned to a dedicated closed loop of logistics; they shall not exit this closed loop. Nevertheless, car manufacturers experience losses and shortages of RIP, mainly because of wrong shipment or misplacement in warehouses. Being able to locate these packaging, particularly when they are empty, is an answer to a real pain point of many car manufacturers.

Trackers, in form of IoT devices fastened to returnable industrial packaging, provide location information, either on a time basis or upon events. Key characteristics for these tracking devices are:

- Low cost and easy installation on RIP (no external battery, no external antenna)
- Battery life of several years to withstand life duration of RIP,
- Harsh environment protection capabilities,
- Location capabilities within car manufacturers and tier-one premises without the addition of a dedicated radio infrastructure, but also in the wild when loaded in trucks.

Figure 1 depicts an implementation of a tracking device on a car manufacturer RIP. Its connectivity uses the Sigfox network [2].

Monitoring of Massive Flow of Valuable Goods

The previous example is about tracking and tracing a known number of RIP, moving in a closed loop. The present example is about another types of logistics: massive flows of valuable goods between two or several points in an open loop. Passenger luggage in airports or cargo in shipping containers are examples of such massive flows of goods that mMTC solution can address. Three constrains are peculiar to these types of use cases:

- *Global scale:* Tracking of luggage in air-flights or in shipping containers is at continent-scale and very often overseas. Technical solutions must be global and

Fig. 2 Example of a tracking tag

compliant with local spectrum regulations in each country where the valuable goods are expected to be shipped.

- *No impact on existing process:* Logistics of cargo or luggage did not wait for the IoT for streamlining their process and interactions of all the actors of its value chain. Improvement from IoT technologies should be implemented as an add-on, with only a side impact.
- *Low return on investment:* IoT technologies will work on side optimization of already existing processes of massive flow logistics. Cost of IoT must be marginal and aligned with expected benefits from IoT solutions.

Figure 2 is an example of tracking tag for personal luggage, connected to Sigfox network. The low power consumption permits an implementation in a form factor of a credit card.

Natural Environment Monitoring and Smart Agriculture
mMTC connectivity applies also to wildlife, natural environment, and smart agriculture for data collection from sensors. The vast majority of these sensors must be battery powered, because electrical power is not natively available in the vicinity of such devices. To overcome maintenance issues, battery operation of several years is a must-have feature for these use cases.

The Challenges of mMTC Communication
Three examples of mMTC for industrial IoT exhibit similar characteristics in their connectivity requirements, as follows:

- *Low amount of data:* Connected devices transmit information such as sensed data, status, index, or alarms and receive commands or configuration parameters. Compared to machines that may be complex and may require cellular M2M communication, IoT devices connect the Internet mainly for a single function, which communicates infrequent small application packets [3].
- *Massive number of connected devices:* Thanks to low cost and easy-to-use IoT communication function, it is possible to connect many types of devices, resulting in a much higher density of connections per square km compared to

the cellular systems. In urban areas, the density of connected devices may be over 50 k per square kilometer [4].

- *Low power in connected devices:* As IoT objects may not require electrical energy for their primary use, communication must be designed to operate with standalone source of energy. A battery-operated sensor with a lifetime up to several years is a prerequisite for many IoT applications. Moreover, battery operation allows non-intrusive installation of connected sensors.
- *Global coverage:* Connectivity for mMTC must be global, especially for mobile IoT devices and compliant with local spectrum regulations and with local frequency sharing rules.

Small amount of data per device and large number of devices per area are a new paradigm for radiocommunication engineers. Although overall volume of data per base station may be quite small compared to those experienced in base stations of cellular system, different traffic models require different solution. In this paper, we address this new paradigm with an innovative approach, named *pervasive listening*, combined with ultra-narrow band modulation.

The rest of this chapter is organized as follows. Pervasive listening approach for radio design of mMTC connectivity is introduced in Sect. 2. Sections 3 and 4 deal with three technologies, that have been renewed to leverage pervasive listening network for IIoT: ultra-narrow band modulation and cognitive algorithms in software-defined radios (SDR) of base stations. Section 5 investigates possible optimizations at network level.

2 Pervasive Listening: A Disruptive Approach in Network Design

2.1 Medium Sharing with Pervasive Listening

When base stations and devices share a common radio resource, a protocol layer for medium access control is commonly used by all parties to organize the random use of the communication resource. For example, base stations in cellular networks allocate the available radio resource with a grant-based random access. In local area network, time-slotted access is another approach for improving random access efficiency. The implementation of such protocols induces various levels of time and frequency coordination between communicating parties. This coordination is obtained by using a medium access protocol layer that exchanges signaling.

Signaling induces, in turn, communication overhead and processing complexity in device and base station. They are affordable in systems with large volume of application data. But they drain excessive amount of energy in IoT devices, that send only infrequent small application messages.

Pervasive listening answers this constraint by removing any coordination between devices and radio access network. Pervasive listening is the ability of a radio access network with multi-base stations deployment to receive a radio burst wherever and whenever it is transmitted, and whatever its carrier center frequency is. Most of the time, base stations in a pervasive listening are in receive mode, ready to detect, demodulate, and decode radio bursts transmitted by devices at random. Devices willing to send an application message have to carry out only minimal operations, such as building a radio packet, selecting a center frequency within the operating frequency band and sending the corresponding radio bursts over the air. With this approach, design complexity in devices is drastically reduced. Nothing is needed to be sent over the air before the actual uplink transmission: devices simply broadcast their radio packets.

2.2 Advantages of Pervasive Listening

As mentioned above, the main advantage of pervasive listening is the absence of frequency and time synchronization in devices. No coordination, thus no power drain is needed in devices for signaling exchange with the radio access network. From the radio access network perspective, the benefits of pervasive listening are single frequency band and cooperative reception, as detailed hereunder.

Frequency Usage

Cellular networks build a global coverage and network capacity with careful radio planning and frequency reuse patterns. On the contrary, pervasive listening builds a radio access network with all base stations on the same frequency band. Devices have to know only one frequency band: the one used by the network they want to connect. This approach is counter-intuitive in usual cellular networks that have a significant downlink traffic, but it is a natural solution for uplink-oriented networks, such as the one of Sigfox. The other benefits are, as follows:

- Base station deployment is much simpler because it does not need frequency planning; all base stations listen for the same frequency range,
- Overlaps of base station coverage is not an issue; on the contrary, it brings multiple receptions of messages sent by devices which improves quality of service at no cost for the IoT devices (see next section),
- When moving, mobile devices do not drain extra power from their battery because they do not maintain attachment to the network. A device has nothing to do to be received by its nearby base stations than just emitting its radio bursts.

Cooperative Reception

With pervasive listening, all base stations are on the same frequency. This results in large overlaps of base stations coverage. Whereas overlaps are kept to the minimum in cellular systems (see part (a) in Fig. 3), they are beneficial in pervasive listening networks because they bring spatial diversity (see part (b) in Fig. 3). Several base stations may receive the same radio packet. This is not an issue because multiple received packets are de-duplicated in core network (see Fig. 8), before transmission

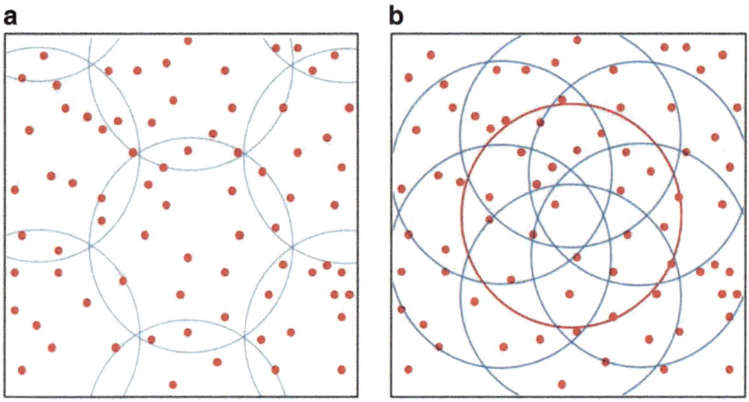

Fig. 3 overlaps of base stations in cellular deployment (**a**), and in a pervasive listening network (**b**)

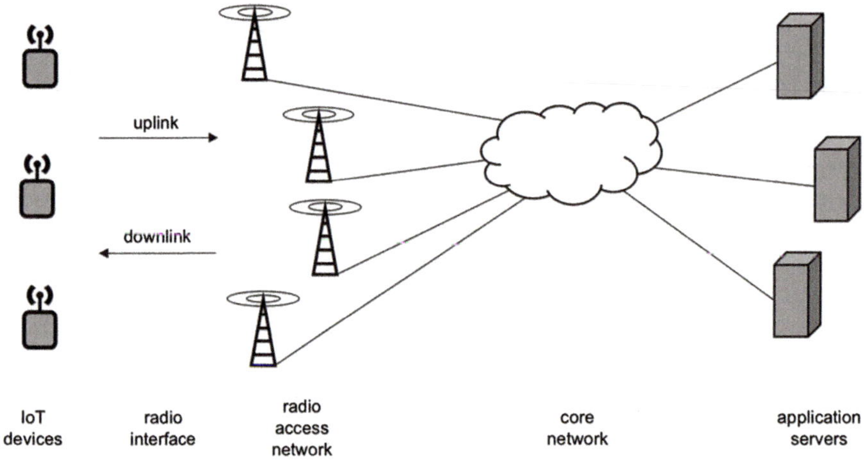

Fig. 4 Overall architecture of an IoT network using pervasive listening

to application server. This feature is named *cooperative reception*. It is a benefit of pervasive listening. It improves the overall quality of service without extra complexity in IoT devices: everything is done in the core network (Fig. 4).

Note 1 Along with deduplication of application messages, the core network runs other services, such as security check, IoT device registration, and billing. Network monitoring is also performed in the core network.

Note 2 The rest of this chapter uses common naming for communication direction: "uplink" is from IoT devices to base stations; "downlink" is from base stations to IoT devices.

2.3 *Disadvantages of Pervasive Listening*

The main drawback of pervasive listening is the collisions of uplink radio packets that occur in the time and frequency domain. Aloha protocol shows that pure random access is effective as long as the offered load is under a few percent [5]. A radio access network with pervasive listening can mitigate collision rate in three ways:

- Ultra-narrow band modulation, which adds a new dimension in the random access (see Sect. 3),
- Software-defined radio processing in base stations, which allows parallel processing for radio packets received simultaneously (see Sect. 4.1),
- Densification of deployment with less-sensitive base stations: with such base stations, the effective coverage area is reduced. With less devices in visibility, a base station experiences less collision.

A less-important disadvantage of pervasive listening is the added complexity needed in base stations for radio processing and in the core network for deduplication of uplink messages. Nevertheless, this added complexity can be easily managed in base stations equipped with industrial-grade PC boards.

3 Ultra-Narrow Band Modulation

UNB, an Old Technique Refreshed with Microelectronics

Ultra-narrow band (UNB) has been known since World War II, with single side band analog modulations. More recently, the progress in microelectronics enables integrated fractional-N PLLs, which may have a frequency synthesis step as low as 1 Hz and a maximum center frequency over the gigahertz. Available in off-the-shelf integrated radio chipsets, these features permit digital modulation rates as low as a few tens of symbols per second, even with a carrier center frequency up to a few gigahertz.

This is what UNB is about: a modulation scheme with an occupied bandwidth smaller than 1 part per million (i.e. 1 ppm) of the carrier center frequency. As an example, the D-BPSK modulation at 100 baud, used in the Sigfox network in Europe [2], is a UNB modulation, because its carrier center frequency is 868.130 MHz.

UNB may be implemented in several ways, resulting in various protocols with specific characteristics [6]. The Sigfox network implements a radio interface named 3D-UNB. 3D stands for triple diversity in time, in frequency, and in space. It is fully specified in [2]. When combined with pervasive listening, 3D-UNB exhibits two significant benefits, as follows.

Improved Link Budget for Less Base Station Density

One benefit of UNB signals is a link budget improvement. As the transmit power is concentrated in a small bandwidth, it results in a less noise values in UNB receivers. Table 1 is an example of the maximum coupling loss (MCL) evaluation for a UNB

Table 1 Maximum coupling loss for 3D-UNB under European country regulations and legacy GPRS uplink

Factor name	Values for 3D-UNB	Values for legacy GPRS uplink
Transmitter		
(1) Tx power (dBm)	+14	+33
Receiver		
(2) Thermal noise density (dBm/Hz)	−174	−174
(3) Receiver noise figure (dB)	4	3
(4) Interference margin (dB)	0	0
(5) Occupied channel bandwidth (Hz)	100	180 k
(6) Effective noise power (dBm) = (2) + (3) + (4) + 10 log((5))	−150	−118.4
(7) Required SINR (dB)	7	12.4
(8) Receiver sensitivity = (6) + (7) (dBm)	−143	−106
(9) Rx processing gain	0	5
(10) MCL = (1)–(8) + (9) (dB)	+157	144

transmission in European sub-gigahertz unlicensed band, where the transmit power of devices is limited to 25 mW (i.e. 14 dBm). The maximum coupling loss of a legacy GPRS uplink is given for reference (from [7]), as GPRS is the first wide-area cellular technology for machine type communication.

A higher MCL means a better link budget for the communication between devices and base stations. In a given area, the same quality of service can be obtained with fewer base stations.

Note The high link budget of UNB modulation is obtained at the cost of low data rate. As the laws of physics are the same for all technologies, a low data rate is the counter part of a high link budget, where local regulations limit the transmit power. For example, the chirp spread spectrum technique of LoRa [8] gives equivalent link budget, but with a more complex implementation in the devices. In licensed spectrum, LTE-M and NB-IoT solutions exhibit MCL over 155 dB, but with the use of complex HARQ mechanism [9].

Medium Access Techniques Renewed with 3D-UNB

In sub-gigahertz spectrum, the unlicensed spectrum is limited to a few hundreds of kilohertz. The conventional narrow band systems implement only a couple of communication channels, each of them having an occupied bandwidth ranging from a few kilohertz up to a couple of tens of kilohertz. On the contrary, the small footprint of UNB modulation (i.e. a bandwidth less than 1 ppm of the carrier center frequency) allows to pack thousands of simultaneous UNB signals in the sub-gigahertz unlicensed spectrum. These UNB signals are named *quasi-tones* in 3D-UNB protocol. With a transmission bandwidth about 100 Hz, UNB brings about 6000 quasi-tones of 100 Hz width, in the 25 mW unlicensed frequency band in

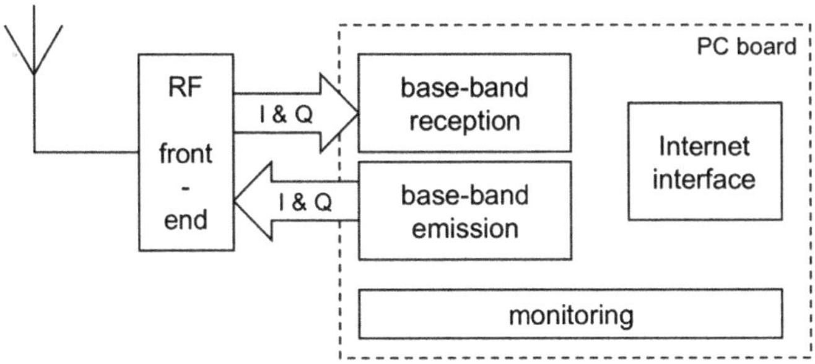

Fig. 5 Base station overall architecture

Europe [10], which is from 868.0 MHz to 868.6 MHz. This large amount of quasi-tones is a real game changer for accessing the medium. Instead of implementing mechanism to share a scarce frequency resource from a central point, there are so many different quasi-tones that a device may select randomly its carrier center frequency, without experiencing excessive collision rate.

Plethora of quasi-tones make time coordination useless. Even if two devices select the same value for their subcarrier center frequency, the probability to have a collision in time is kept under acceptable level, thanks to limited coverage of base stations. Detailed analysis of time and frequency random access can be found in [11] and [12].

4 Cognitive Radio in the Base Station

The use of UNB modulation in devices and pervasive listening in base stations is interesting to meet the requirements of IoT connectivity. As said above, devices select their transmission time and the carrier center frequency of their uplink transmissions randomly. As a consequence, base stations have to be ready for reception at any time and on any frequency. To achieve this capability, 3D-UNB base stations use an innovative design for the software define radio in base stations. This approach brings flexibility and capacity to base stations, at a cost of a limited added complexity, as explained hereafter.

Base Station Functional Blocks Figure 5 depicts 3D-UNB base station architecture, designed for pervasive listening networks. Radiofrequency (RF) front-end implements high dynamic and very linear chains for emission and reception. For uplink reception, base-band functional block implements software-defined radio processing, such as signature detection, demodulation, and decoding. For downlink transmission, base-band processes downlink emissions in cooperation with the core network (see Sect. 5).

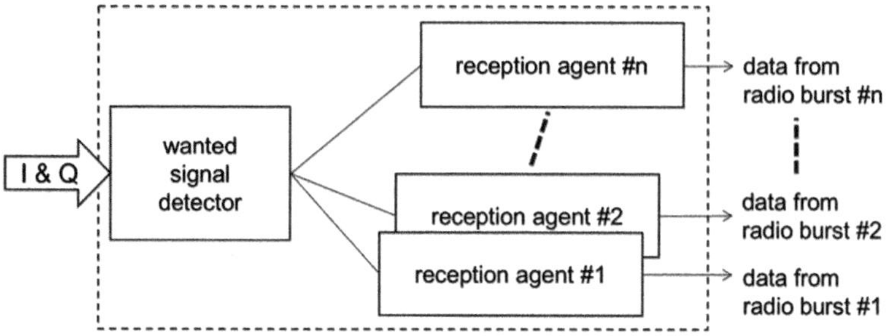

Fig. 6 Instantiation of multiple reception agents to cope with uplink load

As an example, the frequency range for pervasive listening is 200 kHz wide in Sigfox' base stations, at time of printing. The base-band processing runs on an industrial PC board equipped with a Linux operating system. The base stations connection to the Sigfox' core network uses secured IP links (i.e. VPN) over any type of bearer, e.g. ADSL box, LANs, cellular, or satellite.

On-Demand Reception Agent

Base-band processing in reception uses an innovative SDR implementation, where random access is managed by on-demand agents (see Fig. 6).

First process in base-band reception is detection over the full received bandwidth. This process analyses I&Q samples and detects 3D-UNB-like signals, thanks to their frequency and time signature. Once a potential 3D-UNB signal is detected, an instantiation of a tracking and demodulation agent is created and allocated for precisely tracking center frequency of this 3D-UNB signal. Then reception agent demodulates received symbols according to a defined modulation scheme and decodes received packed for extracting the higher layer data.

Next two subsections point out flexibility of this SDR architecture for both capacity and protocol evolutions.

Uplink Reception Capability

Once a new UNB signal is detected by a base station, the wanted signal detector process has to instantiate a new reception agent. The maximum number of messages, that can be received simultaneously, is directly related to the ability of the wanted signal detector to instantiate reception agents, as many as needed. Limiting factors for total number of reception agents are processing power and memory space available on base station PC board. As an example, a base station listening 200 kHz band should be capable of about 2000 instantiations of 3D-UNB reception agents at the same time.

Multi-protocol Capability
It is worth noting that this base station architecture is not limited to a single radio protocol. Several protocols can be processed in parallel, provided the following constraints are fulfilled:

- Frequency and time signature of each protocol can be clearly recognized;
- Reception agents of various type have enough processing power and memory space for running properly.

As the two function blocks "wanted signal detector " and "reception agent" are purely software, changing the behavior of a pool of base stations, or upgrading protocol decoding, can be made remotely with just a software upgrade.

Other Benefits of SDR Implementation
Thanks to software-defined radio implementation, it is possible to use more sophisticated reception agents.

First example is about successive interference cancellation (SIC) algorithms [13]. As 3D-UNB implements a 2D-Aloha random access, collision rate created by offered load has an immediate impact on quality of service at application level. SIC implementation in 3D-UNB base stations benefit from already existing high dynamic range and good linearity in RF front-end of base stations, that are necessary for near-far cases, anyhow.

Second example is about combining algorithms, that may be implemented locally or globally. 3D-UNB communication rules allow transmission of up to three radio bursts for the same message, each radio being encoded with a different convolutional code. Multiple transmissions give an opportunity to have combining algorithms in base stations (local combining) or in the core network (remote combining). Local combining benefits from frequency diversity and convolution codes, whereas remote combining adds spatial diversity to recombination process [11].

5 Network-Level Benefits of Pervasive Listening

From a global perspective, pervasive listening brings benefits at network level, because it allows global optimization.

Base Stations Listen at All Times and Transmit on Demand
3D-UNB uses two separate sub-bands for uplink and downlink transmissions. This frequency arrangement is quite common in large-scale bidirectional system where interface can be full-duplex or half-duplex. 3D-UNB radio interface is half-duplex with a 1 MHz approx. frequency gap between uplink and downlink, for the sake of simplicity in base station design.

To avoid unnecessary period of time when base stations are unable to receive, downlink emissions occur only once triggered by a device request, sent in an uplink message (see Fig. 7). Furthermore, available power in bases stations (i.e. 500 mW

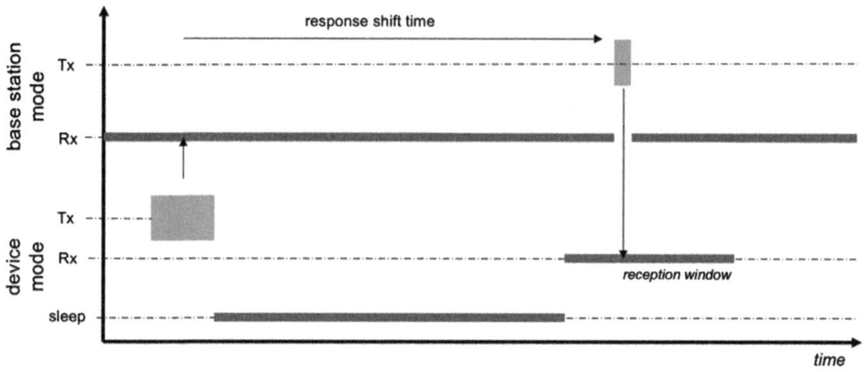

Fig. 7 Half-duplexing in Sigfox base stations

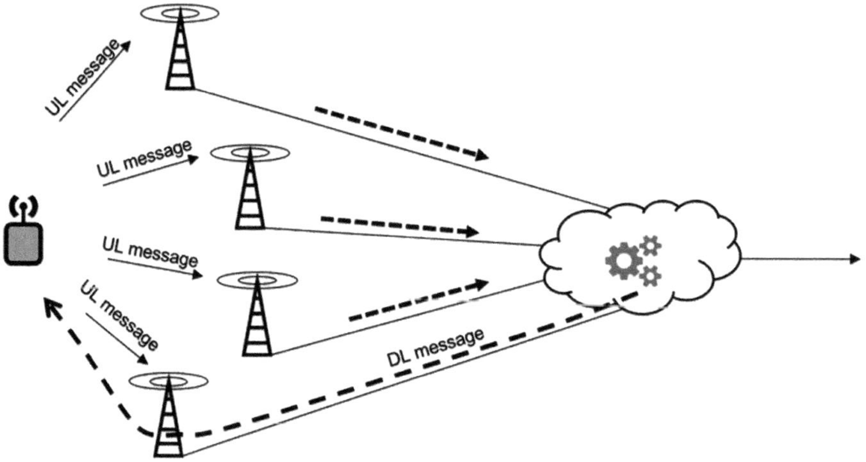

Fig. 8 Example of BS selection for downlink after multiple reception in uplink

in European regulations) allow several downlink emissions to be sent together, each one answering a different IoT device.

Downlink Cognitive Optimization

Base-stations are in reception mode by default. If one base station is unable to receive (because it is in transmit mode), at least one of its surrounding base stations can take over the reception of device transmissions. Toggling between reception mode and transmission mode is scheduled by the core network, which has a global view of the uplink and downlink loads. The scheduling process of downlink transmission is cognitive (see Fig. 8). Central optimization algorithm takes into account self-interference level and noise levels of each base stations for its selection process.

Downlink scheduling is also a multidimensional optimization problem, which is handled by 3D-UNB core network. Variables for optimization are, for example, noise level, downlink traffic load in each base station, uplink traffic load, time delay after uplink transmission, and selectivity in IoT device receivers. As the core network is cloud based, necessary processing power is easily available. Downlink transmission in the selected base station is under full control of 3D-UNB core network.

6 Conclusion

This chapter introduces a new approach in radio communications named pervasive listening. Invented by Sigfox, pervasive listening changes drastically the design and engineering of radio networks for the mMTC. By using software-defined radios and cognitive algorithms in the network along with UNB modulation, it addresses technical constrains, such as low power in devices, high capacity in base stations, and low volume of data per IoT device. Pervasive listening permits low complexity IoT devices by pushing the communication complexity back to the network.

Pervasive listening opens many research topics: interference cancellation, local combining or large-scale MIMO, downlink capacity optimization, etc. Pervasive listening and frugality in IoT devices are challenges of a new era in radio communications.

Acknowledgements The authors would like to thank Laurence Sellier and Gilles Mahé for their technical inputs and their review of this document.

References

1. 3GPP; Release 16 status, updated October 2, 2019; https://www.3gpp.org/release-16 ; consulted on February 2020
2. Sigfox connected objects; Radio specifications; February 2019; https://build.sigfox.com/sigfox-device-radio-specifications; consulted on October 2019
3. Short range devices; Low Throughput Networks (LTN) Architecture; LTN Architecture. ETSI TS 103 358 V1.1.1. 2018–06
4. Low Throughput Networks (LTN); Use cases and system characteristics;. ETSI TR 103 249 V1.1.1. 2017-10
5. Abramson N (1973) The ALOHA system. In: Abramson, Kuo (eds) Computer-communication networks. Prentice-Hall Publishing Company, pp. 501–518
6. Short rang devices; Low Throughput Network (LTN); protocols for radio interface A; technical specification ETSI TS 103 357, revision v1.1.1, pp 2018–06
7. 3GPP, Cellular system support for ultra low complexity and low throughput Internet of Things, Technical report, 3GPP TR 45-820 v2.0.0 (2015–08)
8. Semtech Corporation, LoRa modulations basics, AN 1200.22, revision 2, 2015, May

9. Eric Wang Y-P, Lin X, Adhikary A, Grövlen A, Sui Y, Blankenship Y, Bergman J, Razaghi HS (2017) A primer on 3GPP narrowband Internet of Things. IEEE Commun Mag 55(3):117–123. https://doi.org/10.1109/MCOM.2017.1600510CM
10. (2019) Decision EU 2019/1345 of 2 August 2019 amending decision 2006/771/EC updating technical conditions in the area of radio spectrum use for short range devices. Off J Eur Union
11. Song Q, Lagrange X, Nuaymi L (2017) Evaluation of macro diversity gain in long range ALOHA networks. IEEE Commun Lett 21(11):2472–2475. https://doi.org/10.1109/LCOMM.2017.2732984
12. Zozor S, Zhuocheng L, Lampin Q, Brossier JM (2016) Time-frequency ALOHA-like random access: a scalability study of low power wide area networks of IoT using stochastic geometry, CoRR abs/1606.04791
13. Mo Y, Goursaud C, Gorce J (2018) Uplink multiple base stations diversity for UNB based IoT networks, 2018 IEEE conference on antenna measurements & applications (CAMA), Vasteras, pp. 1–4

Information-Centric Networking for the Industrial Internet of Things

Cenk Gündoğan, Peter Kietzmann, Thomas C. Schmidt, and Matthias Wählisch

1 Introduction

The Internet of Things (IoT) is evolving, and an increasing number of controllers in the field are augmented with network interfaces. Current deployments often are part of larger systems (e.g., a heating) or attached to infrastructure (e.g., smart city lighting). Such devices connect to power, use common broadband links, and adopt the old MQTT protocol [7] for publishing IoT data to a remote cloud. The prevalent use case forecasted for the IoT, however, consists of billions of constrained sensors and actuators mainly not cabled to power, but connected via low power lossy wireless links. A significant portion of this constrained Internet of Things will relate to massive machine type communication (mMTC) for sensing, actuating, and monitoring devices in the industrial domain. The primary objective of the Industrial Internet of Things (IIoT) is content, i.e., the access to small, confined data chunks generated by its mass constituents, which will be tiny, cheap *things* that are severely challenged by the current way of connecting to the Internet.

In view of the network dedication to data units, doubts arose whether host-to-host sessions are the appropriate approach in these disruption-prone environments of (wireless) things, and the data-centric nature at the Internet edge called for rethinking the current IoT architecture [45]. ICN networks [3] have been identified as promising candidates to meet the new challenges of the future IIoT. Name-based routing and in-network caching as contributed by Named Data Networking

C. Gündoğan (✉) · P. Kietzmann · T. C. Schmidt
Institute of Computer Science, HAW Hamburg, Hamburg, Germany
e-mail: cenk.guendogan@haw-hamburg.de; peter.kietzmann@haw-hamburg.de; t.schmidt@haw-hamburg.de

M. Wählisch
Institute of Computer Science, Freie Universität Berlin, Berlin, Germany
e-mail: m.waehlisch@fu-berlin.de

N. H. Mahmood et al. (eds.), *Wireless Networks and Industrial IoT*,
https://doi.org/10.1007/978-3-030-51473-0_9

(NDN) [25, 58] bear the potential to increase robustness of application scenarios in regimes of low reliability and reduced infrastructure (e.g., without DNS). Following initial concepts [36] and early experimental work [5], the adaptation, analysis, and deployment of NDN for the IoT became an active research area that advocated the IoT as a candidate of early NDN adoption. Still open problems persist, namely, naming, routing, forwarding [55], and data push [28], as Shang et al. [48] recently reminded.

Common use cases of the IIoT are mission—sometimes safety—critical, and operative environments are often harsh and challenging. Mobile equipment may be in place, as well as intermittently connected devices. In addition, side channel traffic may be initiated by co-located systems from different manufacturers. In spite of deployment challenges, IIoT network communication faces high quality demands. The most important requirement is *sub-second latency* for alarm messages from detectors of safety application, or for control instructions to actuators. Lost messages may lead to inconsistent conditions in industrial control systems or undetected monitoring state, and hence, a *high reliability* is crucial. From an operational perspective, the network architecture should allow for the deployment of a flexible ecosystem, which enables private as well as open networks.

In this chapter, we review key features and benefits of Information Centric Networking (ICN) for the IIoT in Sect. 2, followed by a real-world system approach to IIoT networking with RIOT (Sect. 3). Section 4 presents an overview about recent advances and achievements of information-centric solutions for the IoT edge including selected performance measurements. We summarize with a prospect on future research directions in Sect. 5.

2 The Case for ICN in the IoT

Information-centric networking is one major outcome of the global future Internet initiatives undertaken in the present millennium. More than a decade of research has created a variety of ICN flavours [3, 57], which essentially have three principles in common [13]: *decoupling of named content from hosts*, *universal caching*, and *content object security*. The approaches liberate content access from physical infrastructure, allow for unhindered content replication and validation, and thereby reduce infrastructure dependency of the content-aware network layer. Some protocols such as NDN even elide network addresses of machine interfaces to rise the barriers for denial of service attacks.

The Internet of Things, which is mainly composed of constrained nodes at network edges, is a clear beneficiary of these directions as they reduce the burden of maintaining dedicated server infrastructure, end-to-end communication channels, and DDoS mitigation. In addition, seamless content replication and caching open the realm to multilateral support of energy preservation and improved transport resilience.

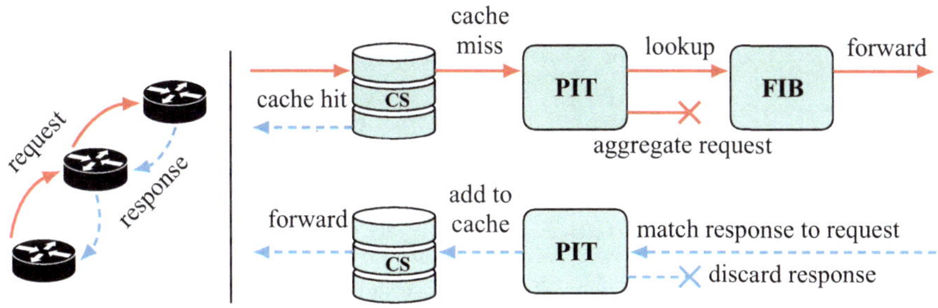

Fig. 1 Stateful forwarding plane of CCNx and NDN: Requests (aggregable Interests) generate reverse path forwarding entries and are served from content caches

> Information-centric networks (ICNs) bear the potential to master the challenges of low reliability and reduced infrastructure support in error-prone wireless regimes. Hop-wise transfer and in-network caching can compensate for link failures, mobility, and intermittent connectivity on the network layer.

2.1 *The CCNx and NDN Architectures*

Content-Centric Networking (CCNx) [32, 33] and Named Data Networking (NDN) [25, 58] enjoy the largest popularity among the information-centric networking schemes. Several open source projects evolved around the NDN concept with an increasing community of mainly academic nature. Prior CCNx work at PARC has been transferred to CISCO and transformed into the Hybrid-ICN architecture [9] with open source software support.

In contrast to the stateless packet processing of the Internet, CCNx and NDN utilize a stateful, name-based forwarding fabric to achieve a decoupling of content objects from their origins. Request and response operations are modeled on the network layer using two distinct message types that are differently treated by the forwarding state machine.

Figure 1 displays the forwarding logic and its data structures. At every hop, an incoming request (Interest) first triggers a cache lookup at the Content Store (CS). On a cache hit, a response containing the requested content is immediately returned to the incoming interface. A cache miss leads to a query of the Pending Interest Table (PIT) for a possibly existing PIT entry with the same content name. If this exists, the new request is aggregated, i.e., the incoming interface is recorded alongside the existing PIT entry, and a subsequent forwarding is suppressed. When a request for a particular content traverses a node for the first time, a PIT entry is initially created to record the content name and the incoming interface. The request

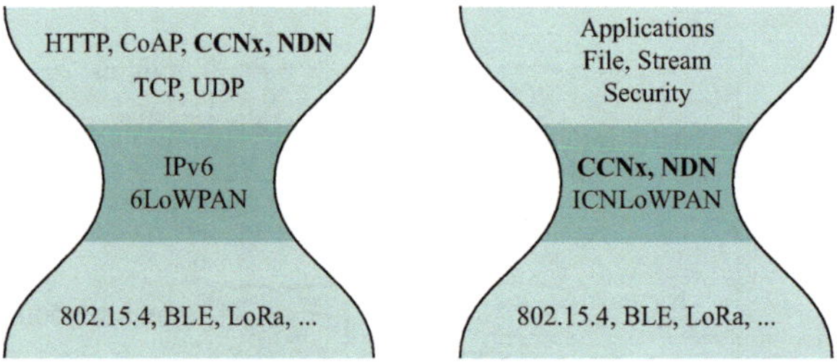

(i) ICN overlay on IP (ii) Native ICN deployment

Fig. 2 Two modes of deployment exist for an information-centric IoT system. Either ICN operates as an overlay on top of IP or ICN is configured natively on top of the data links

is then forwarded toward one or several outgoing interfaces in accordance with the Forwarding Information Base (FIB). Returning responses match against the PIT to identify pending requests. If no requests are pending, then the response is discarded. Otherwise, the response is cached in the CS and forwarded toward the recorded interfaces in the PIT.

2.2 Information-Centric IoT Deployments

Two modes of deployment are commonly considered for information-centric IoT system. Figure 2(i) illustrates the first option, in which an ICN implementation is configured as an overlay of the existing IP infrastructure using TCP or UDP encapsulation. In the second illustration (Fig. 2(ii)), ICN takes the role of the network layer, and IP is completely replaced. The encapsulation mode is generally discouraged, because of its memory overhead and software complexity of hosting two separate network stacks. Further, an overlay deployment prevents the beneficial properties of hop-wise ICN transport with in-network caching. The native deployment of ICN on the network layer is thus the primarily viable approach for challenged IoT networks. It allows for frugal network stacks of reduced complexity and sets free all potentials of an information-centric IoT system.

As low power lossy links and tiny frame sizes are prevalent in the constrained IoT, ICN native deployment requires convergence support analogously to the IPv6 6LowPAN convergence layer. ICNLoWPAN [17] is such a convergence layer that adapts CCNx and NDN primitives to IoT link-layer technologies of low bandwidth, high latency, and small MTU sizes. It introduces hop-by-hop header compression, link fragmentation schemes, and an ICN-specific name eliding. Figure 3 illustrates the protocol composition and shows the positioning of 6LoWPAN and ICNLoW-PAN in the stacks. Each request and response traverses through the convergence layers that deflate or inflate messages accordingly.

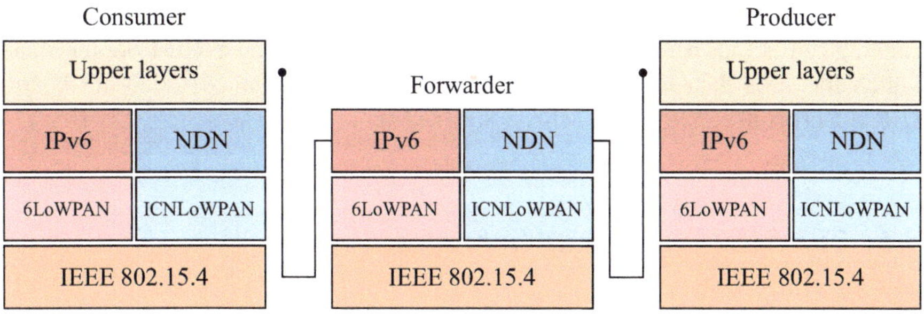

Fig. 3 Stack comparison for constrained IoT networks: IPv6 with 6LoWPAN adaptation versus NDN with ICNLoWPAN

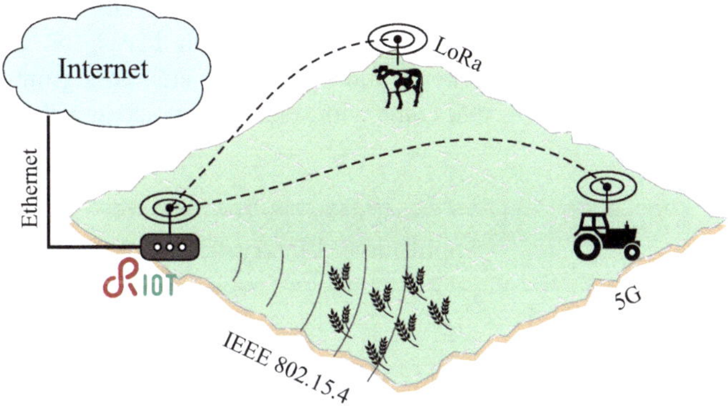

Fig. 4 Sensing and forwarding—a deployment use case of heterogeneous IoT networking

3 A Practical ICN Deployment for the IoT

Bringing IoT networking into the real world is still a challenging, pioneering task. A typical IoT field deployment is displayed in Fig. 4. It consists of sensors attached to elements in the field—a tractor, grain, and a cow in our visualization—as well as a data collection and aggregation system. Embedded nodes are deployed as both, the wireless field sensors and the data collectors—the latter may be connected to some (wired) infrastructure. As the complexity of software for embedded devices has increased over the last decade, it has become state-of-the-art to use operating systems even on memory and CPU constrained machines. The network subsystem is the key component for building the Internet of Things, and the support by an operating system is a key enabler for deploying new networking protocols.

Communication technologies in the IoT are very heterogeneous and need support of an operating system such as RIOT [4, 6]. Figure 4 shows a sample deployment that consists of two RIOT sensors, one equipped with LoRa and one with a 4G/5G radio such as Narrowband IoT (NB-IoT), as well as one RIOT router providing an 802.15.4 LoWPAN wireless interface in addition to both radio links and a wired

Ethernet uplink. In a lab experiment, we can demonstrate this with both sensor nodes running continuous packet streams of single sensor values and a RIOT border router which forwards data upstream. All nodes are very constrained iotlab-m3 boards (ARM Cortex M3, 32-bits, 72 Mhz, 64kB RAM).

3.1 The RIOT Networking Subsystem

The RIOT networking subsystem displays two interfaces to its externals as illustrated in Fig. 5: the application programming interface *sock* and the device driver API *netdev*. Internal to stacks, protocol layers interact via the unified interface *netapi*, thereby defining a recursive layering of a single concept that enables interaction between various building blocks: 6lo with MAC, IP with routing protocols, transport layers with application protocols, etc. This grants enhanced flexibility for network devices that come with stacks integrated at different levels.

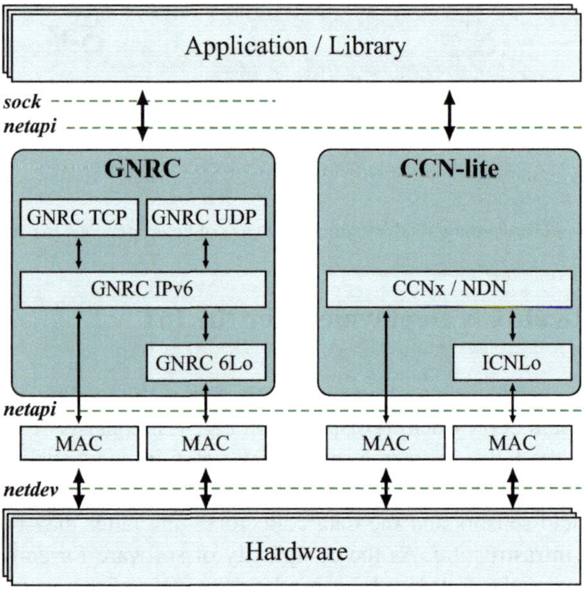

Fig. 5 The RIOT networking subsystem: A recursive layering architecture built around the recurrent protocol interface *netapi*. The generic driver abstraction *netdev* and the application-bound *sock* API allow for flexible composition of network stacks with easy protocol exchange

RIOT—the friendly operating system for the Internet of Things—supports a large variety of hardware platforms, flexible network capabilities closely aligned to standards, and thrives on a global grassroots community composed of academia, companies, and volunteers. Its community-driven 3-month release cycles foster an agile IoT ecosystem.

The Device Driver API: *netdev* RIOT networking abstracts individual devices via *netdev* that allows stacks to access network interfaces in a common, portable way. Unlike common solutions in Linux or other IoT operating systems (e.g., Zephyr, FreeRTOS, Contiki), *netdev* remains fully neutral with respect to the link layer and the network technology. It exchanges full frames including link layer headers in a shared buffer provided by the calling stack. The interface does not enforce implementation details concerning memory allocation, data flattening, and threading, but leaves these decisions explicitly to the user of the interface. With only six function pointers, *netdev* keeps a very low memory profile. The *netdev* interface decomposes into three functional parts: handling of (i) network data, (ii) configuration and initialization, and (iii) events. The combination of these three aspects makes the interface complete with respect to network device functionality and allows for full control of these devices.

The Internal Protocol Interface: *netapi* The RIOT network architecture defines typed message passing between network layers or compound modules with the help of *netapi*. This interface was designed to be as simple and versatile as possible, so that even rather exotic network protocols can be implemented against it. The design also allows for conveniently modular implementations and makes them both easily extensible and testable.

The User Programming API: *sock* POSIX sockets are the typical north-bound interface of an IP network stack. RIOT supports POSIX sockets; however, they require dynamic memory allocation. To cater for plainly static memory access, RIOT offers the optimized API *sock*—a collection of transport-specific network access calls designed to match the constraints of low-power embedded devices.

ICN Integration with RIOT RIOT provides lean mechanisms for integrating third-party software and libraries as packages. Regular CCNx and NDN implementations, though, are designed for general purpose machines and too demanding for the constrained IoT. Two alternatives exist that provide the core ICN feature set for constrained IoT devices: CCN-lite [54] and NDN-lite [47]. Both network stacks integrate into the RIOT operating system by implementing the *netapi* interface. They can thus be plugged-in at any network layer and support the deployment modes of transport encapsulation and native networking. These solutions open a large variety of supported hardware platforms and network technologies to ICN deployment in the wider IoT.

4 Recent ICN Advances for the IoT

The content-specific interface to networking must be considered a significant advantage of NDN/CCNx—consumer and producer applications can access content objects without any intermediary. Still, applications may require dedicated schemes to integrate network access and distribution, for instance, a publish-subscribe function for replacing MQTT. We will now discuss corresponding extensions and also performance comparisons of ICN with traditional protocols such as MQTT [7] and CoAP [49].

Real-world deployment of the Information Centric Network layer demands for additional functions such as service differentiation and publish-subscribe. Foremost, ICN needs to show widespread evidence of a performance superior to traditional approaches.

4.1 Publish-Subscribe

Many IoT scenarios target a very loose coupling between nodes that often run on battery with long sleep cycles and connect via lossy wireless links. Today's default deployment in industrial environments for this follows a publish-subscribe approach using the old MQTT protocol. Information-centric networking decouples content provisioning from data producers in space—additional decoupling in time and synchronization is desirable and attainable by additional publish-subscribe functions.

Early publish-subscribe schemes based on NDN such as Content-based pub/sub [10] and COPSS [11] violate the loose coupling principle in their use of name-based routing or forwarding. Nichols [35] proposes broadcasting for pub-sub, which wastes energy and does not scale. Hop-and-Pull (HoPP) [16] takes up the challenge and seeks for an information-centric IoT networking solution that satisfies all challenges of real-world sensor-actuator networks and allows for an easy deployment. It makes the common assumption that nodes form a stub network and connect to the outside by one or several gateways as displayed in Fig. 6; by exploiting the lean routing protocol PANINI [44], HoPP can built on prefix-specific default routes instead of broadcasting.

For the interior routing, nodes are grouped according to one or several sub-network prefixes (e.g., `/lighting`). One or several distinguished nodes serve as Content Proxies (CPs). CPs are typically more stable and more powerful such as gateways or other infrastructural entities. These Content Proxies take the role of data caches and persistent access points. They will be reachable throughout the network by default routes, unless temporary partitioning occurs. A CP can serve several local

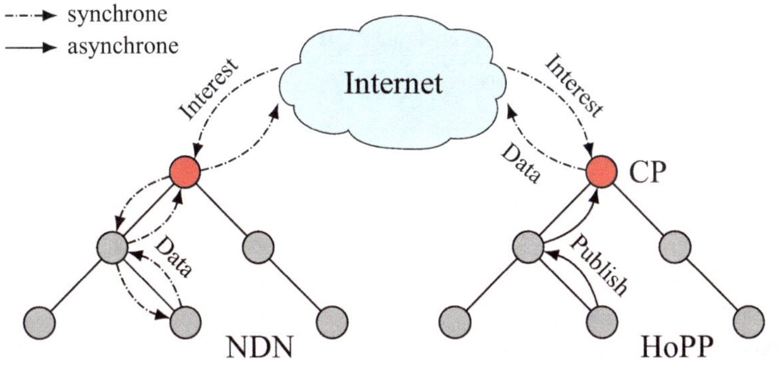

Fig. 6 Communication flow with standard NDN (left) and Pub-Sub approach HoPP (right)

prefixes, but a local prefix may also belong to several CPs. The latter scenario will lead to replicated caching with higher and faster data availability. HoPP is composed of three core primitives:

1. Establishing and maintaining the routing system
2. Publishing content to the CPs
3. Subscribing content from the CPs

This protocol definition strictly complies with the design principles: (a) minimal FIBs that only contain default routes, (b) no push primitive or polling, (c) no broadcast or flooding on the data plane. The HoPP protocol transparently manages consumer and *producer mobility* [15].

An announcement of accumulated names to an exterior network outside the IoT domain is handled by companion protocols. The DNS-like Name Service for NDN (NDNS) [1] and Name Resolution Service (NRS) [23, 24] are two approaches that operate in NDN deployments. NDNSSEC [51] conceptualizes a namespace management for hierarchichal names and defines bindings to the DNS protocol, which allows for an integration into existing DNSSEC equipped DNS deployments.

4.2 Comparing IoT Protocols

Feasibility and benefits of the information-centric protocols need a quantitative evaluation in comparison with traditional Internet protocols such as MQTT and CoAP. Key properties of the three protocol families NDN, CoAP, and MQTT and their variants are compared in Table 1. Variants contain different deployment options for CoAP using the method codes `PUT` and `GET` with an optional resource observation configuration for the latter method. CoAP is further distinguished by its two reliability modes non-confirmable (n) and confirmable (c). MQTT differentiates between its TCP version and its UDP version MQTT-SN. QoS levels Q1 and Q2

Table 1 Comparison of CoAP, MQTT, and ICN protocols. CoAP and MQTT support reliability only in confirmable mode (c) and QoS levels 1 and 2 (Q1, Q2)

	Current IoT protocols					ICN protocols	
	CoAP [49]						
	PUT	GET	Observe	MQTT	MQTT-SN	NDN	HoPP
Transport	UDP	UDP	UDP	TCP	UDP	n/a	n/a
Pub/Sub	✘	✘	✓	✓	✓	✘	✓
Push	✓	✘	✓	✓	✓	✘	✘
Pull	✘	✓	✘	✘	✘	✓	✓
Flow control	✘	✘	✘	✓	✘	✓	✓
Reliability	(c)	(c)	✘	(Q1, Q2)	(Q1, Q2)	✓	✓

provide a reliability layer, where Q1 adds a simple retransmission mechanism and Q2 follows a two-step acknowledgement process to guarantee an exactly once delivery. Specialized properties of the different approaches become apparent: Every protocol variant features distinct capabilities. Notably in the IoT, where TCP (aka generic MQTT) is unavailable, the pull-based NDN and HoPP are the only protocols admitting flow control and reliability as a generic service.

An extensive study [14] performed comparative evaluations of all the protocols in Table 1 on a large-scale IoT testbed with varying amounts of network stress configured. The analyses include *memory consumption* on nodes and effective *network utilization* by control and data traffic including *protocol overhead* and *link stress* caused by retransmissions. The actual performance of data transmission is measured in *data loss*, *goodput*, and *content arrival time* which represents the delay between issuing a transaction and data arrival at the sink. The study also considers the *data flows* and its *energy consumption*. Figures 7, 8, and 9 summarize the goodput analysis of this study for NDN, HoPP, MQTT, and CoAP, respectively. The different experimental results of the data goodput are displayed in box plots and compared to the theoretical optimum (lines). Time series of data goodput are further revealing the flow behavior as displayed in the lower row of this figure. HoPP clearly admits the most evenly balanced flows and shows nearly optimal goodput values, closely followed by NDN. Flow performances for MQTT and CoAP fluctuate with some tendency of instability when approaching its full transmissions speed.

Overall results of this study clearly show a smoother operation in challenged networks with the ICN variants due to in-network caches and the hop-wise nature of (re-)transmissions.

4.3 Caching Strategies

Caching is a delicate subject in the IoT for the pronounced reasons that it reduces latency as well as forwarding load, improves data availability in networks without perpetual connectivity, and lessens overall energy expenditures. Since caching

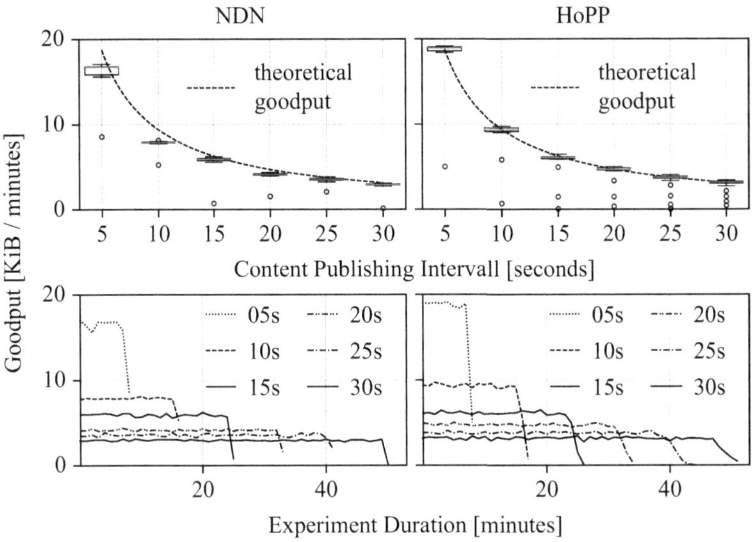

Fig. 7 Goodput summary and flow evolution for NDN and HoPP protocols at different publishing intervals

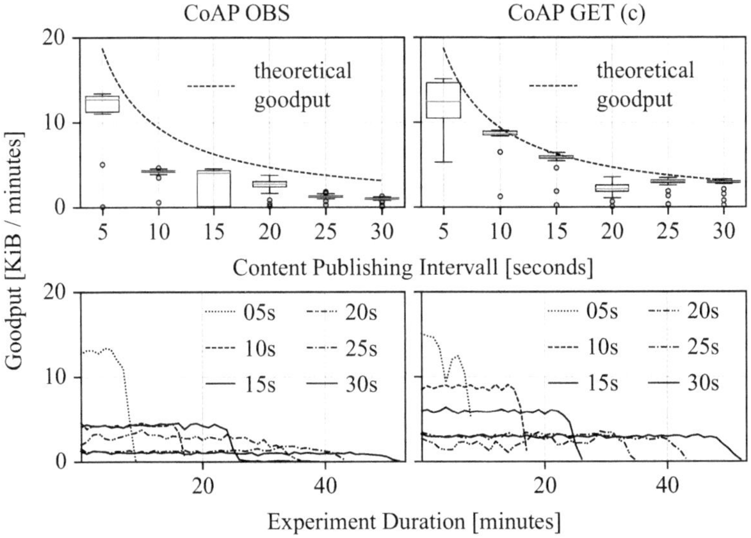

Fig. 8 Goodput summary and flow evolution for CoAP OBS and GET protocols at different publishing intervals

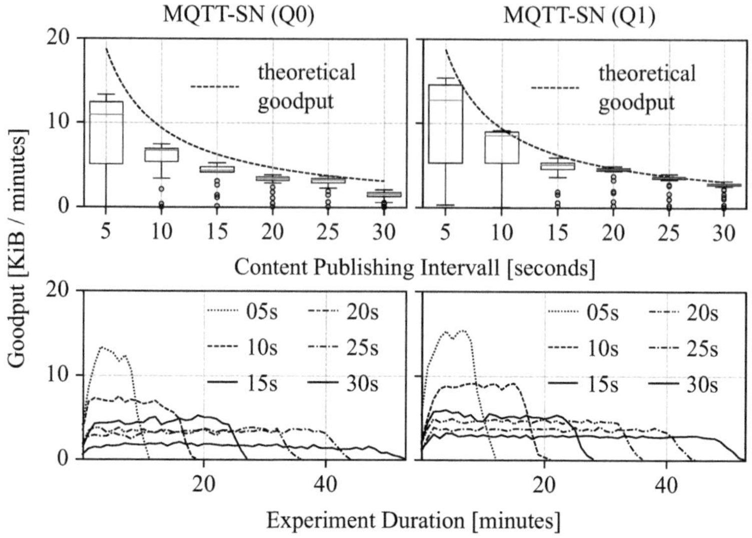

Fig. 9 Goodput summary and flow evolution for MQTT-SN protocols at different publishing intervals

resources on a single node are not available in abundance, a high cache diversity throughout the network is imperative to achieve the caching benefits. Caching policies that promote a high cache diversity can be categorized into different families.

Probabilistic caching, commonly known as $Prob(p)$, uses a static probability p on each forwarding hop to decide whether a node is eligible to cache incoming data. Several research groups [22, 42, 59] confirm that $Prob(p)$ yields a higher cache diversity, especially for lower p values, compared to the default policy of leaving a copy on every node.

Conversely to statically choosing a value for p, a policy may consider external hints to dynamically calculate values for each node, or even for distinct data objects. Examples for this class of policies include *ProbCache* [42], which computes p for data objects based on the hop distance between consumers and producers, and *pCASTING* [22], which considers content freshness, nodal battery levels, and cache saturation to calculate p. A recent study [39] shows that these approaches can increase the cache hit ratio, lower the number of hop traversals, and reduce cache evictions in resource-constrained regimes.

Another class of caching policies takes topological information into account to benefit from knowledge about the global cache utilization. A serious drawback of these approaches is their significant cost originating from long convergence times, increased signaling overhead, and susceptibility to topological changes [40].

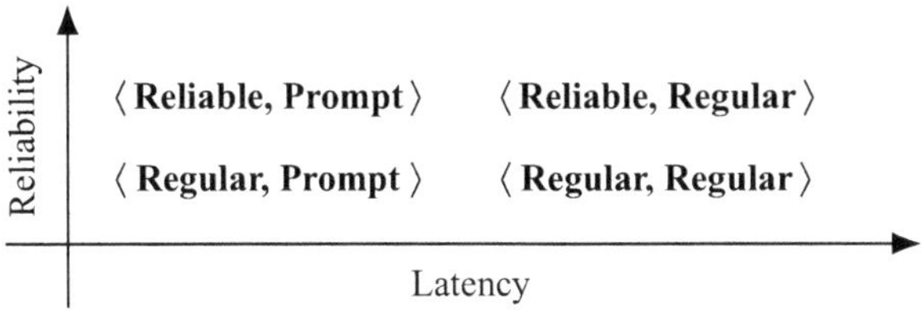

Fig. 10 QoS service levels along the two QoS dimensions *reliability* and *latency*

4.4 Quality of Service

Implementing service differentiation and assurance in a network raises the challenge of managing distributed resources without sacrificing them. In-network caches in ICN enrich the field of manageable resources. Caches reduce latency and forwarding load and often take the role of a (large, delay-tolerant) retransmission buffer. With CCNx and NDN, additional resources come into play in the form of Pending Interest Tables (PITs) that govern stateful forwarding. Capacities in forwarding, caching, and pending Interest state may be largely heterogeneous. A wirespeed forwarder may supply negligible cache memory compared to its transmission capacity, for example. In the IoT the opposite is often true in that flash memory is normally shipped in "infinite" sizes when compared to the main memory (PIT) or the wireless data rate. A beneficial resource management faces the problem of how to carefully balance these resources and arrive at an overall optimized network performance [28].

Resource complexity, however, extends beyond a single system. The impact of distributed resources is easily flawed if management cannot jointly coordinate contributions. Neighboring caches, for instance, are less effective if filled with identical data. A more delicate problem arises from PIT state management. If neighboring PITs diverge and no longer represent common forwarding paths, all data flows terminate, and forwarding resources are wasted. This problem of state decorrelation was first reported in [55].

QoS extensions for ICN have recently attracted attention and generated various efforts within the IRTF ICN research group [19, 26, 31, 37, 38]. Based on this effort, a QoS management for constrained CCNx and NDN networks emerged [18] that defines a lightweight, prefix-based flow classification mechanism and resource management rules to orchestrate manageable resources based on the QoS service levels displayed in Fig. 10. The resource management rules are grouped into three categories:

1. Locally isolated decisions that have no interactions with other mechanisms.
2. Local resource correlations that entail interactions between mechanisms.
3. Distributed resource coordination that affects resources across multiple devices.

An experimental evaluation of this QoS management scheme in a 31-node IoT deployment as depicted in Fig. 11 shows surprising results [18]. These experiments

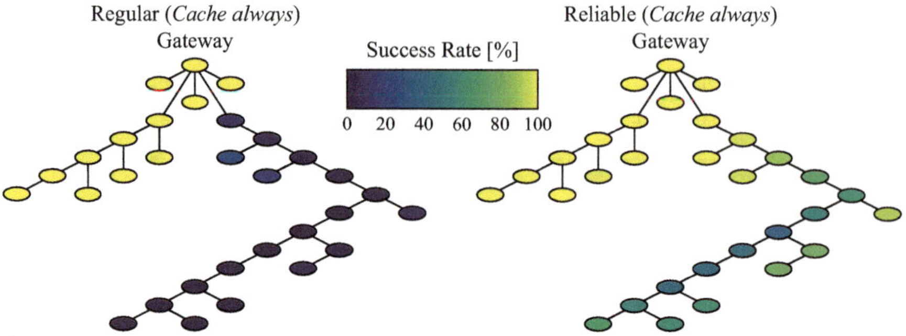

Fig. 11 Nodal success rates using regular traffic (left) and reliable actuator traffic (right)

run on the open-access FIT IoTLab testbed using the open source software platforms RIOT and CCN-lite with QoS extensions. The deployments consist of a gateway node, sensors, and actuators that periodically request sensor readings and control commands. The left-hand side of Fig. 11 shows very poor performances for nodes that are positioned far away from the gateway or forwarder nodes that belong to paths of high hop counts. The right-hand side shows the same setup and deployment but enables QoS features [18]. A great improvement of nodal performances is apparent. Nodes that formerly showed success rates of 0–10% now exhibit success rates of 40–100%.

4.5 *Network Deployment and Security Considerations*

Industrial safety and control systems are increasingly interconnected to interchange operational conditions locally and to report their status updates to external observers. A typical deployment scenario consists of IoT stub networks that are often wireless and confined to the production plant, together with gateways that uplink to an Internet service provider. Current initial deployment scenarios further involve a (private) cloud which a dedicated group of trustees can access. Typical stakeholders are the operators of the systems. All parties rely on secure communication channels established between the network endpoints and the cloud. This scenario builds closed data silos for a preselected, confined group. It is visualized on the left-hand side of Fig. 12.

Already today it becomes apparent that the number of stakeholders in emerging scenarios will widen—plant operators, emergency teams, equipment vendors, and supervisory authorities may retrieve information about current safety conditions, intermediate operational statistics, as well as long-term reports. Furthermore, even a wider public may legitimately require civil participation in affairs of common impact, as is developing from many open urban sensing initiatives [8], as well as participatory European laws. Following this demand, data silos need to break up in favor of a flexible, distributed data access that cannot easily rely on preconfigured

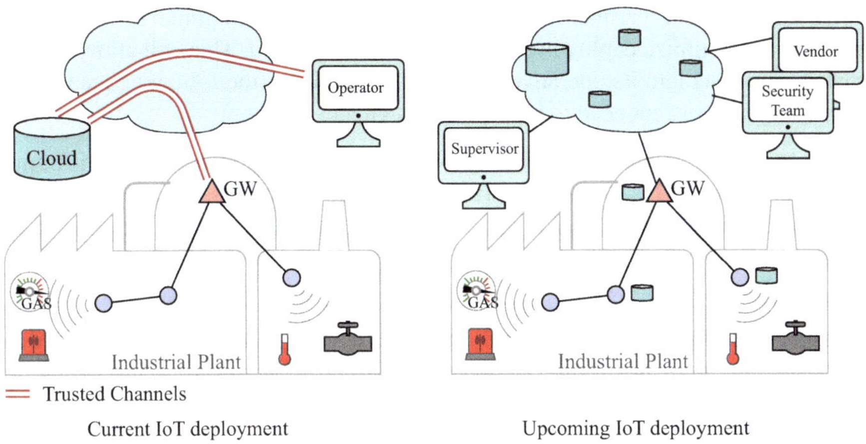

Fig. 12 Current and future deployment scenarios of the industrial Internet

trusted channels. Still, data might not be uniformly public, but continue to require protection. Protecting the data itself instead of the transmission channels paves the way to transparent data replication and caching—an efficient method for diversifying today's silos. Such a heterogeneous environment built from several independent stakeholders is visualized on the right-hand side of Fig. 12.

Taken from real-world deployment, studies [12, 20, 21] make the case for a distributed, multi-stakeholder environment and identify three major objectives for the networking layer:

1. Allow for ubiquitous multiparty data access without pre-established secure data channels or VPNs in the constrained IoT.
2. Provide a robustly secured networking infrastructure that is resilient to varying link conditions and mobility with the ability to recover locally from intermittent impairments.
3. Raise the barriers for DDoS attacks of constrained devices and confine the attack surface of unwanted traffic to local links.

The studies show how the NDN approaches to Information Centric Networking can significantly contribute to these goals. They also assess the shortcomings of current IoT solutions such as MQTT and CoAP over transport layer security as well as OSCORE [46], which brings object security to CoAP.

5 Conclusions and Future Research

In this chapter, we reviewed the state of the art of Information Centric Networking (ICN) in the Internet of Things (IoT), dedicating special focus to practical, ready-to-use solutions for the constrained IoT. We emphasized the benefits in robustness, resilience, and security of ICN and surveyed key performance results.

In the future, experimentally driven research should continue to thoroughly analyze and optimize deployments of ICN in the IoT world. This will allow to gain realistic insights into its operative potentials. In addition, the following areas have seen initial efforts, but need to be explored in greater depths.

Experimental exploration of 5G information-centric network slicing Initial work integrates ICN in the 5G core nextgen architecture [43], or contributes a framework to enable Internet Services over ICN in 5G LAN environments [53]. 5G core infrastructure, though, has not fully developed, and many experiences have to follow in future stages.

Named functions for in-network data processing at the network edge Named Function Networking (NFN) [50] and Remote Method Invocation in ICN (RICE) [27] are two notable approaches to in-network computing at the (mobile) edge. The IRTF COIN Research Group is in the process of formalizing the directions and requirements of the field [29, 30], which will eventually open up new networking opportunities.

Coexistence of information-centric and host-centric networking Tunneling ICN traffic over IP is common. The converse, however, is also an option [52, 56] and considered to be advantageous in mobile and IoT environments. Alternatively, hICN [9, 34] promises a closer integration with ICN packets mapped to the IPv6 protocol. hICN traffic can therefore be sent over the existing infrastructure—this deployment option still requires practical demonstration.

Economic and social implications of networking autonomous content objects Content object security as immanent to ICN facilitates a pluralistic networking approach, in which content can freely replicate and diffuse the Internet. This acts in contrast to end-to-end data encryption, which fosters service monopolies that are capable of controlling applications involved at one or both ends. This fundamental difference in networking bears the potential to alter the economics of the Internet, as well as its principles of topology building. Introductory studies on the economics and social contexts exist [2, 41], but wide-ranging case studies that elaborate on the implications are still missing.

References

1. Afanasyev A, Jiang X, Yu Y, Tan J, Xia Y, Mankin A, Zhang L (2017) NDNS: a DNS-like name service for NDN. In: Proceedings of the 26th IEEE International Conference on Computer Communications and Networks (ICCCN'17). IEEE, Piscataway, pp 1–9, https://doi.org/10.1109/ICCCN.2017.8038461
2. Agyapong PK, Sirbu M (2012) Economic incentives in information-centric networking: implications for protocol design and public policy. IEEE Commun Mag 50(12):18–26
3. Ahlgren B, Dannewitz C, Imbrenda C, Kutscher D, Ohlman B (2012) A survey of information-centric networking. IEEE Commun Mag 50(7):26–36

4. Baccelli E, Hahm O, Günes M, Wählisch M, Schmidt TC (2013) RIOT OS: towards an OS for the internet of things. In: Proceedings of the 32nd IEEE INFOCOM, Poster. IEEE Press, Piscataway, pp 79–80
5. Baccelli E, Mehlis C, Hahm O, Schmidt TC, Wählisch M (2014) Information centric networking in the IoT: experiments with NDN in the wild. In: Proceedings of 1st ACM Conference on Information-Centric Networking (ICN-2014). ACM, New York, pp 77–86. https://doi.org/10.1145/2660129.2660144
6. Baccelli E, Gündogan C, Hahm O, Kietzmann P, Lenders M, Petersen H, Schleiser K, Schmidt TC, Wählisch M (2018) RIOT: an open source operating system for low-end embedded devices in the IoT. IEEE Internet Things J 5(6):4428–4440. https://doi.org/10.1109/JIOT.2018.2815038
7. Banks A, Gupta R (eds) (2014) MQTT Version 3.1.1. Oasis standard, OASIS. http://docs.oasis-open.org/mqtt/mqtt/v3.1.1/os/mqtt-v3.1.1-os.html
8. Bornholdt H, Jost D, Kisters P, Rottleuthner M, Bade D, Lamersdorf WH, Schmidt TC, Fischer M (2019) SANE: Smart networks for urban citizen participation. In: 2019 26th International Conference on Telecommunications (ICT) (ICT 2019). IEEE Press, Piscataway, pp 496–500. https://doi.org/10.1109/ICT.2019.8798771
9. Carofiglio G, Muscariello L, Augé J, Papalini M, Sardara M, Compagno A (2019) Enabling ICN in the internet protocol: analysis and evaluation of the hybrid-ICN architecture. In: Proceedings of the 6th ACM Conference on Information-Centric Networking, ICN '19. ACM, New York, pp 55–66
10. Carzaniga A, Papalini M, Wolf AL (2011) Content-based publish/subscribe networking and information-centric networking. In: Proceedings of the ACM SIGCOMM WS on Information-centric Networking (ICN '11). ACM, New York, pp 56–61
11. Chen J, Arumaithurai M, Jiao L, Fu X, Ramakrishnan K (2011) COPSS: an efficient content oriented publish/subscribe system. In: ACM/IEEE Symposium on Architectures for Networking and Communications Systems (ANCS'11). IEEE Computer Society, Los Alamitos, pp 99–110
12. Frey M, Gündogan C, Kietzmann P, Lenders M, Petersen H, Schmidt TC, Shzu-Juraschek F, Wählisch M (2019) Security for the industrial IoT: the case for information-centric networking. In: 2019 IEEE 5th World Forum on Internet of Things (WF-IoT) (WF-IoT 2019). IEEE Press, Piscataway, pp 424–429. https://doi.org/10.1109/WF-IoT.2019.8767183
13. Ghodsi A, Shenker S, Koponen T, Singla A, Raghavan B, Wilcox J (2011) Information-centric networking: seeing the forest for the trees. In: Proceedings of the 10th ACM HotNets Workshop. ACM, New York, NY, USA, HotNets-X
14. Gündogan C, Kietzmann P, Lenders M, Petersen H, Schmidt TC, Wählisch M (2018) NDN, CoAP, and MQTT: a comparative measurement study in the IoT. In: Proceedings of 5th ACM Conference on Information-Centric Networking (ICN). ACM, New York, pp 159–171. https://doi.org/10.1145/3267955.3267967
15. Gündogan C, Kietzmann P, Schmidt TC, Lenders M, Petersen H, Wählisch M, Frey M, Shzu-Juraschek F (2018) Demo: seamless producer mobility for the industrial information-centric internet. In: Proceedings of 16th ACM International Conference on Mobile Systems, Applications (MobiSys), Demo Session, ACM, New York, best demo award
16. Gündogan C, Kietzmann P, Schmidt TC, Wählisch M (2018) HoPP: robust and resilient publish-subscribe for an information-centric internet of things. In: Proceedings of the 43rd IEEE Conference on Local Computer Networks (LCN). IEEE Press, Piscataway, pp 331–334. https://doi.org/10.1109/LCN.2018.8638030
17. Gündogan C, Kietzmann P, Schmidt TC, Wählisch M (2019) ICNLoWPAN – named-data networking in low power IoT networks. In: Proceedings of 18th IFIP Networking Conference. IEEE Press, pp 1–9. https://doi.org/10.23919/IFIPNetworking.2019.8816850

18. Gündogan C, Pfender J, Frey M, Schmidt TC, Shzu-Juraschek F, Wählisch M (2019) Gain more for less: the surprising benefits of QoS management in constrained NDN networks. In: Proceedings of 6th ACM Conference on Information-Centric Networking (ICN). ACM, New York, pp 141–152. https://doi.org/10.1145/3357150.3357404
19. Gündogan C, Schmidt TC, Wählisch M, Frey M, Shzu-Juraschek F, Pfender J (2019) Quality of service for ICN in the IoT. IRTF Internet Draft – work in progress 01, IRTF. http://tools.ietf.org/html/draft-gundogan-icnrg-iotqos
20. Gündogan C, Amsüss C, Schmidt TC, Wählisch M (2020) IoT content object security with OSCORE and NDN: a first experimental comparison. In: Proceedings of 19th IFIP Networking Conference. IEEE Press, pp 19–27. Best paper award
21. Gündogan C, Amsüss C, Schmidt TC, Wählisch M (2020) Toward a RESTful Information-centric Web of Things: a deeper look at data orientation in CoAP. In: Proceedings of 7th ACM Conference on Information-Centric Networking (ICN). ACM, New York
22. Hail MA, Amadeo M, Molinaro A, Fischer S (2015) Caching in named data networking for the wireless internet of things. In: International Conference on Recent Advances in Internet of Things (RIoT). IEEE, pp 1–6
23. Hong J, You T, Hong YG, Dong L, Westphal C, Ohlman B (2019) Design guidelines for name resolution service in ICN. Internet-Draft – work in progress 03, IETF
24. Hong J, You T, Hong YG, Kafle V (2019) Architectural considerations of ICN using name resolution service. Internet-Draft – work in progress 03, IETF
25. Jacobson V, Smetters DK, Thornton JD, Plass MF (2009) Networking named content. In: 5th International Conference on emerging Networking Experiments and Technologies (ACM CoNEXT'09). ACM, New York, pp 1–12
26. Jangam A, suthar P, Stolic M (2019) QoS treatments in ICN using disaggregated name components. Internet-Draft – work in progress 01, IETF
27. Król M, Habak K, Oran D, Kutscher D, Psaras I (2018) RICE: remote method invocation in ICN. In: Proceedings of the 5th ACM Conference on Information-Centric Networking, ICN'18. ACM, New York, p 1–11
28. Kutscher D, Eum S, Pentikousis K, Psaras I, Corujo D, Saucez D, Schmidt T, Waehlisch M (2016) Information-Centric Networking (ICN) Research Challenges. RFC 7927, IETF
29. Kutscher D, Karkkainen T, Ott J (2019) Directions for computing in the network. Internet-Draft – work in progress 01, IETF
30. Liu P, Geng L (2019) Requirement of computing in network. Internet-Draft – work in progress 01, IETF
31. Moiseenko I, Oran D (2020) Flow classification in information centric networking. Internet-Draft – work in progress 05, IETF
32. Mosko M, Solis I, Wood C (2019) Content-centric networking (CCNx) messages in TLV format. RFC 8609, IETF
33. Mosko M, Solis I, Wood C (2019) Content-centric networking (CCNx) semantics. RFC 8569, IETF
34. Muscariello L, Carofiglio G, Auge J, Papalini M (2019) Hybrid information-centric networking. Internet-Draft – work in progress 03, IETF
35. Nichols K (2019) Lessons learned building a secure network measurement framework using basic NDN. In: Proceedings of the 6th ACM Conference on Information-Centric Networking, ICN '19. ACM, New York, pp 112–122
36. Oh SY, Lau D, Gerla M (2010) Content Centric Networking in tactical and emergency MANETs. In: 2010 IFIP Wireless Days. IEEE, Piscataway, pp 1–5
37. Oran D (2019) Considerations in the development of a QoS architecture for CCNx-like ICN protocols. Internet-Draft – work in progress 03, IETF
38. Oran D (2020) Maintaining CCNx or NDN flow balance with highly variable data object sizes. Internet-Draft – work in progress 02, IETF
39. Pfender J, Valera A, Seah WKG (2018) Performance comparison of caching strategies for information-centric IoT. In: Proceedings of the 5th ACM Conference on Information-Centric Networking, ICN '18. ACM, New York, pp 43–53

40. Pfender J, Valera A, Seah WKG (2019) Content delivery latency of caching strategies for information-centric IoT. arXiv:190501011 [cs]. http://arxiv.org/abs/1905.01011, arXiv: 1905.01011
41. Piro G, Signorello S, Palattella MR, Grieco LA, Boggia G, Engel T (2017) Understanding the social impact of ICN: between myth and reality. AI SOCIETY 32(3):401–419
42. Psaras I, Chai WK, Pavlou G (2012) Probabilistic in-network caching for information-centric networks. In: Proceedings of the Second Edition of the ICN Workshop on Information-Centric Networking, Helsinki, pp 55–60
43. Ravindran R, suthar P, Trossen D, Wang C, White G (2020) Enabling ICN in 3GPP's 5G NextGen core architecture. Internet-Draft – work in progress 02, IETF
44. Schmidt TC, Wölke S, Berg N, Wählisch M (2016) Let's collect names: how PANINI limits FIB tables in name based routing. In: Proceedings of 15th IFIP Networking Conference. IEEE Press, Piscataway, pp 458–466
45. Schooler EM, Zage D, Sedayao J, Moustafa H, Brown A, Ambrosin M (2017) An architectural vision for a data-centric IoT: rethinking things, trust and clouds. In: IEEE 37th International Conference on Distributed Computing Systems (ICDCS). IEEE, Piscataway, pp 1717–1728
46. Selander G, Mattsson J, Palombini F, Seitz L (2019) Object security for constrained RESTful environments (OSCORE). RFC 8613, IETF
47. Shang W, Afanasyev A, Zhang L (2016) The design and implementation of the NDN protocol stack for RIOT-OS. In: Proceedings of IEEE GLOBECOM 2016. IEEE, Washington, DC, pp 1–6
48. Shang W, Bannis A, Liang T, Wang Z, Yu Y, Afanasyev A, Thompson J, Burke J, Zhang B, Zhang L (2016) Named data networking of things (Invited Paper). In: Proceedings of IEEE International Conference on Internet-of-Things Design and Implementation (IoTDI). IEEE Computer Society, Los Alamitos, pp 117–128
49. Shelby Z, Hartke K, Bormann C (2014) The constrained application protocol (CoAP). RFC 7252, IETF
50. Sifalakis M, Kohler B, Scherb C, Tschudin C (2014) An information centric network for computing the distribution of computations. In: Proceedings of the 1st ACM Conference on Information-Centric Networking, ICN'14. ACM, New York, pp 137–146
51. Tehrani PF, Osterweil E, Schiller J, Schmidt TC, Wählisch M (2019) The missing piece: on namespace management in NDN and how DNSSEC might help. In: Proceedings of 6th ACM Conference on Information-Centric Networking (ICN). ACM, New York, pp 37–43. https://doi.org/10.1145/3357150.3357401
52. Trossen D, Reed MJ, Riihijärvi J, Georgiades M, Fotiou N, Xylomenos G (2015) IP over ICN – the better IP? In: 2015 European Conference on Networks and Communications (EuCNC). IEEE Computer Society, Los Alamitos, pp 413–417
53. Trossen D, Wang C, Robitzsch S, Reed M, AL-Naday M, Riihijarvi J (2020) Internet services over ICN in 5G LAN environments. Internet-Draft – work in progress 01, IETF
54. Tschudin C, Scherb C, et al. (2018) CCN Lite: lightweight implementation of the content centric networking protocol. http://ccn-lite.net
55. Wählisch M, Schmidt TC, Vahlenkamp M (2013) Backscatter from the data plane – threats to stability and security in information-centric network infrastructure. Comput Netw 57(16):3192–3206. https://doi.org/10.1016/j.comnet.2013.07.009
56. White G, Shannigrahi S, Fan C (2020) Internet protocol tunneling over content centric mobile networks. Internet-Draft – work in progress 01, IETF
57. Xylomenos G, Ververidis CN, Siris VA, Fotiou N, Tsilopoulos C, Vasilakos X, Katsaros KV, Polyzos GC (2014) A survey of information-centric networking research. IEEE Commun Surv Tutor 16(2):1024–1049
58. Zhang L, Afanasyev A, Burke J, Jacobson V, Claffy K, Crowley P, Papadopoulos C, Wang L, Zhang B (2014) Named data networking. SIGCOMM Comput Commun Rev 44(3):66–73
59. Zhang M, Luo H, Zhang H (2015) A survey of caching mechanisms in information-centric networking. IEEE Commun Surv Tutor 17(3):1473–1499

Part III
Enabling Technologies for Industrial IoT

Security Challenges for Industrial IoT

Lehlogonolo P. I. Ledwaba and Gerhard P. Hancke

1 Introduction

The concept of the Industrial Internet represents the incorporation of the Internet of Things (IoT), machinery control and operational techniques, information and communications technology (ICT) and people within a larger Industrial Internet of Things (IIoT) to realise the use of advanced data analytics to improve business outcome [1]. This joining of "global industrial sectors, advanced computing and manufacturing, pervasive sensing and ubiquitous network connectivity" [1] results in a single, cohesive system. This also serves in connecting previously isolated, simple, physical operations to the cyber world for smarter, self-aware independent actuation [1]. Industrial systems connected using the Industrial Internet typically operate in mission-critical environments and have higher standards of safety, security, availability and resilience for all components than general consumer and commercial sectors [1]. In the industrial context, safety is defined as the condition in which "the system is able to operate without unacceptable risk of physical damage or damage to the health of the people directly or indirectly in contact with the system as a result of damage to system property or the system environment" [1], security as the "operating condition of the system which does not allow for the unintended or unauthorised access, change or destruction of the system, its data and the information it encompasses" [1] and resilience as the "system condition that is capable of avoiding, absorbing or dynamically managing adversarial conditions while in the process of completing assigned missions or reconstructing operational capabilities after suffering casualties within the system" [1]. For the Industrial Internet to be considered effective, significant increases should

L. P. I. Ledwaba (✉) · G. P. Hancke
Department of Computer Science, City University of Hong Kong, Kowloon, Hong Kong
e-mail: lpledwaba2-c@my.cityu.edu.hk; gp.hancke@cityu.edu.hk

N. H. Mahmood et al. (eds.), *Wireless Networks and Industrial IoT*,
https://doi.org/10.1007/978-3-030-51473-0_10

be seen in the overall system performance, scalability, efficiency and compatibility, enabling interoperability for a wide variety of open standards, frameworks and architectures [1].

The addition of computing capability to industrial processes brings with it a variety of challenges. The vulnerability of IIoT to malicious attacks is a growing concern as more "smart" deployments are established globally. Standardisation in the production of IIoT devices, their communication protocols and the degree of security that the devices are capable of providing is essential for deployment into industrial processes with strict operational guidelines. The scale required of IIoT deployments means that future solutions should be highly scalable and interoperable to avoid vendor lock-in [2]. The availability and integrity of the IIoT network should always be preserved to be able to meet strict, real-time deadlines and to prevent cascading failures which could result in physical harm [2]. The constraint of resources such as available power, processing and memory and long operational periods means that developed IIoT solutions should be able to support low power operation and utilise a small portion of the memory and processor resources [2].

The challenges seen with IIoT devices also extend into the domain of security. IIoT devices are vulnerable to physical attacks such as tampering and theft as large-scale deployments are often unmonitored [2]. The devices are also subject to eavesdropping, man-in-the-middle, denial-of-service and masquerade attacks as a result of the peer-to-peer, wireless broadcast network which currently implements little to no mechanisms to verify the identity of communicating nodes and authenticity of data received [2, 3]. Implementing traditional IT security techniques fails to secure these devices as the added delays often compromise the availability of the system. Security solutions for the IIoT context therefore need to be capable of securing networks while minimising trade-offs in power consumption, processing capacity and memory footprint.

2 Security Standards for the Industrial IoT

Security standards can be used to define what security is expected for an IIoT network, the depth at which security services should be implemented, and to validate the security mechanisms and solutions designed to secure IIoT. In an effort towards standardising how IIoT networks are developed and deployed, improving and accelerating the move towards the IIoT, the Industrial Internet Consortium (IIC) was formed by businesses and academic institutions. As part of their work, the IIC developed a reference architecture and security framework detailing a standardised method for designing secure IIoT networks with the aim of making the Industrial Internet easily understandable and supported by "widely applicable, standard-based, open architecture frameworks and reference architectures" [1]. The vendor-agnostic reference architecture details the interactions and interoperability of the various viewpoints within the Industrial Internet and provides guidelines for the development and deployment of future network solutions and application architectures [1].

The security framework [4] details the security techniques and technologies which are to be employed within the various areas and stack levels of the network architecture to guarantee safe, secure and resilient operation throughout the effective life span of IIoT deployment. The top layer comprises four (4) foundations, namely, "endpoint protection, communication and connectivity protection, security monitoring and analysis and security configuration management" [4]. When used in conjunction with supplementary documents such as *Endpoint Security Best Practices* [5], the IIC provides a comprehensive pool of resources that allows developers to build in appropriate security services at design time.

The OpenFog Reference Architecture for Fog Computing, also known as standard IEEE 1934–2018, was developed in respect of the need for an open, fog computing architecture capable of ensuring interoperable and secure systems and one that is independent of, but fully supported by, the wider vendor space [6]. In the Industrial Internet, fog computing architectures are used to "selectively move comput[ing], storage, communication, control and decision making closer to the network edge where data is being generated in order to solve the limitations in current infrastructure to enable mission-critical, data-dense use cases" [6]. This allows for the computing resources at the edge of the IIoT network to interface with wider cloud services with reduced latency as fog computing maintains the benefits of a cloud computing scheme [6]. The reference architecture defines eight main pillars – "security, scalability, openness, autonomy, RAS (reliability-availability and serviceability), agility, hierarchy and programmability" [6] – as well as the relevant stakeholders and their roles in the wider fog value chain. These include silicon manufacturers, application developers, operating systems, etc. [6].

The security pillar describes the functions and mechanisms that could be applied to secure a fog node, from the silicon utilised in the node design to the software applications used on and with the node. Privacy, anonymity, integrity, trust, attestation, verification and measurement are identified by the architecture as key security attributes which should be guaranteed on a node to the best of one's ability [6]. As a basis for a secure design, a secure node must provide an immutable root of trust, preferably hardware-based. The root of trust should then be attestable by the software agents running within and throughout the fog infrastructure. Edge nodes should provide the first point of access control and encryption within the wider network in addition to providing contextual integrity, isolation and control aggregation of privacy-sensitive data prior to their departure from the network edge. Should there be any network components that cannot be attestable, they should be prevented from participating within and with the fog nodes and should be deemed to provide data that is not fully trustworthy [6].

Comparing the architectures developed for the Industrial Internet and fog computing, one can see that they are complementary in their recommendations made for node security. The IIC Reference and Security frameworks serve to provide a guideline on what functions should be included and the objectives that they should meet, while the OpenFog Reference Architecture provides a recommendation as to which mechanisms and technologies could be used to provide those functions. Combining the two architectures gives a solid, standard base design, as the uncertainty associated with the required functions for node security and the tools that are to be used in order to meet the objectives set for the functions have been removed.

In addition to the newer standards and guidelines, existing standards may also be applied to the design of IIoT networks. The Federal Information Processing Standard (FIPS) 140-2 Security Requirements for Cryptographic Modules standard and the Common Criteria (CC) Protection Profiles (PP) [7] define the various levels of security which can be established across a module implementing cryptographic processes and can subsequently be used for designing secure IIoT endpoints. Industry-specific standards, such as the National Institute of Standards and Technology (NIST)'s Guidelines for Smart Grid Cybersecurity, will also provide guidelines for allowable tolerances in latency, jitter and availability that can serve to influence the design of the IIoT communications network.

3 Requirements and Trade-Offs for Industrial IoT Security

With the establishment of any security services in a network comes trade-offs that occur as a result of allocating additional resources towards protecting devices from malicious activities. In the context of IIoT devices, these trade-offs need to be given due consideration given the limitation on available resources. Adding security capability has the potential to deplete the endpoint resources or introduce delays such that the device becomes unsuitable for the real-time, mission-critical contexts in which it is required to operate. To be able to secure the IIoT, it is important to identify where compromise will be seen and to choose security solutions where a trade-off should not negatively impact the network's usefulness to the application for which it is intended. By considering the trade-offs given in line with the industrial standards of safety and security, secure IIoT deployments can be designed in compliance with the different industry regulations.

Another important consideration for IIoT security is the timing of when security mechanisms are to be included into the design of devices. By the nature of some security solutions, their inclusion would need to be considered in earlier design stages to ensure the most effective protection. Considering Fig. 1, one can see that security mechanisms that would affect the physical configuration of the device would need to be considered earlier in the device design stages, while those that are achievable through firmware could be considered in later design stages. The design timeline together with the associated trade-offs of the security solution would allow designers to be able to choose future upgradeable solutions early, thus preventing the need for intensive and expensive physical redesigns.

In the following sections, the requirements and trade-offs for the IIoT security mechanisms introduced in Fig. 1 are discussed in further detail. A brief summary of the main points are presented in Table 1.

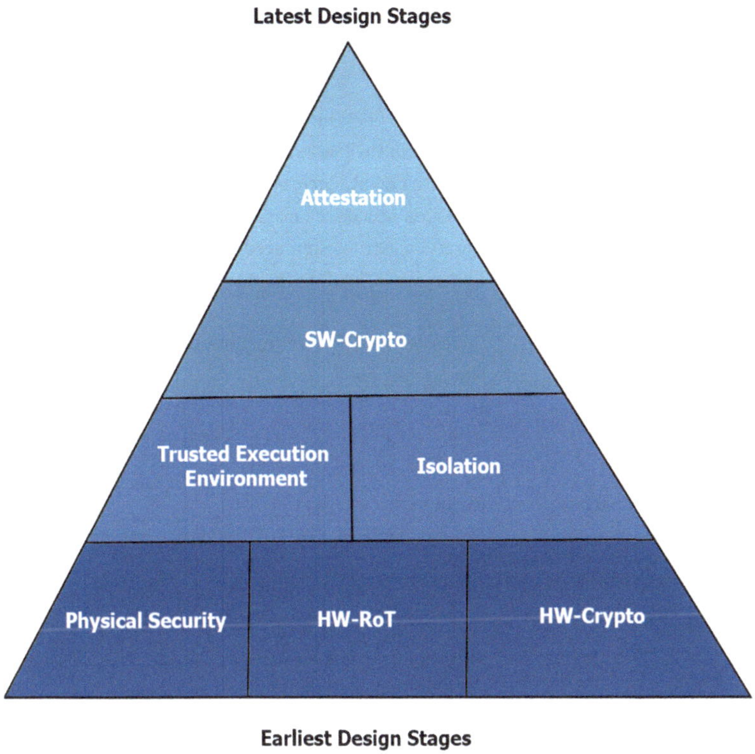

Fig. 1 Inclusion stages for the incorporation of security into IIoT device design

3.1 Physical Security

Devices in the IIoT are vulnerable to four main types of attacks – invasive, non-invasive, fault injection and software attacks – which arise as a result of compromised physical security [8]. Invasive attacks require the physical capture of the endpoint and often involve physical intrusion at device level, where physical intrusion occurs to the product enclosure, or at chip level, where intrusion occurs to the chip packaging [8, 9]. Non-invasive attacks do not include physical intrusion or damage to the endpoint device but are the result of observing the behaviour of the endpoint as security operations are carried out [8]. Side-channel attacks such as timing analysis attacks, electromagnetic analysis and power analysis attacks are examples of common endpoint non-invasive attacks [8]. Fault injection attacks occur when the attacker alters the environment or operating conditions of the IIoT endpoint in order to initiate a malfunction that compromises device security [8]. Over- or under-voltage attacks, over- or under-temperature attacks and timing attacks are common examples of fault injection attacks [8]. Software attacks are

Table 1 Summary of security solutions and trade-offs for the IIoT

Physical security	
Existing solution	Trade-off
Enclosure monitoring sensors	Increase in enclosure size to accommodate tamper sensors
Electromagnetic leakage shields	Increase in IIoT mote size to accommodate shields
Physical unclonable functions	Increased delay, decrease in available ROM and RAM
Anti-tamper mesh	Inclusion needed at design phase
	Careful pattern design needed
	Expensive/difficult to include on legacy devices
Secure and trusted execution	
Hardware security modules	Increased device power requirements
	Not upgradable in future
	Increased PCB size to accommodate new IC
	Added delay to transmit encrypted data
Isolation	
TEEs and ARM TrustZone	Requires use of ARM MCU
	Not independently tested for security compliance because of NDA
Attestation	
Commercial solutions	Remain focused on single-prover attestation
	Still subject to a wide variety of shortcomings and lack of consensus on methodology
Academic solutions	Would still need to be verified and tested against industrial standards
Cryptography	
Software implementations	Large increase in memory occupation owing to large code sizes
	Long computation delays introduced into network
	Increased power consumption by endpoints
	Need to use standard cryptographic algorithms and constantly check for algorithm deprecations
Hardware crypto accelerators	Difficult to upgrade if algorithm is deprecated

typically launched through the communication interfaces of the device such as debug interfaces, programming interfaces and communication interfaces [8].

The vastness of IIoT network deployments means that it is highly infeasible to completely prevent node capture [9]. Therefore, as the first building block towards securing the entire IIoT network, tamper protection mechanisms need to be employed to improve the physical security of isolated network devices. Complete physical security solutions require the inclusion of tamper detection, tamper response, tamper resistance, if possible, and tamper evidence logging [8]. Standardisation, licensing or certification specifications are mechanisms which can be used as a guideline in the design of a security solution and to test for compliance for physical security. The FIPS 140-2 standard [7] defines four requirement levels for physical security, while IBM defines six levels of physical security protection [10].

While it is vital that physical security measures be designed and included from the design stages of a secure IIoT endpoint (also known as a secure mote), they come at a variety costs which also need to be factored into the design of the mote. External tamper sensing protections such as enclosure monitoring sensors and electromagnetic leakage shields will need to be provided with sufficient space and ventilation, leading to possible increases in enclosure sizes. Should the size increase not be constrained, situations will arise in which the enclosure size becomes a limitation in the application areas in which the mote is used. Other considerations for using external tamper sensing protection includes:

- identifying appropriate power sources for the sensing circuitry,
- establishing the impact the additional drain tamper detection circuits may have on the lifetime of the mote's power source,
- ensuring that the installed tamper protections allow for maintenance and upgrade work,
- developing maintenance and upgrade policies such that exploitable weak points (back doors) are not introduced by the maintenance process.

Physical security measures for the mote processor, such as anti-tamper mesh and physical unclonable functions, require careful design in order to properly disguise the signal and wiring patterns that are of interest to malicious attackers while not impacting the performance of the processor. These measures need to be implemented during the design phase of the mote, making their inclusion on legacy devices expensive or very difficult to achieve.

3.2 Secure and Trusted Execution

In Industrial Internet applications, it is essential to define the levels of trust allocated to network components, communications and maintenance installations. This trust can be identified as being either static or dynamic. Static trust is based on "evaluations against a specific set of security requirements" such as international standards for security [11]. Dynamic trust is highly dependent on the continued running state of the system under consideration and is measured throughout the system life cycle. Fundamentally, dynamic trust is determined through the existence of a secure and reliable means within the system capable of providing evidence that the trust state is unchanged and that the system remains in an expected, secure state [11]. The IIC framework recommends implementing a root of trust (RoT) from which mechanisms for identification and integrity checking can be derived, thereby establishing dynamic trust. The root of trust is to provide initial confidence in the system operations by validating that the entities requesting network access are both authorised to access network resources and cannot access resources for which they do not have access permission [4]. The root of trust also aids with establishing network integrity by providing a baseline for identifying and preventing unauthorised access attempts [4].

After having established trust in the network operation, establishing trust in network users is the next challenge to be handled. The use of credentials to verify the identity of the various devices communicating within the network could establish varying levels of trust and, consequently, varying levels of access privileges [4]. Choosing an appropriate credential scheme to be applied to endpoints however is highly dependent on the credential's uniqueness and strength, and the context in which the endpoint will be operating [4]. Care needs to be taken to ensure that credentials offer sufficient uniqueness and strength – to prevent the falsification of a device's identity – while also allowing for new devices to be easily and securely added to the growing network space [4]. ISO/IEC 24760-1 [12] provides detailed guidelines in determining the three levels of trust – identity, unique identity and secure identity – for endpoint identities, and the Industry 4.0 documentation [13] provides additional information on the requirements of a secure identity technology that is to be used in industrial contexts.

Hardware security modules (HSMs) may be used to implement a root of trust however they bring with a variety of trade-offs in terms of the power consumption and upgradability of IIoT devices. The use of hardware security chips as a security device could serve to shorten the security lifetime of the secure mote. As encryption and security standards are continually updated, one may find that the standard version implemented on the HSM employed to provide a RoT may be superseded within by the newer version sooner than expected, decreasing the level of trust that the secure mote provides. Given that these chips are hard soldered into the design, they would be difficult to replace. With large IIoT network deployments, such an operation would be highly expensive and infeasible. The use of a separate hardware module could also lead to an increase in the power consumption for IIoT devices both while active and while asleep. Appropriate testing would need to be conducted in order to determine the added power drain and the new effective lifetime of the IIoT device power source. The addition of a separate chip also serves to increase the printed circuit board size and could introduce delay in the MCU start-up and processing times, as communication would need to be routed through to the security module and back. Again, tests would need to be conducted to determine the added delay time and adjust the network operations to accommodate it within the application area requirements.

3.3 Isolation

Isolation techniques can be used to shelter parts of the IIoT network or device in order to prevent the cascade of undesirable effects caused by a failure in other areas [4]. As a result, a minimum operational baseline can be guaranteed even during the event of a malicious attack. Physical isolation techniques may also be used to provide security services separately from normal operations by employing the use of a dedicated chip, device or execution environment. One such example is the use of a dedicated gateway to provide security services for older, legacy devices. Often, the firmware cannot be upgraded on these devices to accommodate the updated security policies owing to insufficient resources or a lack of legacy support in the new security firmware [4]. Traffic flowing to and from these devices would be filtered through the gateway, where security operations would be subsequently handled. This allows for the provision of adequate coverage in vulnerable areas of the attack space while trying to minimise the impact on network operations.

Generally, isolation can be achieved through the operating system to isolate business and operational processes from security processes (process isolation); through boundaries determined by hardware, software or a hybrid implementation (container isolation); or through a hypervisor configured to isolate each running instance on an IIoT device (virtual isolation) [4]. Already, isolation practices can be seen in some existing security solutions. HSMs provide physical isolation of security processes by implementing security functions on a separate, physical device. Security modes, such as those implemented by a trusted execution environment (TEE), provide a form of virtual isolation through the separation of security processes and resources by making them unavailable to normal operations operating outside of the secure world. Current hypervisor and container-based technologies remain heavily focused on securing traditional ICT technologies and operating systems; however solutions for the IIoT are slowly emerging, with implementations focusing on the development of container technologies for IoT cloud services or Linux-based embedded operating systems designed to support gateway functions.

Therefore, the main problem facing the use of isolation techniques with the IIoT is the lack of appropriate solutions given that hypervisor use is still primarily seen within tradition ICT systems. Although forms of isolation are provided within the ARM TrustZone TEE, the use of TrustZone is currently limited to ARM MCU solutions whose architecture is TrustZone capable. Another trade-off with the use of TrustZone, and vendor-specific isolation solutions, is that the lack of independent compliance testing by unaffiliated developers as a result of non-disclosure agreements. One is limited to trusting a manufacturer's claims of compliance to security standards.

3.4 Attestation

Assuring the integrity of IIoT data is often achieved by using a digital signature. The signing key is protected in secure storage using a RoT, and signing operations would be conducted in a trusted execution environment such as within a trusted platform module (TPM) [4]. In using a digital signature, an IIoT device would be able to validate the integrity of firmware updates prior to installation while configuration and log files could be signed to ensure their integrity for further network uses [4].

Attestation is another technique that is utilised towards the assurance of integrity. The basis of attestation is that "the entity that is to be tested, called the prover, sends a status report of its current configuration to another party, called the verifier, to demonstrate that it is in a known and thus trustworthy state" [14, 15]. To provide attestation, a trusted third party needs to be provided along with a mechanism to provide provable information fields that can be bound together with a digital signature, called an attest [16]. A variety of attestation methods have been previously used to provide trust and integrity within IIoT networks, each with varying degrees of success and shortcomings.

Remote attestation schemes assume that the prover is provided with a trusted mechanism, such as a TPM, with integrity measurements being taken and securely stored during the secure boot process [15]. When conducting the attestation, the verifier sends a request for the device configuration measurements, and the prover retrieves and signs the measurements, through the use of a digital signature algorithm or a digital certificate from a trusted third party, before sending them to the verifier [15]. The verifier then verifies the signature and compares the measurements against expected measurements for that device configuration [15]. Various shortcomings have been seen with the remote attestation scheme when applied to an IoT configuration. Firstly, as it is best suited for single-prover settings, it is infeasible for the verifier to know every possible device configuration in the network, especially given large-scale IIoT deployments [15, 17]. Secondly, with IIoT devices being left largely unattended and in remote deployments, the assumption about no physical attacks occurring on the devices can no longer be considered valid [16].

Software-based attestation was typically targeted for the resource-constrained devices at the edge of a wireless sensor network (WSN). Differing from the RoT-based remote attestation, software attestation uses challenge-response techniques which allow for the verifier to check the integrity of the prover's memory contents against modification, relying on checking the computation time of the prover in responding to the attestation challenge as an indicator of whether the device has been compromised [14]. Traditionally, the technique is heavily reliant on the assumption that an attacker is not actively attacking the network during the attestation period [14]. Again, previous implementations of software-based attestation focused on single-prover scenarios, making existing commercial attestation solutions unsuitable for use in WSN/IoT applications.

As with isolation, the use of attestation in the IIoT lacks appropriate solutions that can be implemented as part of a security policy design. Commercially available solutions for attestation remain primarily focused on single-prover methods, which are inappropriate for the peer-to-peer nature of IIoT network deployments. Academic solutions for attestation attempt at designing multi-prover methods. However, these are still subject to shortcomings that are to be handled as future work and lack of consensus on methodology. In addition, academic solutions would need to be taken into a lengthy, commercial development cycle in which verification and testing against industry standards would still be required.

3.5 Cryptography

Under the guidelines given in [4], IIoT devices should use standard cryptographic algorithms with regularly maintained and updated libraries [4]. The framework recommends the use of hardware random number generators (RNG) to ensure the randomness and uniqueness of cryptographic keys and a key revocation scheme should the invalidation of a key be required prior to its expiration [4].

Performing cryptographic operations on IoT endpoint devices has been a continuous challenge owing to their resource-constrained nature and the intensive mathematical processing required of encryption and decryption operations (especially in asymmetric solutions). In such cases, hardware accelerators are often employed to enable cryptographic operations. More recently, IIoT devices are being fitted with 32-bit central processing units (CPU), which provide more processing capability, but the random access memory (RAM) and read-only memory (ROM) available on these devices are still far less than what can be found on a traditional personal computer (PC). Existing studies provide a good indication of the capability of older-generation sensor nodes to handle cryptographic algorithms; however, of the algorithms often tested, many may not be the most appropriate to use towards safeguarding an IIoT endpoint given their age, and subsequent deprecation as a standard, or lack of standardisation or openness. More recent studies showcase the ability of new-generation IIoT processors in running unmodified, standard cryptographic algorithms, but it can be seen that the available processing capabilities are not yet sufficient to adequately handle public key cryptography techniques [18].

A number of trade-offs arise from the use of cryptographic solutions. Updates by standard bodies would need to be monitored to ensure that cryptographic algorithms are still appropriate to use for industrial and commercial applications and are still considered secure. As with the HSM, a hardware crypto accelerator would be difficult to upgrade in the event of the provided algorithm's deprecation as a standard. Additionally, care would need to be taken to protect the communication paths between the MCU and the crypto accelerator to ensure that no security information is leaked.

With the use of software cryptographic algorithm implementations, previous studies have shown large increases in memory occupation, computation delays and increased power consumption which were observed when implemented on older-generation devices [19–22]. Although these observed performances may improve with the use of new-generation IoT processors, software implementations of cryptography are unsuitable for use on legacy devices. This would then either require a replacement of all legacy devices with newer, more future-proof solutions, deployment of security gateways in areas where legacy devices are in use, or result in a network with a mixture of secure and insecure devices, which fails to adequately address the security requirements of the network. The use of a security gateway may be able to provide cryptographic ability for communications originating from legacy but would result in an increase in the overall network size and would require a large deployment effort with an associated cost. Additionally, care would need to be taken to adjust the network with appropriate routing protocols in order to prevent communication delays, as a result of message queuing, or instances of message dropping should multiple devices try communicating with the gateway at once.

4 Conclusion

Throughout the course of this chapter, it has been shown that security for the IIoT needs to be implemented from the design stages of application technologies in order to maximise the attack space covered and the effective lifetime of the security protection. The frameworks proposed by the IIC and OpenFog foundation have aided in identifying the standard security features needed to properly secure IIoT, supported by established industry standards. By conducting a detailed analysis of the identified security features, appropriate security technologies were found to provide security for the IIoT, including pre-designed secure MCUs. In addition, the need for open, standard security solutions was highlighted as a mechanism to ensure and enforce vendor compliance to industrial security regulations.

It was also seen that the inclusion of security mechanisms into an IIoT network would come with added trade-offs – some of which included increased device size, increased power consumption, additional memory requirements and increases in monetary cost. Also identified were gaps in IIoT security implementations for areas such as data loss prevention, device monitoring, attestation and isolation, illustrating that a complete security solution is yet to be readily available for the IIoT. As a result, a collaborative, in-depth research effort is needed across the academic, industrial, private and public sectors to be able to support multi-layer solution-development.

Acknowledgments This research has been supported by the Council for Scientific and Industrial Research (CSIR), South Africa, funding under project number 05400 054AT KC9EICF.

References

1. Industrial Internet Consortium (2015) Industrial Internet Reference Architecture, p 100, version 1.7. [Online]. Available: http://www.iiconsortium.org/IIRA-1-7-ajs.pdf
2. Sadeghi AR, Wachsmann C, Waidner M (2015) Security and Privacy Challenges in Industrial Internet of Things, San Francisco, June 2015, pp. 1–6, ID: doc:58de387be4b0cc37dc282eef. [Online]. Available: http://ieeexplore.ieee.org/document/7167238/
3. Gollmann D, Krotofil M (2016) Cyber physical system security, 1st edn, ser. The new codebreakers. Springer, Berlin/Heidelberg, pp 195–204, presentation. [Online]. Available: https://link.springer.com/chapter/10.1007/978-3-662-49301-4_14
4. Industrial Internet Consortium (2016) Industrial Internet Security Framework Volume G4, p 173, volume G4. [Online]. Available: http://www.iiconsortium.org/pdf/IIC_PUB_G4_V1.00_PB-3.pdf

5. Hanna S, Kumar S, Weber D (2018) IIC endpoint security best practices, Mar 2018. [Online]. Available: https://www.iiconsortium.org/pdf/Endpoint_Security_Best_Practices_Final_Mar_2018.pdf
6. OpenFog Consortium (2017) OpenFog reference architecture for fog computing, Feb 2017, reference architecture. [Online]. Available: https://www.openfogconsortium.org/wp-content/uploads/OpenFog_Reference_Architecture_2_09_17-FINAL.pdf
7. National Institute of Standards and Technology (2001) Security requirements for cryptographic modules, Federal information processing standards (FIPS), Technical report, FIPS 140-2. [Online]. Available: http://nvlpubs.nist.gov/nistpubs/FIPS/NIST.FIPS.140-2.pdf
8. Nisarga B, Peeters E (2016) System-level tamper protection using MSP MCUs, Aug 2016, application report SLAA715. [Online]. Available: http://www.ti.com/lit/an/slaa715/slaa715.pdf
9. Yussoff YM, Hashim H, Rosli R, Baba MD (2012) A review of physical attacks and trusted platforms in wireless sensor networks. Proc Eng 41:580–587, Jan 2012. [Online]. Available: http://www.sciencedirect.com/science/article/pii/S187770581202615X
10. Skorobogatov S (2012) Physical attacks and tamper resistance, 1st edn. Ser. Introduction to hardware security and trust. Springer, New York, pp 143–173. [Online]. Available: https://link.springer.com/chapter/10.1007/978-1-4419-8080-9_7
11. Sabt M, Achemlal M, Bouabdallah A (2015) Trusted execution environment: what it is, and what it is not. In: 2015 IEEE Trustcom/BigDataSE/ISPA, vol 1, Helsinki, Aug 2015, pp 57–64. [Online]. Available: http://ieeexplore.ieee.org/document/7345265/
12. International Organisation for Standardisation (2011) SANS ISO/IEC 24760-1:2011: information technology. Security techniques. A framework for identity management. Terminology and concepts, SABS standards division, Technical report, Dec 2011, ISO/IEC standard. [Online]. Available: https://www.iso.org/obp/ui/#iso:std:iso-iec:24760:-1:ed-1:v1:en
13. Plattform Industrie 4.0 (2016) Technical overview: secure identities, working paper. [Online]. Available: https://www.plattform-i40.de/I40/Redaktion/EN/Downloads/Publikation/secure-identities.pdf?__blob=publicationFile&v=7
14. Asokan N, Brasser F, Ibrahim A, Sadeghi A-R, Schunter M, Tsudik G, Wachsmann C (2015) SEDA: scalable embedded device attestation. In: Proceedings of the 22nd ACM SIGSAC Conference on Computer and Communications Security, ser. CCS'15. Association for Computing Machinery, New York, pp 964–975. [Online]. Available: https://doi.org/10.1145/2810103.2813670
15. Valente J, Barreto C, Cardenas AA (2014) Cyber-physical systems attestation. In: 2014 IEEE International Conference on Distributed Computing in Sensor Systems, Marina Del Rey, May 2014, pp 354–357. [Online]. Available: http://ieeexplore.ieee.org/document/6846189/
16. Fongen A, Mancini F (2015) Integrity attestation in military IoT. In: 2015 IEEE 2nd World Forum on Internet of Things (WF-IoT), Milan, Dec 2015, pp 484–489. [Online]. Available: http://ieeexplore.ieee.org/document/7389102/
17. Ibrahim A, Sadeghi A-R, Tsudik G, Zeitouni S (2016) DARPA: device attestation resilient to physical attacks. In: Proceedings of the 9th ACM Conference on Security & Privacy in Wireless and Mobile Networks. ACM, Darmstadt, July 2016, pp 171–182. [Online]. Available: http://doi.acm.org/10.1145/2939918.2939938
18. Ledwaba LPI, Hancke GP, Venter HS, Isaac SJ (2018) Performance costs of software cryptography in securing new-generation internet of energy endpoint devices. IEEE Access 6:9303–9323
19. Antonopoulos CP, Petropoulos C, Antonopoulos K, Triantafyllou V, Voros NS (2012) The effect of symmetric block ciphers on WSN performance and behaviour. In: IEEE 8th International Conference on Wireless and Mobile Computing, Networking and Communications (WiMob), Barcelona, Oct 2012, pp 799–806. [Online]. Available: http://ieeexplore.ieee.org/document/6379167/

20. Chang CC, Muftic S, Nagel DJ (2007) Measurement of energy costs of security in wireless sensor nodes. In: 16th International Conference on Computer Communications and Networks, Honolulu, Aug 2007, pp 95–102. [Online]. Available: http://ieeexplore.ieee.org/document/4317803/
21. Guimaraes G, Souto E, Sadok D, Kelner J (2005) Evaluation of security mechanisms in wireless sensor networks. In: 2005 Systems Communications (ICW'05, ICHSN'05, ICMCS'05, SENET'05), Montreal, Aug 2005, pp 428–433. [Online]. Available: http://ieeexplore.ieee.org/document/1515560/
22. Trad A, Bahattab AA, Othman SB (2014) Performance trade-offs of encryption algorithms for wireless sensor networks. In: 2014 World Congress on Computer Applications and Information Systems (WCCAIS), Hammamet, Jan 2014, pp 1–6. [Online]. Available: http://ieeexplore.ieee.org/document/6916625/

Machine Learning/AI as IoT Enablers

Yue Wang, Maziar Nekovee, Emil J. Khatib, and Raquel Barco

1 Introduction

Recent years have evidenced a rapid growth in the application of advanced Artificial Intelligence (AI) technologies in numerous fields, such as industry, healthcare, transportation, and domestic appliances. AI is a form of computing that allows a machine to perform cognitive functions, such as adapting their behaviour and modifying their decisions according to changing environment and conditions. Machine learning (ML) is an application of AI that provides a system the ability to automatically learn and adapt to the environment through experience. In particular, machines and tools that support AI are designed to react and learn from data collected from the environment, and the knowledge and insights created from them, through data analytics. Data analytics discovers new knowledge and creates new value through the exchange, selection, integration, and analysis of massive data. It provides a technology that reveals the knowledge and correlation in systems that may not be discovered or fully described with conventional mathematical models. The properties and problems of data analytics vary when the volume, generation velocity, and variability of the collected data grow above a certain threshold, entering into the Big Data analytics realm. To support these conditions, novel technologies have entered the market, such as cloud computing and NoSQL databases. Big Data analytics, combined with the underlying AI technologies, have

Y. Wang (✉)
Samsung Research UK, Staines, UK
e-mail: yue2.wang@samsung.com

M. Nekovee
University of Sussex and Quantrom Technologies Ltd, Brighton, UK

E. J. Khatib · R. Barco
University of Málaga, Málaga, Spain

N. H. Mahmood et al. (eds.), *Wireless Networks and Industrial IoT*,
https://doi.org/10.1007/978-3-030-51473-0_11

found their applications in all aspects of business, society, and life, which are reshaping our future technology landscape.

Industry 4.0 is one of the main consequences of this data-centric revolution. Information on the processes, consumer demands, supply chain, etc. become a necessity to achieve the flexibility and agility required in Industry 4.0. To obtain this information, data must be intensively collected by different kinds of IoT sensors (product tracking, environmental monitoring, etc.) and processes (online shopping trends, machine status information, network traffic information, etc.). The collected information is then stored and processed using the aforementioned Big Data storage and analytics technologies, etc. Wireless connectivity plays an even more important role in industrial environments, due to the ease of deployment, low maintenance costs, and the high flexibility they offer.

Future wireless networks are data-intensive and service-driven. The adoption of wireless technologies has enabled a new paradigm in connectivity and computation, where machines have access to the Internet to autonomously send data and receive instructions. These machine-to-machine (M2M) communications, which have varying characteristics and requirements, have enabled a rich set of novel applications, and, combined with mobile computing devices, shaped the Internet-of-Things (IoT). In IoT, novel applications have appeared, such as smart wearables, smart mobility, smart utility management, eHealth, virtual/augmented reality, ultra-high definition (UHD) video, driverless cars, etc. It has been predicted that around 25 billion IoT devices will be connected by 2025 [1]. Specifically, cellular technologies are seeing a great adoption by the IoT market, thanks to the ubiquitous connectivity they offer, plus their ease of use and maintenance from the point of view of the clients. 5G technologies, with their capability of providing high data rate, low latency, and guaranteed services through network slicing, are designed to cater for the needs of different IoT applications. The connection of the massive numbers of devices will generate a huge amount of data, gathered by individual devices, and shared over the IoT network in near real time.

An important point to take into account in industrial IoT networks are the particularities of the scenarios where connectivity occurs. Industrial environments such as factories or distribution centres are especially harsh for radio propagation, due to the presence of large metallic structures that cause shadowing and a large number of transmitters that produce interference. Therefore, a key point in deploying intelligent connectivity in industry, and a major differential factor with respect to the general use cases, is to use the appropriate Radio Access Network technologies and be especially careful with their dimensioning.

The accumulation and sharing of massive amounts of data and knowledge will in turn facilitate AI and ML, enabling the so-called intelligent connectivity – a vision of future network empowered by the combination of emerging technologies, including 5G, AI, ML, Big Data, and IoT [2]. Underpinned by ubiquitous hyperconnectivity, as well as real-time decision making with collective intelligence, intelligent connectivity is foreseen to transform industries such as energy, transportation, and manufacturing, as well as every aspect of our daily lives.

To fully unleash the potential of intelligent connectivity, there are some challenging topics that must be addressed, for instance, data and intelligence sharing, scalability of the existing solutions, security, and the underlying transformation in infrastructure. In particular, data analytics and AI face some unique challenges when applied to IoT networks. Firstly, a great variety of device types exist, and data collected from different devices may follow different format with different data types. It is a huge task to harmonize the collected raw data into a universal language where insights and knowledge can be shared. Secondly, the constantly changing network conditions and surrounding environment needs to be detected and the connectivity methods adapted by the AI algorithms, ideally in real time. This implies a fast exchange of information and knowledge, as well as a need for selecting what data to pass on and what data to retain locally in the device. In addition, there is a problem of model applicability. The majority of the current AI models deployed in IoT networks are based on exhaustive experimenting over available data, so these models are highly adapted to the existing datasets. There is a significant problem of reusing and scaling the existing AI models extracted in one scenario to a different scenario, or a different part of the network for the same application. For example, an AI model extracted from a specific industrial process in a small factory cannot be easily scaled to larger factories.

This chapter provides an overview of the current and future applications enabled by the merging of AI, 5G, and IoT, and their future looking technologies.

2 The Role of AI and Big Data

AI and Big Data, as the key enablers of intelligent connectivity, have been evolving hand in hand with emerging IoT technologies, where the most significant sources of data are generated. Considerable interest from the industry and research efforts have been attracted to this field. In the developments both from academia and industry, data mining and ML are used to extract the insights and knowledge from the data collected by IoT networks.

AI is a set of techniques and algorithms that are meant to perform actions that usually would require human intervention. AI algorithms are ultimately functions that, given a certain input, return a corresponding output through a non-linear relation. The inputs are usually a set of complex observations, which may require a pre-processing with operations such as quantification or normalization. The outputs are dependent on the application where the algorithm is used and the nature of the algorithm. The type of output defines a taxonomy where AI algorithms can be grouped as classifiers, regressors, etc. The non-linear relation between the input and the output is shaped by a set of parameters that are commonly complex and hard to adjust. Some examples of AI algorithms are Recurrent Neural Networks, Fuzzy Logic Controllers, and Bayesian Networks. Figure 1 summarizes how AI methods are used in intelligent connectivity, and their relation with ML and Big Data Analytics. In intelligent connectivity, AI algorithms take as inputs the data

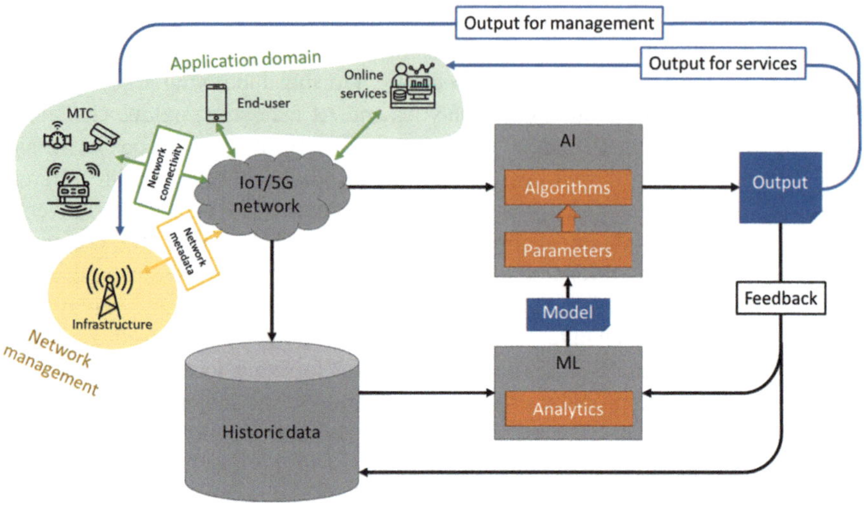

Fig. 1 General scheme for intelligent connectivity

collected by the IoT devices in the network, along with data from other sources such as the network infrastructure or external online services; and produce outputs that can be used to interact with the IoT applications and services, or to improve the connectivity by modifying network configuration.

AI algorithms have a large number of configuration parameters that must be fine-tuned to a specific scenario to work correctly. Although there are AI systems where these parameters are adjusted manually, ML is more often used to do this. In these kinds of setup, ML algorithms take as input large historic datasets similar to those that the AI algorithm will process once deployed and return as output optimal sets of configuration parameters.

- *Supervised learning:* The ML algorithm has access to sets of input variables of the AI and the expected output (labeled data). In this case, the ML needs to configure the AI so that it imitates the process that generated the training samples in the first place. Although supervised learning usually produces AI systems that need less post-processing and that have higher accuracy, one major issue is the availability of training data. Some common examples of supervised learning algorithms are Deep Learning and Support Vector Machines.
- *Unsupervised learning:* The ML algorithm only has access to sets of input data (unlabeled data). In this case, the ML will search for patterns and train the AI to find them. These systems will usually produce less accurate AIs, but accessibility of unlabeled data is much easier. Some examples of unsupervised learning are clustering and anomaly detection.
- *Reinforcement learning:* In the third kind, which is sometimes classified as a kind of supervised learning, the ML algorithm has access to the input data of the AI and, although it cannot access the expected output, there is a certain

feedback that indicates whether the output produced by a trained AI is correct or not. In this case, the ML will train the AI algorithm in a case-by-case fashion. Q-learning is an example of reinforcement learning.

AI and ML algorithms are based on the processing of large amounts of data. Both the storage and the processing of these data consumes a lot of resources. In fact, above a certain threshold, traditional computing techniques are insufficient for the successful execution of some AI/ML systems. This is where Big Data technologies come into play. Three features determine whether a problem can be considered part of this Big Data domain [3]:

- High volume: A very large amount of sources produce data. This is true for IoT networks, where a very high number of devices produces large amounts of data.
- High variability: The data from different data sources come in different formats that requires harmonization. In an IoT network, devices of different models, manufacturers, and purposes operate, producing data in many different formats (numerical records, audio/video files, etc.).
- High velocity: Data is generated quickly, that is, faster than it can be processed by traditional methods. In applications where speed is important (such as self-driving cars), processing the data fast is critical; and this is difficult when the maximum allowed delay is close to the minimal processing time.

In the case of intelligent connectivity, the collected and processed data has all the three features. Big Data techniques help to overcome these challenges by offering special storage and processing methods. NoSQL databases improve the storage and retrieval of data with high variability (i.e. data that may have different formats at different moments). Cloud computing is a set of Big Data technologies for improving the speed of processing. In cloud computing, tasks are divided into many parallel processes, reducing the overall computing time. Schemes such as Map-Reduce [4] and the Lambda architecture [5] are examples of Big Data processing techniques.

3 Use Cases of AI-Enabled Intelligent Connectivity

So far, the mainstream applications of AI in technologies include computer vision, natural language processing, voice recognition, and prediction. These technologies can be widely used by end consumers and businesses. Table 1 below gives an overview of the applications of AI in different technologies and their application scenarios for consumers and enterprise customers. In Fig. 2, we provide an overview of some of the use cases on AI-enabled intelligent connectivity. Next, we describe a few use cases of AI-enabled intelligent connectivity.

Table 1 AI algorithms and their applications

Technology	Overview	Application Scenarios
Computer vision	Computer replaces human vision to recognize, follow and measure the objects	Smart home AR, VR Shopping via image searching Intelligent home security 3D analytics
Natural language processing	Interpret meanings of texts and extract abstracts from articles	Search engine Recommendations and advertisement Machine translate
Voice recognition	Translate human instructions to texts and commands to machines	Smart TV Call centre Voice assistant Smart home assistant
Enterprise applications	AI applications for third-party, business customers	Network management Stock exchange Production planning

Fig. 2 Overview of intelligent connectivity use cases

3.1 Smart Manufacturing

In the last years, market trends have driven to a demand of highly customized manufacturing goods. To efficiently serve this new customized market, where production volumes of a single product are low, but total sales keep increasing, factories need to adopt agility as a basis for their operation. Agility is achieved with a vast set of novel technologies collected under the umbrella of Industry 4.0 [6]. Wireless connectivity, Big Data, robotics, and sensors are the four pillars of Industry 4.0. intelligent connectivity, as a combination of Wireless connectivity and Big Data, plays a major role in many Industry 4.0 applications. In this section, two of these applications will be described: predictive maintenance and hazard detection.

Predictive Maintenance As costs of production grow, the need for cutting expenses is an ever-increasing need in industry. In industrial machinery, there are two sources of expense: waste of unspoiled elements and machine breakdowns. To be more specific, some industrial machinery require a periodic maintenance, which can be done proactively or reactively. The first approach implies that some wear items (such as metallic pieces that are subject to stress, or parts that perform abrasive processes) may be changed before their lifespan is consumed, increasing the expense in replacements. On the other hand, the reactive approach consists of only replacing parts once they wear off and cause a malfunction. Although this means that the expendable elements are fully used, they may cause machine breakdowns that increase the cost with the need of repairs. Therefore, there is a need to optimize the scheduling of predictive maintenance so that the wear elements are fully used without causing breakdowns.

Sensors are one of the key technologies in Industry 4.0. In recent years their cost has dropped, making vast deployments affordable from an economic standpoint. Novel IoT technologies, such as cloud-based platforms for collecting and processing sensor data, greatly simplify the process of sensorization at large scale. Information from many kinds of sensors can be collected, and models of the monitored processes can be extracted with ML processes. These models can then be used to perform predictions.

This scheme of heavy monitoring, modelling, and prediction can be used for predictive maintenance. Depending on the specifications of the machine, magnitudes such as vibrations, flow of fluids, electric current, conductivity, and thickness of certain pieces can be measured. Supervised ML using the collected data from reactive maintenance, can determine which variables contain information on an immediate breakdown, leading to an interruption in the operation of the machine next time maintenance is required. This achieves both objectives of fully utilizing wear elements and preventing breakdowns.

Hazard Detection Factories, having large and powerful machines, are dangerous places. Some dangers to personnel are fires, accidents with machinery, accidents with vehicles, toxic fumes, falling objects, and incorrectly isolated electric lines. According to Eurostat, over 3000 deaths per year are registered in the EU in the sector of agriculture, construction, and manufacturing. Strict regulations are in place to minimize these fatalities, determining how machines and buildings must be built in order to prevent hazards and facilitate danger mitigation measures. Active safety systems, such as automatic fire extinction systems, also help reduce the risk and incidence of accidents. A critical aspect of safety in factories is to detect hazards early, before they cause accidents or irreversible situations. These hazards may be varied in nature, determining the mechanisms that can be used for their detection. Smoke detectors, radiation detectors, or temperature sensors are some common examples.

Video analysis is a function that can be performed with intelligent connectivity. Processes such as object recognition or movement detection can be performed in the cloud with AI using video feeds from connected cameras. To train the analysis AI algorithms, videos of known activities can be used to feed supervised learning processes to train a model. Another option is to model normal behaviour and train the AI to recognize when abnormal activity occurs. The output of video analysis can then be fed to other systems to perform certain activities, such as raising alarms or activating actuators through integrated IoT platforms.

Video analysis can be used to detect hazards such as smoke, fire, or even sabotage in factories. Since surveillance cameras are usually deployed in factories, video analysis can be deployed over their feeds to increase the coverage of hazard detection, reducing the reaction time for active safety systems.

3.2 Connected Cars

As novel technologies arrive to the mobile communications market, an increasing demand to integrate them into vehicles is growing [8]. Services such as video streaming are starting to be part of on-board entertainment systems. But, beyond entertainment for passengers, mobile communications can offer some very interesting services to assist drivers: collision avoidance systems, navigation, predictive maintenance, etc. Mobile communications will also play a major role in the future of self-driving cars, where it is expected that autonomous cars will communicate among each other to create self-organizing traffic patterns. Two examples where intelligent connectivity has a central role are the transmission of traffic-related warnings to drivers and remote driving.

Connected cars also has the potential to transform logistics, which is a key aspect of supply chain management in industries.

Traffic-Related Warning Transmission Although fully autonomous self-driving cars are quickly becoming a realistic possibility, there is yet a long way to go in its social and legal aspects. In the first place, the psychological implications of not having any control over a self-driving vehicle cause a general rejection over the wider public. Also, issues like the coexistence between autonomous and non-autonomous traffic, the ethical dilemmas on what decisions should AI systems take in case of emergency, the liabilities in case of accident, etc. are debates that are still open to resolution. Therefore, the implantation of these technologies is taking place in a gradual manner; starting with technologies that assist drivers and provide information for better decision making. Information on traffic, road conditions, tolls, accidents, or weather all help drivers to plan their routes or be especially cautious at certain moments.

One of the main use cases of intelligent connectivity is the obtention of information from sensors deployed over large geographic areas. The two basic building blocks of this use case are the sensors themselves, and the wireless technologies that provide connectivity. The reduction of the price in sensors and the availability of low power systems in the last decades has made the deployment of massive amounts of sensors economically feasible. But thanks to AI and ML, other devices, such as cameras, smartphones or network access points, can be used to extract additional information after some processing. Image recognition, location analysis or network traffic modelling are just some of the processes that can be used to obtain rich information from these devices. Regarding wireless networks, wireless access network (WAN) technologies, such as 5G or 6LoWPAN, enable low cost and low power connectivity both for sensors and for users of the information.

For traffic information, intelligent connectivity can be used to gather and centralize all the useful information. This information may come from sources such as road cameras, which can be used to measure the traffic and detect jams through image recognition. Other incidents, such as oil stains over the asphalt, can be recognized either by cameras or by sensors in the cars that upload this data to the cloud. Collision detection systems, which are currently being installed in modern vehicles and are being enforced by legislation, can report accidents. All this information can then be curated and customized for each driver based on the analysis of their trajectory, which can be obtained inspecting the geolocation information of smartphones. As a result, drivers will have an updated and simple newsfeed on their dashboards, that warns them of any important event they might encounter in the near future.

Remote Driving Another intermediate steps towards full automation is remote driving. In this stage, although there is still a human making the decisions, the driver is in a remote location, so there is a delay in the feedback that, if not appropriately dealt with, may cause accidents due to late decisions.

Mixed-Criticality Systems model devices where several different information paths coexist. Some are more proprietary, therefore they are processed earlier, having to wait less in queues, and receiving more resources (such as increased CPU frequency) when needed. The wireless network can also establish different policies for different kinds of traffic. Currently, 5G networks consider three main traffic types: Enhanced Mobile Broadband (eMBB), Massive Machine Type Communications (mMTC), and Ultra-Reliable Low Latency Communications (URLLC). This differentiation allows to adapt the resources available in the Radio Access Network carrier and the Core Network connections to better serve the needs of each type of message.

In remote driving, video feeds are a very important data source, allowing the driver to visualize the environment. To transmit a high-resolution view, eMBB connections are required in order to provide the required bandwidth. On the other hand, when a sudden obstacle, such as an animal, appears in view, eMBB may introduce a high latency, so a URLLC message showing the danger would be required. To differentiate when this warning must be sent, image recognition must be running at all times in the car's CPU using a high-priority process. Collision avoidance between cars must also be dealt with high priority. In this case, geolocation information must be sent regularly by vehicles to the network, and a collision prediction must be run in the network edge, where the information of neighbouring vehicles can be aggregated. Once a potential collision is detected, a warning can be sent to both drivers using URLLC.

3.3 *Next-Generation Healthcare*

Applications of IoT in healthcare seem to be endless: from remote monitoring and personal healthcare to smart sensors and medical device integration, as well as the pharmaceutical industry, healthcare insurance, healthcare building facilities, robotics, smart pills, and even treatments of diseases [7]. It has the potential to not only keep patients safe and healthy, but also to improve how physicians deliver care. In the following we will focus on a few prominent IoT use cases in health with the greatest potential from AI.

Remote Patient Monitoring Personal health and medical data are collected from an individual and transmitted to a provider for use in care and related support. In this way the provider can track healthcare data for a patient once released to home or a care facility, reducing readmission rates. Healthcare devices as insulin pumps, defibrillators, scales, continuous positive airway pressure machines, cardiac monitoring devices, and oxygen tanks are now connected in the IoT to ensure remote monitoring, providing patients and their caregivers valuable real-time information.

IoT-supported healthcare services can provide better and more efficient treatment to patients while also inducing cost saving for the providers. On the other hand, interconnectivity can provide for easy data collection, asset management, Over-the-Air updates, and device remote control and monitoring.

Assisted Living Demographics, public policy, and the labour market are driving an emerging market for IoT to deliver elder care services. By 2029, 20 percent of the U.S. population will be over the age of 65 and 70 percent of those individuals will need some form of assisted care, according to recent research [16].

AI and the IoT have the potential to shape a new collection of technologies to improve the quality and availability of elder care while helping to control its costs. Ambient intelligence, which combines AI and IoT, will provide real-time monitoring of an environment and event-driven response to changes in that environment. Sensors designed to detect changes in sound, motion, physiological signals, as well as more generalized image processing are core components of an ambient intelligent environment.

Ambient intelligence thus is poised to serve a range of functions with regards to elder care, but most applications will address three broad functions: maintaining routine activities and social connectedness, enhancing safety, and monitoring health. Routine activities and social support are especially suited for elders suffering cognitive decline. These systems detect changes in patients' location or environment and provide verbal assistance as needed, or if needed, notify caregivers. Safety-enhancing sensors are often wearable and provide early warnings of potentially threatening situations, such as falls. Health monitoring systems may combine wearable and stationary sensors to monitor blood pressure, pulse, and movement of the patient as well as environmental data, such as ambient temperature.

Unlike IoT applications that function primarily to monitor and control devices or environmental conditions, ambient intelligence systems are designed to monitor and support humans, creating an additional dimension of complexity. Developers of ambient intelligence systems face challenges common to IoT as well as some specific to this domain. Real-time processing, quality control, and data integration are especially important when making decisions about the physical well-being of a patient.

4 Architecture for AI-Enabled IoT

Intelligent connectivity encompasses a wide set of ML and AI algorithms for a very wide array of solutions applied over a great variety of use cases. To implement these solutions in practice, the first question to resolve is its architecture, that is, what elements will be used, and at which location in the IoT system they will be set up. Figure 1 provides an overview of such an architecture. In this section, we will delve into the details of the elements of this architecture.

4.1 IoT Network

The IoT network has a main role as a gatherer of information. By providing connectivity to IoT devices, it collects all the data and redirects them to the services they are connected to. It also plays the reverse role, that is, to send commands and responses from the services to the devices.

In intelligent connectivity, the IoT network adapts itself to the connectivity needs of the devices, ensuring that they have the resources they need for their operation. This means that the network is reactive to external changes, and in some cases even proactive, in the sense that it uses predictions to adapt to these changes beforehand. Therefore, a secondary role of the network is as a client of the AI services. For this, the IoT network must share its configuration parameters and performance indicators with the AI/ML blocks, and use their outputs for self-configuration.

This adaptability functionality is especially important in harsh environments such as those found in Industry 4.0. Shadowing and interference are major problems, as earlier stated. Intelligent connectivity solutions can help in tasks such as the detection of coverage holes (i.e. zones in an industrial premise where no wireless connectivity is present), interference mitigation and load balancing. AI techniques can perform these functions, and even do it in a predictive manner.

Although the network is a central component to intelligent connectivity, it is also the component over which the least control is usually feasible by the industrial installation owners. Large deployments are usually undertaken and operated by network operators, while often the applications are demanded and developed by external entities with very specific needs. There is a need, therefore, for coordination among the different entities.

4.2 Databases

In intelligent connectivity, the IoT network "knows" where all the information from all the devices is located. This knowledge can be modeled as a single, huge database, where the AI/ML blocks can query specific data. Since the central database is actually a set of disperse network services, common formats [9] such as XML or JSON, and normalized interfaces such as REST [10] or Graphql [11], are key technologies to retrieve the information when required. Technologies such as NoSQL databases are used in online services to store very large amounts of schema-less data, which are common when the data sources (IoT devices, smartphones, etc.) are from different vendors. This technology can also be used by the IoT network to centralize the data from different services once they are queried by the AI/ML block.

Some important features of databases in industrial applications are their reliability, their performance and their security. Databases play a central role in data-centric applications, therefore, it is important that they are accessible at all times, information is not corrupted easily, since otherwise the outcomes of ML/AI processes would be affected and the cost due to errors would escalate. Performant databases are the basis of performant ML/AI algorithms that can cater for large data-centric applications and processes with very high throughput. Security is key to avoid industrial information theft and sabotage; and it is also a major selling point for owners, which ultimately helps in the expansion of the intelligent connectivity market.

4.3 AI and ML Components

The elements that will be used for the AI block depend highly on the application, which imposes a certain output, for example, a classification label, or the prediction of a time series. Two boundary conditions must be set based on the requirements of the application:

- Selection of input data: to decide the datasets that will first train the AI algorithm with ML and afterwards be used as input of the trained AI algorithms, the main criteria is the availability of the information within the data. In other words, the first step to take is to assess what is the base dataset that contains the target information. The base dataset determines also the physical data sources that must be used (e.g. databases, file systems, devices, etc.), and the flow of data throughout the network. These are aspects that must be taken into account to ensure that requirements such as latency and reliability are fulfilled.
- Selection of the type of output: this decision depends on what the objective of the intelligent connectivity solution is. It defines what information will be extracted from the input data. There are many kinds of output; for instance, class labels, that classify a certain input dataset into one of a finite number of classes; or predicted values, that provide a value for a variable in the near future based on past values of that same variable or others. The output can also consist in model parameters, such as statistic indicators (averages, quantiles, etc.).

Once these boundary conditions are adjusted, the set of AI algorithms that can be used is narrowed down to those that can provide the expected output with the selected inputs. In some cases, some algorithms (such as Artificial Neural Networks) can be used for different kinds of applications (prediction and classification), but the mode of working with them, the set of selected inputs and their roles varies widely for each case.

Once the AI algorithm is fixed, the ML method that will train it must be selected. If the data available for training includes examples of the output, supervised learning can be used; otherwise, unsupervised learning must be chosen. In a system where ML is done online, that is, when the output of the AI is validated by an external factor and fed back to the ML algorithm, reinforcement learning can also be used.

Selecting the datasets and the ML/AI algorithm are the base of the intelligent connectivity solution design; but to actually implement the system other decisions must also be taken. Specifically these decisions affect where each of the functions composing the solution are implemented:

- Physical computing element: aspects such as the dimension of data or the complexity of the operations determine the required computing power. Also, the cost of such computing power must be taken into account.
- Centralized/decentralized architecture: this aspect of the architecture determines whether the algorithm is implemented in a single server (centralized) or aggregating the results of instances running in a distributed set of devices (decentralized).

These two architectural decisions comprehend a very large set of technologies that exist in the market nowadays. Also, the very broad range of ML/AI algorithms that exists in the market gives place to a very heterogeneous set of requirements. There are some algorithms that are very lightweight while others are demanding; some that are parallelizable while others are not; some that need a global view of the use case (for instance, they need to access data coming from many sensors as well as data saved in the cloud) while others only need a local view (for instance, only on the sensing device). All these considerations define the boundary conditions of the selected location for the implementation of the algorithm.

Considering the physical computing element and the centralized/decentralized decisions as a single issue, there are three main options for the implementation location:

- Local device: For algorithms that only need local visibility and are simple enough so that they can run within the computational resources offered by the device, Implementation in the same device is a possibility. The main advantage of this location is that the latency is very low. On the other hand for energy constrained devices this implies a higher consumption. This is an example of a decentralized implementation option. Different instances running in different devices can communicate among themselves using peer to peer communications.
- Remote device: The traditional client server scheme also has its place in intelligent connectivity. In this case the algorithms will not be running the devices but in a remote location. The devices only act as information collectors and actuators. Also data from other data sources (such as databases or the Internet) can be used in this scheme. Applications that require a global view must be implemented using this scheme. This is a traditionally centralized architecture, where a network connection is established between the devices (clients) and one remote computer (server). Nowadays this scheme is a little bit more complex but also more flexible, thanks to technologies such as virtualization (that allows to run virtualized servers and share physical resources between them) and cloud computing (that allows running a flexible number of parallel instances of a specific algorithm). These technologies combined allow the access to a high computing power with a low cost.
- Edge computing: In the last years, the separation between the communications infrastructure and the computing platform has started to vanish. In edge computing the implementation of the algorithms is done over computing elements located in the access network nodes (e.g. gNBs in 5G). This combines the advantages of a local implementation (low latency) with the advantages of a centralized implementation (global visibility, energy saving, computing power, etc.)

In IoT, one important development of the last years is the emergence of integral platforms, such as OneM2M [12], Fiware [13] or OpenStack [14] that offer premade solutions including data storage, processing, computation resource management, edge computing, security, etc. These platforms offer a scalable starting point for any new intelligent connectivity solution.

5 Future Outlook

The next few years are going to see the merging of the emerging technologies, including the convergence of big data, AI, and IoT. In particular, industrial IoT will harness the power of AI for optimized manufacturing process, including predictive maintenance and root cause analysis.

5.1 Digital Twins

As "things" becoming connected and with increased capability of producing data through sensing, virtual replicas of physical entities and processes can be produced to run simulation, before actual entities are built and deployed. Such virtual replicas are referred to as 'digital twins'. In essence, a digital twin is a computer program that takes real-world data and contexts about a physical system and process and reproduces how the system or the process will react to these inputs. Digital twins have been applied to manufacturing industry to facilitate production and proactive maintenance, and can include large items such as buildings, factories, and even cities.

Digital twin is a perfect example as the merging of emerging technologies including big data, AI, and IoT. The technology has been made possible due to the massive number of IoT sensors. In particular, construction of digital twins requires inputs from massive sensors gathering all relevant features—in the form of big data—of its physical counterpart, such that its digital twin can represent the physical entity, and reactions of these data can be simulated in real time. Representing a complicated physical entity (e.g., a factory, a bridge) may rely on the underlying features of the material and structure of the physical entity, and conventional method of modelling such an entity may not be sufficient. AI can serve as an effective tool in this case to reflect the underlying features of the physical entity, offering recommendations and insights to performance validation, with or without a specific modelling. It can also effectively react to the dynamic contexts of the twin, and provide enhancement in real time, according to the contexts. In many cases, a digital twin could serve as a prototype of the physical entity, before it is physically deployed in practice.

5.2 *Next-Generation IoT*

The next-generation IoT (NG-IoT) technologies and applications [8] will be human-centric. A human-centric IoT environment requires tackling new technological trends and challenges. The next-generation IoT development, including human-centred approaches, is interlinked with the evolution of enabling technologies (AI, connectivity, security, virtualization) that require strengthening trustworthiness with electronic identity services, services and data portability across applications and IoT platforms. This ensures evolution to platforms with better efficiency, scalability, end-to-end security, privacy, and resilience. The virtualization of functions and rule-base policies will allow for free, fair flow and sharing of data and knowledge, while protecting the integrity and privacy of data.

Intelligent/cognitive IoT networks provide multiple functionalities, including physical connectivity that supports transfer of information and adaptive features that adapt to user needs. These networks can efficiently exploit network-generated data and functionality in real time and can be dynamically instantiated close to where data are generated and needed. The dynamically instantiated functions are based on (artificial) intelligent algorithms that enable the network to adapt and evolve to meet changing requirements and scenarios and to provide context and content suitable services to users. The AI embedded in the network allows the functions of IoT platforms to be embedded within the network infrastructure.

Advanced technologies are required for the NG-IoT to provide energy-efficient, intelligent, scalable, and high-connectivity performance, with intelligent and dynamically adaptive infrastructure to provide high quality experience that can be developed by humans and things. In this context, the connectivity networks provide energy efficiency and high performance as well as the edge-network intelligence infrastructure using AI, ML, Deep Learning, Neural Networks, and other techniques of decentralized and automated network management, adaptive analytics, and shared context and knowledge.

The development of AI and IoT combined in NG-IoT enables new ways of interacting with connected objects through voice or gesture, while augmented reality (AR) and virtual reality (VR) are powered by the data generated by IoT. Furthermore, sensors and actuator technologies together with AI and connectivity will push the development of tactile IoT based on convergence of these technologies, where the boundaries between virtual and physical worlds blur.

Bibliography

1. Ericsson (2019) Ericsson mobility report
2. GSMA (2019) Intelligent Connectivity, how the combination of 5G, AI, Big Data and IoT is set to change everything
3. Russom P (2011) TDWI best practices report, fourth quarter: big data analytics. TDWI

4. Dean J, Ghemawat S (2004) MapReduce: simplified data processing on large clusters. Google Inc
5. Marz N, Warren J (2015) Big Data: principles and best practices of scalable realtime data systems. Manning Publications Co
6. Bundesministerium für Bildung und Forschung, «Industrie 4.0. Innovationen für die Produktion von morgen,» 2017
7. Zhu H et al (2019) Smart healthcare in the era of Internet-of-Things. IEEE Consumer Electronics Magazine 8(5):26
8. Everis (2014) everis Connected Car Report
9. Di Martino B, Esposito A, Nacchia S, Maisto SA (2018) Towards an integrated internet of things: current approaches and challenges. In: Internet of everything. Springer, Singapore, pp. 13–33
10. Fielding RT (2000) Architectural styles and the design of network-based software architectures. University of California Irvine, Irvine
11. GraphQL, «GraphQL specification,» June 2018. [En línea]. Available: https://spec.graphql.org/June2018/
12. «OneM2M,» [En línea]. Available: www.onem2m.org
13. «Fiware,» [En línea]. Available: www.fiware.org
14. «OpenStack,» [En línea]. Available: www.openstack.org
15. Vermesan O, Bacquet J (2018) The next generation Internet of Things – distributed intelligence at the edge and human machine-to-machine cooperation. River Publishing
16. Population Reference Bureu of the United States, fact sheet: aging in the United States, July 15 2019, https://www.prb.org/aging-unitedstates-fact-sheet/

Edge Computing for Industrial IoT: Challenges and Solutions

Erkki Harjula, Alexander Artemenko, and Stefan Forsström

1 Introduction

Today, we can observe large global trends in the digitalization of many aspects of our everyday life. In particular, we see applications that can utilize information from sensors attached to things that can also communicate among each other over the Internet. This concept is commonly referred to as the Internet of Things (IoT) and provides us with services that are more personalized, automated, and have more intelligent behavior. Related to this, we can also see trends in IoT Cloud Computing (CC) for large-scale data storage, big data analysis on a massive amount of gathered data from IoT sources, and incorporation of Cyber-Physical Systems (CPS) into machine to machine (M2M) systems. Concurrent to this development, much work is being done in the Industry 4.0 initiative, including smart cities, smart industry, factories of the future, and smart manufacturing, hence, forming the concept of Industrial IoT (IIoT) [1]. At the same time, the deployment of the 5G wireless communication technology is also increasing everyday around the world [2], enabling a new magnitude in speed and low latency wireless communication with ultra-high reliability and availability.

E. Harjula (✉)
Centre for Wireless Communication, University of Oulu, Oulu, Finland
e-mail: erkki.harjula@oulu.fi

A. Artemenko
Corporate Sector Research and Advance Engineering, Robert Bosch GmbH, Renningen, Germany
e-mail: alexander.artemenko@de.bosch.com

S. Forsström
Institution of Information Systems and Technology, Mid Sweden University, Sundsvall, Sweden
e-mail: stefan.forsstrom@miun.se

N. H. Mahmood et al. (eds.), *Wireless Networks and Industrial IoT*,
https://doi.org/10.1007/978-3-030-51473-0_12

Edge Computing (EC) enables services to exploit the proximity of devices by providing computational resources closer to end nodes, therefore enabling ultra-low latency and high data rate communication. At the same time, it provides a means for controlling and limiting the propagation of sensitive data. Multi-access Edge Computing (MEC) is a standard by the European Telecommunications Standards Institute (ETSI) for 5G networks, among others, to offload processing and data storage from mobile and IoT devices to the edge of mobile networks instead of passing all of the data and computation to data centers or handling them locally [3]. Fog Computing (FC) and Mist Computing (MC) are closely related to both EC and CC, as they can be interpreted as low flying cloud computing near the edge [4]. We make the distinction that EC mainly refers to the computational edge infrastructure, FC mainly refers to the logical architectures enabling distributed virtualized services on the edge architecture utilizing the hardware capacity of EC nodes, and MC to the extreme edge of the networks and local edge computations. FC typically covers caching, data processing, and analytics occurring near the source of the data that improve the performance at the edges of the network, reduces the burden on data centers and core networks, and improves the resilience against networking problems [3]. Figure 1 shows an overview of how cloud, fog, mist, and edge computing fit together in a layered structure, including the scale of each layer and typical operations.

IIoT is one of the most important application areas of IoT, and therefore, it is vital for defining the requirements for EC systems. In recent years, the CC paradigm has found its way into the manufacturing industry addressing the need to process vast data originating from a massive number of sensor devices. It offers centralized resources to perform computationally intensive operations. Here, a

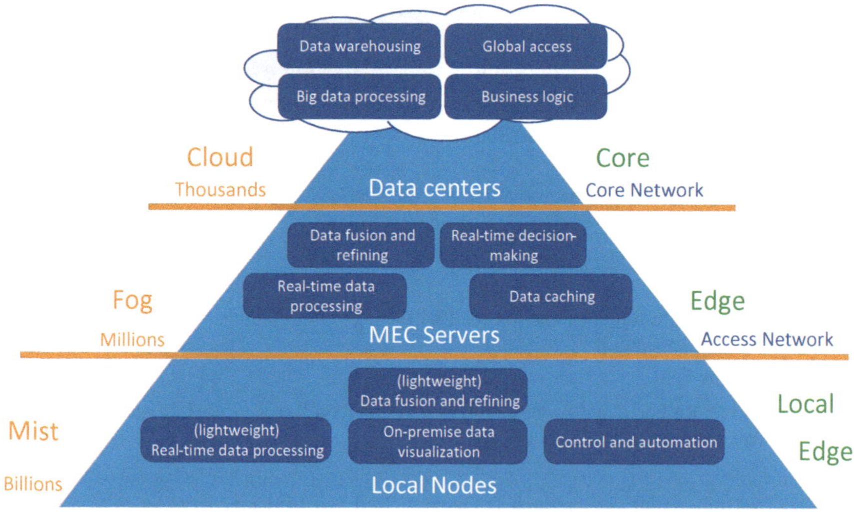

Fig. 1 A simplified view of Cloud, Fog, Mist, and Edge computing layers

representative example is predictive maintenance, which detects conditions leading to malfunctions, and therefore enables flexible manufacturing that increases the reconfigurability level of production systems enabling batch-size-one products. Due to improved connectivity with guaranteed Quality of Service (QoS), machines do not need to rely on their own dedicated computing hardware anymore but can rather use connectivity to access the cloud resources [5]. While the use of a centralized CC entity presents several vulnerabilities like single point of failure, backbone congestion, security, and data privacy, EC and MEC introduce computational resources, storage, and services at the edge of a network [6]. Applying EC, FC, and MEC for industrial manufacturing can solve most of the weaknesses of traditional cloud computing. However, several challenges still remain that will be addressed in this chapter.

In order to highlight the advantages, disadvantages, and future research in the intersection of the IIoT and EC, this chapter will give an introduction to the applications, challenges, and solutions of MEC. The remainder of this chapter is organized as follows: Sect. 2 goes further into the state of the art of EC and MEC for industrial use, the standardization, applications, and challenges. Section 3 focuses on solutions and future development potential, the next steps, and research directions. It includes three-tier edge cloud architectures, microservice architectures, SDN and NFV integration, security/privacy management, and the use of artificial intelligence (AI) for MEC. Finally, Sect. 4 summarizes and concludes the chapter.

2 State of the Art in Edge Computing for IIoT

This section describes the current state of the art in EC with focus on industrial applications. After the introduction of the EC potential for existing and appealing industrial use cases and standardization activities, we focus on the most crucial EC aspects and highlight their relevant open points and challenges.

2.1 Edge Computing Technology for Industrial Use

CC has, already for some time, been a standard in industry, bringing a vast amount of processing capabilities to analyze data generated by a huge number of already operational IoT devices. Many large industrial companies have taken into use their own in-house clouds (aka private clouds), as well as public clouds to satisfy their production needs [7]. Along with this, the vision of a fully automated and flexible factory of the future gets closer to reality. However, novel concepts, such as factory with zero-downtime, digital twins, flexible production planning, pro-active system surveillance, intelligent technical assistance, and batch-size-one products, still require further improvement of the network performance and services on the factory floor [8]. Nevertheless, EC is considered as one of the enabling technologies to unveil the full potential of the proposed smart factory concepts.

The EC technique can support IIoT devices with limited capabilities (with regard to, e.g., battery, CPU, and GPU) in their ever-increasing computation demands created by various kinds of novel use cases. High-complexity robotic applications, Automated Guided Vehicles (AGV), real-time interactive multi-user co-working, mobile production cells, sensors and actuators connected over wireless communication technologies, augmented reality (AR), and virtual reality (VR) are only a few examples of such industrial use cases [9]. Many of them introduce requirements that differ from those considered in conventional IT systems. Such requirements include ultra-high availability, reliability, predictability, very low latency, and strictly deterministic real-time behavior of all system components. MEC alone cannot satisfy all these requirements. Therefore, a combination of many new enabling technologies is required, e.g., time-sensitive networking, real-time virtualization, software-defined networking, 5G with enhanced mobile broadband, ultra-reliable low-latency and massive machine-type communication, etc. Many of these enabling technologies are covered in this book.

First EC products, known as edge gateways [10], are gaining popularity in the industrial context, connecting thousands of IIoT devices to data processing units at the edge of a network, close to sensors, and hence, avoiding the issues of sending all the data directly to the cloud, which is often not feasible due to cost, privacy, and network issues. Software giants offer first software platforms for deployment and management of edge clouds (e.g., Azure IoT Edge from Microsoft, AWS IoT Greengrass from Amazon, Cloud IoT Core from Google, Bosch IoT Gateway, etc.). Many of these products still present proprietary solutions. To improve this situation, different standardization activities are working on specifications for different EC aspects.

2.2 *MEC Standardization*

Shortly after an introduction of small cloud data centers close to the data source called Cloudlets [6], the Open Edge Computing (OEC) and OpenFog Consortium (OFC) initiatives have been generated to accelerate the standardization and dissemination of the EC technology. Thereafter, multiple committees, working groups, and standardization bodies around the world have been created. According to [11], the most important ones are the following:

- **Multi-access Edge Computing initiative** as an Industry Specification Group within the European Telecommunications Standards Institute (ETSI)
- **MEC in 5G networks** within the 3rd Generation Partnership Project (3GPP)
- **MEC system as service-oriented RAN** within the China Communications Standards Association (CCSA)

The majority of the core partners in all standardization entities come from telecommunications industry. This is reflected in the core activities, as well as the goals of working groups. All bodies perform different conceptual, architectural, and functional work and intend to develop a standardized, open environment that will allow efficient and seamless integration of third-party applications across multi-vendor platforms [11]. From the perspective of the authors of this book, however, ETSI MEC initiative considers the broadest range of applications and architecture scenarios among all EC standardization entities. This is reflected in the aspect that the EC platform is not bound to any access technology, which is reflected in the title of MEC.

ETSI MEC introduces a reference architecture and technical requirements enabling efficient and seamless execution as well as interoperability and deployment of a wide range of EC scenarios that include IIoT. The multi-vendor proof-of-concept projects visualize key aspects of MEC technology and prove it is feasible and valuable. Important aspects like latency, energy efficiency, system resource utilization, network throughput, and quality of service are constantly highlighted.

2.3 MEC Applications: Industrial IoT

As mentioned above, EC excels in application scenarios where there is a need for low latency, high bandwidth, and high resilience computation and communication in order to enable its real-time, intelligent, and autonomous decision-making. This can be required, for example, in different smart appliances, such as smart vacuum cleaners using sensor information available inside the house. But also, edge device video analysis, mobile big data analysis, connected vehicles, smart building control, and safety monitoring present appealing use cases in the IIoT context. A new trend on the factory floor is represented by different kinds of mobile vehicles, e.g., Unmanned Aerial Vehicles (UAV) and Automated Guided Vehicles (AGV), which cooperatively solve certain tasks. Some typical industrial applications include edge services such as industrial production robots, where the low latency and resilience of EC is paramount. Cooperating robotic arms in a production line show a good and appealing example of the robotic cooperation on the factory floor. Here, EC supports production systems by offloading of data analytics and smart data processing in the close proximity to the data sources. Furthermore, smart industrial environment monitoring, grid system controls, and self-organized massive wireless sensor and actuator networks constantly continue to attract manufacturers' attention. Many of the applications, mentioned above, have already been implemented using the current technology base. Some of the use cases are shown in Fig. 2. Many further useful applications present a source for discussions due to the challenges they introduce to the infrastructure.

First products are available in the IIoT market in this context, proving the benefits of the EC paradigm. As an example, Bosch IoT Gateway presents an IIoT solution with support of open APIs, a variety of development tools for creation of edge

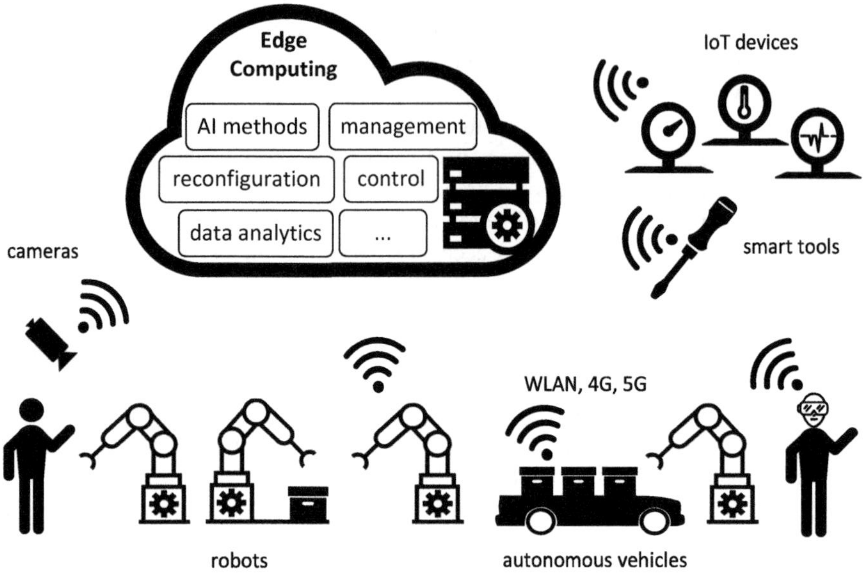

Fig. 2 EC technology in industrial applications

applications, providing autonomy and intelligence at the edge [12]. The product is in use in many scenarios including IoT platforms with EC support for intelligent data processing, optimization of electric vehicle charging, and smart field device connectivity at the edge [13].

2.4 Edge Computing Challenges

IoT systems can greatly benefit from the EC technology, but several challenges still remain, related to, e.g., performance, efficiency, reliability, availability, scalability, security, and privacy [14]. The following sections discuss these challenges in more detail.

Performance and Quality of Service

Novel 5G wireless technologies enable low-latency communication, which is crucial in various IIoT scenarios requiring real-time functionality [15]. As mentioned earlier, MEC (and EC in general) is another of the two main enablers of reliable low-latency wireless services since it minimizes the route length between the local nodes and computing resources [14, 16–18]. EC also helps improving other QoS factors since it is easier to remove and manage performance bottlenecks on short routes compared to longer ones. Most of the challenges related to maintaining high performance of EC/MEC systems concern special situations, such as fast mobility of end nodes and rapid changes in both service demand and density of end devices, which can be the case in many industrial applications or, e.g., during public mass events. Therefore, important research topics are related to, e.g., placement of edge resources and deciding where to deploy computation and data in different scenarios. The ability to rapidly migrate data and computations among MEC servers or between MEC and core network servers to ensure QoS in dynamic scenarios is among the most important research areas of EC.

Reliability and Availability

The resilience against computing and network infrastructure problems is one of the most important research areas of EC/MEC [19]. In many IIoT systems, the processing of sensor data, at least some degree of the decision-making logic and the control of actuators, is beneficial to manage locally on site, because the connection with the access network may occasionally become unreliable or low in performance [18, 20]. Therefore, it is beneficial to bring some EC capacity also within local IoT clusters. This is called "local edge computing." The challenges related to reliability of EC systems concern the ability to adapt to dynamically changing situations, related to, e.g., mobility, network failures, disturbances, and hardware failures. The EC system should automatically manage, analyze, and optimize its operation, including placement of data management and computational tasks, based on the current situation and foreseen changes. With regard to availability, the critical questions are where are the system components located, and who/where are the users? Availability becomes a particular problem in cases where the different stakeholders of the services are logically and geographically distributed.

Scalability and Deployability

In IIoT systems, sensor information is gathered from a high number of devices connected with short-range or novel long-range low-power wireless technologies [1, 21]. Furthermore, advanced sensors, control systems, surveillance video streaming, and still image capturing devices are already producing huge amounts of data to be processed [22]. In traditional systems, all of this data processing and related decision-making logic has been handled at data centers, which is becoming problematic from the viewpoint of scalability, performance, and reliability. In this context, EC helps by providing computational capacity near the source of the data, allowing various data pre-processing, refining, and analysis functions to reduce the amount of data to be sent to cloud servers and therefore reducing the load inflicted to core networks and data centers. The important research challenge in this area is to develop intelligent algorithms for deciding on which tier to manage different functions and prioritizing tasks when limited resources do not allow globally optimal solutions.

Security, Privacy, and Trust

Since the digital world penetrates deeper and deeper in the industry and business processes, as well as our everyday life, a particular concern is related to preserving the privacy and security of networked systems. We are living in a world where the data collected from the users is ruthlessly exploited by different organizations and authorities around the world. Centralized cloud systems are an inherently challenging environment from the viewpoint of security and privacy, since all data need to pass several links and devices, owned by a wide set of stakeholders between end devices and servers, not forgetting the chance for data leaks at public servers [14, 23, 24]. What is even worse, IoT, surrounding us almost everywhere, gives cybersecurity attackers further tools to even threaten our health or life by infiltrating to systems affecting our physical safety [25]. Therefore, the need for technologies allowing local data management and decision-making to limit data propagation toward public networks is obvious. In this context, EC is a centric building block for service providers to guarantee customer data preservation within set boundaries. Regarding security and trust challenges in the EC and IIoT domain, there are still many open and difficult challenges [26], including end device security, protocol and network security, cloud/fog security, end user application security, data protection, malicious attacks as well as identity and authentication management, access control, trust management, intrusion detection systems, privacy, virtualization, and forensics.

Resource Efficiency

Resource efficiency, including energy-efficiency, is a powerful measure for promoting sustainability in technological evolution. Internet and Communications Technology (ICT) is one of the main tools for improving the resource efficiency of the infrastructures around us [27], but its intrinsic resource demand is rising rapidly [28]. In this context, local data pre-processing, refining, and analysis functions enabled by EC help reducing the load inflicted to various components of the cloud systems and therefore promote sustainability through improved energy and resource efficiency. IoT systems include numerous low-power sensors, actuators, and other devices that are resource-constrained in their nature [14, 18]. In order to maintain both the system-level performance and resource efficiency of constrained-capacity nodes, IoT systems need to take into account the limited hardware and energy capacity of the end nodes. One of the main measures for achieving this is to offload computing and data management to higher layers on the IoT architecture. The traditional IoT systems do this by offloading computation to cloud servers. In data-intensive computing, such as video surveillance, this is not optimal from the viewpoint of network utilization, since all data need to pass several communication links along the way from the end node to the server. A more efficient approach would be to handle as much of the data-intensive computing near the source of data as possible. In this context, EC is in a key position to improve resource efficiency.

The challenges related to resource efficiency concern, e.g., how to reliably measure and communicate resource usage, how to minimize resource consumption while still maintaining the availability of nodes in highly interactive scenarios, and how to prioritize resource efficiency with several other constantly changing requirements in complex multi-tier IoT systems.

Being not complete, the list of challenges, presented in this section, gives an overview of open points, which show a potential for further improvements.

3 Solutions and Future Development Potential

In the previous section, we pointed out some important research challenges for EC in IIoT. To address those challenges, in this section we discuss on some of the most relevant research directions and potential solutions.

3.1 Three-Tier IoT Edge Architecture

To deal with the vast amount of data originating from a massive number of sensor devices, the risk of connectivity problems, and to limit the propagation of sensitive data, at least some degree of processing of the sensor data and decision-making/control logic is beneficial to be managed locally. Since it cannot be expected that local IoT clusters include devices with sufficient stability and hardware capacity to accommodate full-functional MEC servers, decentralized solutions become essential to accommodate the local processing, data management, and decision-making. To make this possible, a three-tier IoT Edge model has been proposed by the authors in [18]. In this model, the data and processing can be deployed on three alternative levels of operation: (1) public servers, e.g., in data centers, (2) MEC servers, and (3) local nodes as virtualized functions. Figure 3 illustrates the model and its benefits. The model enables dynamic optimization of service deployments, based on the service requirements, available computational and network capacity, and load. We see high potential in using microservice and serverless architectures, introduced in Sect. 3.2 and AI (Sect. 3.5), in defining the optimal deployments for different types of services in the three-tier architecture.

3.2 Microservices and Serverless Architectures

IoT services have traditionally been designed as monolithic cloud applications associating multiple software components into a single entity. Due to their ponderous maintenance and deployment, the current industry trend is toward microservice paradigm, an architectural style to build, manage, and evolve service architectures

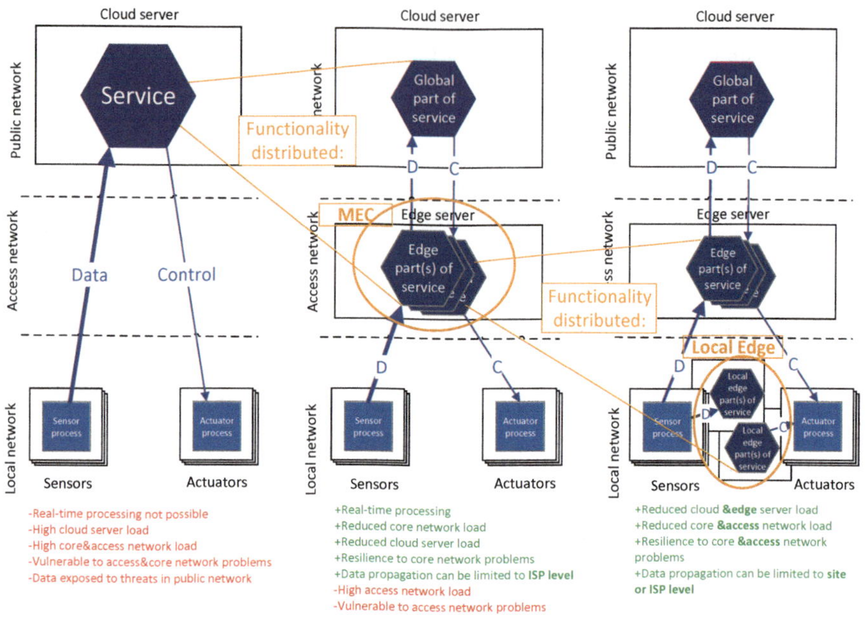

Fig. 3 Three-Tier IoT edge architecture. (**a**) Traditional cloud-based IoT. (**b**) 2-tier edge IoT. (**c**) 3-tier edge IoT

consisting of small, self-contained units, microservices [29, 30], which enable the development of distributed service compilations. Microservices are small and self-acting virtualized components, typically based on container technology (such as Linux or Docker containers), that are relatively easy to develop in isolation and maintain as standalone software components. Microservice architectures enable continuous software evolution, seamless technology integration, optimal runtime performance, horizontal scalability, and reliability through fault tolerance.

A new class of applications is emerging, namely, "serverless applications." They are exemplified by Function-as-a-Service (FaaS) [31, 32] systems that enable the future "service anywhere" architecture. FaaS has been considered as one of the technologies to realize lightweight microservices, also called as FaaS functions or "nanoservices" [18]. Whereas traditional microservices have larger role and are expected to be always available, FaaS-functions are considered as smaller logical units that become alive when needed and then execute and terminate when not needed anymore. Since FaaS functions typically do not run long periods and their size is small, their deployment does not require dedicated servers. Based on this, FaaS functions can be deployed on any device providing sufficient computational capacity, and therefore FaaS is also called as serverless computing. In FaaS-based agile development of microservices, the developers do not need to consider the computing infrastructure, as both resource provisioning and scaling are automated.

Microservices and serverless architectures can be seen as a technology for implementing various types of fog services on the three-tier IoT edge architecture, where service functions can be deployed on the most optimal tier, based on the current conditions of the network and computational environment.

3.3 Integration with SDN and NFV

EC per se represents a distributed approach that combines end devices and processing capabilities of remote servers. The network and its performance become a vital part of the EC paradigm. Network resources require a simple and flexible management to deal with the low latency and reliability requirements in addition to the huge data transmissions involved. Software-defined networking (SDN) is a new networking approach that decouples the network control from the data forwarding hardware. The network intelligence is logically located in software-based controllers (aka control plane) and the network devices become mere packet forwarding entities (aka data plane) [33]. SDN enables a new level of network management including better control, higher flexibility, and scalability. Moreover, SDN introduces fast network reconfiguration and self-healing that address important issues, such as user and application mobility, as well as uninterrupted service provisioning in the case of factory automation. Integrating SDN mechanisms with EC helps to provide the required computation resources and to satisfy the unique quality of service requirements of the applications, which is one of the MEC challenges as mentioned in Sect. 2.4. Having been researched over a long time, SDN has seen many novel approaches for network management, control, fast reconfiguration, healing, network function abstraction, placement, etc. Especially in wireless and mobile scenarios, SDN is able to make networks more controllable and programmable, update routing tables according to the often predictable mobility patterns of the nodes, and thus select the most appropriate paths from or to the end devices (e.g., [34]). Thus, SDN addresses many weak and challenging aspects of EC. However, there are only a few works that take those SDN approaches to the factory floor where they could be highly beneficial.

Another complementary network technology is presented by Network Function Virtualization (NFV). It adds on further hardware abstraction and can be combined with SDN to extend the virtualization approach toward higher layer network functions like load balancing, firewalls, intrusion detection, WAN acceleration, etc. The integration of SDN and NFV with EC brings a lot of new possibilities that improve the overall network and computational performance. In an industrial context, however, the establishment of any hardware abstraction is only feasible if such vital capabilities like availability, reliability, predictability, and deterministic behavior of the resulting system are not harmed. These aspects are still not sufficiently covered in the state of the art and present a research opportunity for both academia and industry.

3.4 Security, Privacy and Trust Management

Many different potential security attack vectors and risks for privacy breaches exist in an IIoT value chain from sensors, via gateways and fog nodes, to data centers, including end-user applications. The remainder of this section will present details of some of the identified open challenges in each of these areas.

The end devices are an integral part of EC systems due to the additional responsibilities that have been given to those. However, securing devices against unauthorized access by a malicious person is extremely difficult. Thus, key management, storing the keys and handling them in a secure way becomes paramount. It is also not uncommon to see hard-coded keys or group-key systems on IIoT devices, where a single compromised device can compromise the whole system security. There are many examples of extracting keys from devices if one has access to a physical device. Examples include physical side channel attacks, tampering, reverse engineering, power/electromagnetic analysis, timing attacks, known fault attacks, and clock glitches. End devices also tend to be the target of malicious software, including trojan horses, spyware, viruses, and other malware.

Network security is also a difficult but integral part of IIoT EC systems. The broad and heterogeneous network architecture with multiple network components using different hardware and software implementations is a challenging environment for security management. Different networks have their own vulnerabilities and weaknesses, for example, Local Area Networks (LAN), Wide Area Networks (WAN), low-power wide-area networks (LPWAN), and industrial networks. Therefore MEC and IIoT systems need to take a broad range of network types into consideration, making this a difficult challenge. Additionally, the wireless communication medium, which is often used in the IIoT, introduces an extra vulnerability and an opening for a wide range of attacks such as eavesdropping and jamming.

Another important challenge is dealing with trust and securing sensitive industrial data. This includes hiding and protecting the sensitive industrial data, such as sensor values, algorithms, and industrial process information, where a data breach can lead to competitors gaining an advantage over them. Therefore, the need for technologies allowing local data management and decision-making to limit data propagation toward public networks is obvious. In this context, EC is a centric building block for providing guarantees for customers to keep their data within set boundaries. However, the systems consisting of functions distributed on computing nodes on several architectural tiers, owned and managed by different stakeholders with their own security policies, are inherently very complex, which requires attention in the future research. There is a clear need for Security as a Service-type components, capable of running in constrained-capacity nodes [35]. Furthermore, building trust between stakeholders of these complex systems, based on, e.g., Blockchain [36], is an interesting avenue for future research.

3.5 *Use of Artificial Intelligence for EC Optimization*

AI has become a very important technique in many different domains. Being now for a long time successfully used in applications like speech and image recognition, strategic game systems (chess and Go), autonomous robots, etc., AI methods can be applied in EC for the optimization of many different aspects. In this section, we want to highlight several approaches of Machine Learning, being an area of AI, applied for task offloading.

The EC servers are usually densely distributed close to end devices to reduce the cost for offloading of computational tasks to these servers through wireless or wired links. Among many benefits, users can observably reduce the experienced delay of applications, energy consumption, and improve the QoS with the help of offloading. However, a list of unresolved questions arise [37]:

- **What** part of an application needs to be offloaded considering the complexity of the application, data to be shared between the user device and the EC server as well as the available network capacity?
- **When** is an optimal time to start the offloading considering the dynamic behavior of the end device, the available network capacity, as well as the dynamic load of the server?
- **Where**, on **which** node and at **which** architectural tier (local node/EC server/cloud server) should the offloaded task be processed considering the CPU and GPU availability on different nodes as well as the distance to these nodes from the user device?
- **How** should the offloading be organized?

In a complex scenario, offloading becomes a multi-objective decision-making problem. Designing an offloading strategy does not have a straightforward solution due to the dynamic behavior of EC systems. Stochastic characteristics of edge environment can make pre-decided offloading strategy impractical. Reinforcement Learning (RL), an area of Machine Learning, can be applied in training an AI agent to observe the current state of the EC system, to make an intelligent offloading decision based on specified criteria, and to learn from the history of such decisions. However, conventional RL algorithms cannot scale well as the number of edge devices increase, since the explosion of state space will make traditional tabular methods of RL infeasible. Another approach from the Machine Learning area is based on Deep Learning (DL), aka Deep Neural Networks (DNN). It operates efficiently with a large number of state spaces. The benefit of using DNN methods in EC is to extract hidden patterns from large and complex data sets of heterogeneous applications. A combined strategy, called Deep Reinforcement Learning (DRL) [38], shows a good offloading performance in various complicated EC scenarios. DRL methods treat the complicated EC system as a black box and interact with it to learn the optimal policies without modeling the system dynamics. Although there are significant advantageous in DRL methods, notable challenges related to dynamic behavior of considered applications remain in applying DRL to solve task-offloading problems in EC.

4 Conclusion

The unveiling of novel 5G and EC technologies will be one of the major driving factors in increasing productivity and therefore key enablers for long-envisaged vertical applications in various sectors including IIoT. In this book chapter, we have given an introduction to the applications, challenges and solutions of EC including an overview of the state of the art in EC for IIoT, different standardization activities, open challenges, and future development potential. Based on this, we believe that EC is an important piece of the IIoT puzzle and a key concept to meet the demands of future industrial services. The open challenges and research directions mentioned in this chapter represent attractive points for improvement and active work in both academia and industry. For example, the solutions of using three-tier IoT edge architecture, microservices and serverless architectures, integration with SDN and NFV, the use of AI for EC optimization, as well as aspects of security, privacy, and trust management have just recently become popular discussion hotspots around EC technology. In each of the mentioned areas, we have highlighted the advantages, disadvantages, and needed future research for the proliferation of the IIoT and EC in particular.

References

1. Xu LD, He W, Li S (2014) Internet of Things in industries: a survey. IEEE Trans Indus Inf 10(4):2233–2243
2. Shafi M, Molisch AF, Smith PJ, Haustein T, Zhu P, De Silva P, Tufvesson F, Benjebbour A, Wunder G (2017) 5G: a tutorial overview of standards, trials, challenges, deployment, and practice. IEEE J Sel Areas Commun 35(6):1201–1221
3. Garcia Lopez P, Montresor A, Epema D, Datta A, Higashino T, Iamnitchi A, Barcellos M, Felber P, Riviere E (2015) Edge-centric computing: vision and challenges. SIGCOMM Comput Commun Rev 45(5):37–42
4. Puliafito C, Mingozzi E, Longo F, Puliafito A, Rana O (2019) Fog computing for the Internet of Things: a survey. ACM Trans Internet Technol 19(2): 18:1–18:41
5. Mohanarajah G, Hunziker D, D'Andrea R, Waibel M (2014) Rapyuta: a cloud robotics platform. IEEE Trans Autom Sci Eng 12(2):481–493
6. Satyanarayanan M, Bahl P, Caceres R, Davies N (2009) The case for VM-based cloudlets in mobile computing. IEEE Pervasive Comput 8(4):14–23
7. Cardoso P, Monteiro J, Semi ao J, Rodrigues J (2019) Harnessing the internet of everything (IoE) for accelerated innovation opportunities. IGI Global, Pennsylvania
8. Chen X, Jiao L, Li W, Fu X (2016) Efficient multi-user computation offloading for mobile-edge cloud computing. IEEE/ACM Trans Netw 24(5):2795–2808

9. Govindaraj K, Grewe D, Artemenko A, Kirstaedter A (2018) Towards zero factory downtime: edge computing and SDN as enabling technologies. In: 2018 14th International Conference on Wireless and Mobile Computing, Networking and Communications (WiMob), pp 285–290, Oct 2018
10. Morabito R, Petrolo R, Loscri V, Mitton N (2018) LEGIoT: a lightweight edge gateway for the Internet of Things. Futur Gener Comput Syst 92:11
11. Ai Y, Peng M, Zhang K (2018) Edge computing technologies for Internet of Things: a primer. Digit Commun Netw 4(2):77–86
12. Edge computing: the perfect complement to the cloud in IoT. https://www.bosch-iot-suite.com/edge-computing/. Accessed 09 Mar 2020
13. Connectivity and intelligence at the edge of IoT. https://developer.bosch-iot-suite.com/service/gateway-software/. Accessed 09 Mar 2020
14. Porambage P, Okwuibe J, Liyanage M, Ylianttila M, Taleb T (2018) Survey on multi-access edge computing for Internet of Things realization. IEEE Commun Surv Tutorials 20(4):2961–2991, Fourthquarter 2018
15. Chettri L, Bera R (2020) A comprehensive survey on Internet of Things (IoT) toward 5G wireless systems. IEEE Internet Things J 7(1):16–32
16. Tran TX, Hajisami A, Pandey P, Pompili D (2017) Collaborative mobile edge computing in 5G networks: new paradigms, scenarios, and challenges. IEEE Commun Mag 55(4):54–61
17. Mach P, Becvar Z (2017) Mobile edge computing: a survey on architecture and computation offloading. IEEE Commun Surv Tutorials 19(3):1628–1656, thirdquarter 2017
18. Harjula E, Karhula P, Islam J, LeppÃd'nen T, Manzoor A, Liyanage M, Chauhan J, Kumar T, Ahmad I, Ylianttila M (2019) Decentralized IoT edge nanoservice architecture for future gadget-free computing. IEEE Access 7:119856–119872
19. Gubbi J, Buyya R, Marusic S, Palaniswami M (2013) Internet of Things (IoT): a vision, architectural elements, and future directions. Futur Gener Comput Syst 29(7):1645–1660. Including special sections: cyber-enabled distributed computing for ubiquitous cloud and network services & cloud computing and scientific applications – big data, scalable analytics, and beyond
20. Rossi F, van Beek P, Walsh T (2006) Handbook of constraint programming (foundations of artificial intelligence). Elsevier Science Inc., Amsterdam/Boston
21. Sezer OB, Dogdu E, Ozbayoglu AM (2018) Context-aware computing, learning, and big data in Internet of Things: a survey. IEEE Internet Things J 5(1):1–27
22. Xu G, Ngai EC, Liu J (2018) Ubiquitous transmission of multimedia sensor data in Internet of Things. IEEE Internet Things J 5(1):403–414
23. Sadeghi A, Wachsmann C, Waidner M (2015) Security and privacy challenges in industrial Internet of Things. In: 2015 52nd ACM/EDAC/IEEE Design Automation Conference (DAC), pp 1–6, June 2015
24. Kasinathan P, Pastrone C, Spirito MA, Vinkovits M (2013) Denial-of-service detection in 6lowpan based Internet of Things. In: 2013 IEEE 9th International Conference on Wireless and Mobile Computing, Networking and Communications (WiMob), pp 600–607, Oct 2013
25. BMWs connected drive feature vulnerable to hackers. https://www.autoblog.com/2015/02/03/bmws-connected-drive-feature-vulnerable-to-hackers. Accessed 26 Jan 2020
26. Roman R, Lopez J, Mambo M (2018) Mobile edge computing, fog et al.: a survey and analysis of security threats and challenges. Futur Gener Comput Syst 78:680–698
27. Kramers A, Hoejer M, Loevehagen N, Wangel J (2014) Smart sustainable cities – exploring ICT solutions for reduced energy use in cities. Environ Modell Softw 56:52–62. Thematic issue on modelling and evaluating the sustainability of smart solutions
28. Schlomann B, Eichhammer W, Stobbe L (2015) Energy saving potential of information and communication technology. Int J Decis Support Syst 1(2):152–163
29. Pahl C, Brogi A, Soldani J, Jamshidi P (2019) Cloud container technologies: a state-of-the-art review. IEEE Trans Cloud Comput 7(3):677–692
30. Rodger R (2018) The tao of microservices. Manning, Shelter Island

31. Kuhlenkamp J, Werner S (2018) Benchmarking FaaS platforms: call for community participation. In: 2018 IEEE/ACM International Conference on Utility and Cloud Computing Companion (UCC Companion), pp 189–194, Dec 2018
32. García López P, Sánchez-Artigas M, París G, Barcelona Pons D, Ruiz Ollobarren Á, Arroyo Pinto D (2018) Comparison of FaaS orchestration systems. In: 2018 IEEE/ACM International Conference on Utility and Cloud Computing Companion (UCC Companion), pp 148–153, Dec 2018
33. Nunes BAA, Mendonca M, Nguyen XN, Obraczka K, Turletti T (2014) A survey of software-defined networking: past, present, and future of programmable networks. IEEE Commun Surv Tutorials 16(3):1617–1634, Third 2014
34. Yang M, Li Y, Jin D, Zeng L, Wu X, Vasilakos AV (2015) Software-defined and virtualized future mobile and wireless networks: a survey. Mobile Netw Appl 20(1):4–18
35. Ranaweere P, Imrith VN, Liyanage M, Jurcut AD (2020) Security as a service platform leveraging multi-access edge computing infrastructure provisions. In: International Conference on Communications (ICC), 2020 IEEE. IEEE
36. Cinque M, Esposito C, Russo S (2018) Trust management in fog/edge computing by means of blockchain technologies. In: 2018 IEEE International Conference on Internet of Things (iThings) and IEEE Green Computing and Communications (GreenCom) and IEEE Cyber, Physical and Social Computing (CPSCom) and IEEE Smart Data (SmartData), pp 1433–1439, July 2018
37. Wu H (2018) Multi-objective decision-making for mobile cloud offloading: a survey. IEEE Access 6:3962–3976
38. Wang J, Hu J, Min G, Zhan W, Ni Q, Georgalas N (2019) Computation offloading in multi-access edge computing using a deep sequential model based on reinforcement learning. IEEE Commun Mag 57(5):64–69

Part IV
Selected Use Cases in Connected Industries

Intelligent Transport System as an Example of a Wireless IoT System

Roshan Sedar, Charalampos Kalalas, Francisco Vázquez-Gallego, and Jesus Alonso-Zarate

1 Introduction

1.1 The Transformation of Mobility

In view of the increased rate of fatal road accidents, resulting in approximately 1.25 million deaths every year [1], the development of Intelligent Transport Systems (ITS) has recently attracted growing interest from the transport industry, expecting to improve transport safety and mobility. Among other definitions, the European Telecommunications Standards Institute (ETSI) defines that ITS use modern communication technologies to improve safety, reliability, efficiency, and quality in transport as long as relevant technologies are integrated into infrastructure and vehicles [2]. In turn, the involvement of a wide set of stakeholders, e.g., network operators, service providers, regulatory entities, road traffic authorities, and automotive original equipment manufacturers (OEMs), can be seen in various aspects related to the development of ITS.

The integration of advanced communication technologies into all modes of passenger and freight transport systems effectively leads to making transport safer, more efficient, and more sustainable; also it reduces the environmental impact. ITS are not only limited to road vehicles but also span across other services being implemented in navigation systems, air traffic systems, and water and rail transport systems. The new mobility paradigm is transforming the landscape of entire industries, which will require advanced and competitive next-generation transport systems in the context of Cooperative, Connected and Automated Mobility (CCAM), e.g., Cooperative-ITS (C-ITS). The C-ITS constitute a subset of ITS

R. Sedar · C. Kalalas · F. Vázquez-Gallego · J. Alonso-Zarate (✉)
Centre Tecnològic de Telecomunicacions de Catalunya (CTTC/CERCA), Barcelona, Spain
e-mail: roshan.sedar@cttc.es; ckalalas@cttc.es; francisco.vazquez@cttc.es; jesus.alonso@cttc.es

N. H. Mahmood et al. (eds.), *Wireless Networks and Industrial IoT*,
https://doi.org/10.1007/978-3-030-51473-0_13

and enable real-time wireless communication and information sharing between ITS stations[1] to achieve coordinated decisions through cooperation, thus extending the capabilities of a vehicle beyond the scope of a typical stand-alone entity.

The C-ITS enable effective data exchange leveraging wireless connectivity where vehicles can communicate with each other, with roadside units (RSUs), and with other road users, e.g., pedestrians, cyclists, etc. These interactions can be categorized into vehicle-to-vehicle (V2V), vehicle-to-infrastructure (V2I), and vehicle-to-network (V2N) communication, often clustered under the generic term vehicle-to-everything (V2X) communication. The ubiquitous V2X communication contributes to the realization of the Internet of Vehicles (IoV) paradigm, a concept which has recently emerged from the Internet of Things (IoT). The IoV enables road users and both road and traffic managers to mutually share information and reach to more informed and coordinated decisions. CCAM services relying on IoV are expected to constitute the key enablers of future C-ITS. Nevertheless, to successfully implement CCAM services and shape the new transport ecosystem, both automotive and telecommunication industries along with relevant stakeholders will have to overcome a number of challenges.

1.2 IoT in Automotive Systems

The large-scale deployment of low-cost wireless sensors with integrated sensing, computing, and storing capabilities offers the potential to significantly improve the efficiency of existing C-ITS toward increased road safety, optimized driving decision-making, and real-time traffic control. The integration of ultra-high-definition cameras, radars, lidars, ultrasonic range finders, and other types of sensors, realizing the concept of Advanced Driving Assistance Systems (ADAS), is progressively turning vehicles into sophisticated computing units able to gather, process, and exchange information, between vehicles and other road users, and with an increasingly intelligent road infrastructure. As the level of driving automation increases,[2] unprecedented volumes of data will be generated per vehicle rendering the automotive services of emerging C-ITS much more demanding in terms of network capacity. In this context, the underlying communication network becomes of utmost importance to support V2X connectivity and fulfill the requirements of emerging V2X use cases in terms of latency, reliability, scalability, coverage, data rate, and positioning accuracy. By leveraging the full potential of artificial intelligence (AI) and virtual/augmented reality capabilities, next-generation communication technologies (e.g., 5G and beyond) will become pervasive across

[1]ITS stations can be categorized as mobile ITS stations (vehicles) or fixed ITS stations (roadside installations) [3].

[2]On the way to fully automated vehicles, six levels of autonomy are identified by the Society of Automotive Engineers (SAE) from complete driver control (level 0) to full autonomy (level 5).

multiple V2X use cases and will radically transform the automotive sector. For example, data-driven AI techniques will allow vehicles acquire a detailed understanding of the driving environment, e.g., traffic mobility patterns and channel conditions of other vehicles, and make real-time decisions with minimum user control. The foundation of emerging C-ITS, therefore, lies at the intersection of the multidisciplinary areas of wireless connectivity, IoT, AI, and data analytics.

The remainder of this chapter is organized as follows: Section 2 provides a categorization of the basic V2X use cases and their performance requirements. Section 3 elaborates on the key features of the two dominant V2X radio access technologies and summarizes their standardization evolution. Section 4 discusses fundamental challenges toward future V2X communication and highlights potential enablers and research directions. Finally, Sect. 5 provides our concluding remarks.

2 Use Case Examples and Key Performance Indicators

From telecom perspective, a variety of vehicular use cases (commonly referred to as V2X use cases) have been mainly identified by the ETSI ITS [4] and The 3rd Generation Partnership Project (3GPP) in Rel-14[3] and Rel-15 [5, 6]. The V2X use cases are typically focused on safety, traffic efficiency, and comfort services (i.e., infotainment). In particular, each of these service categories imposes different quality of service (QoS) requirements in terms of end-to-end latency, data rate, reliability, or communication range. In what follows, we provide an overview for a set of use cases along with their requirements derived from 3GPP Rel-15 [6]. Graphical representations of example V2X scenarios are also illustrated in Fig. 1.

2.1 *Example of V2X Use Case Groups*

V2X use cases are typically classified into groups or families. In the following, some of the most representative V2X use case groups are introduced.

Cooperative Awareness The awareness among networked entities, e.g., road users and the roadside infrastructure, brings benefits in the established connectivity and, essentially, sets the basis for a number of safety and traffic-efficiency ITS applications. Information of mutual interest, e.g., location, speed, trajectory, and type attributes, can be proactively communicated allowing for more informed decisions with enough room of time and space. In the case of Vulnerable Road User (VRU) protection in Fig. 1b, the vehicles are warned about the presence of a VRU, e.g., pedestrian or cyclist, when there is a dangerous situation. In turn, when a

[3]Each 3GPP Release is henceforth abbreviated as Rel-*No*.

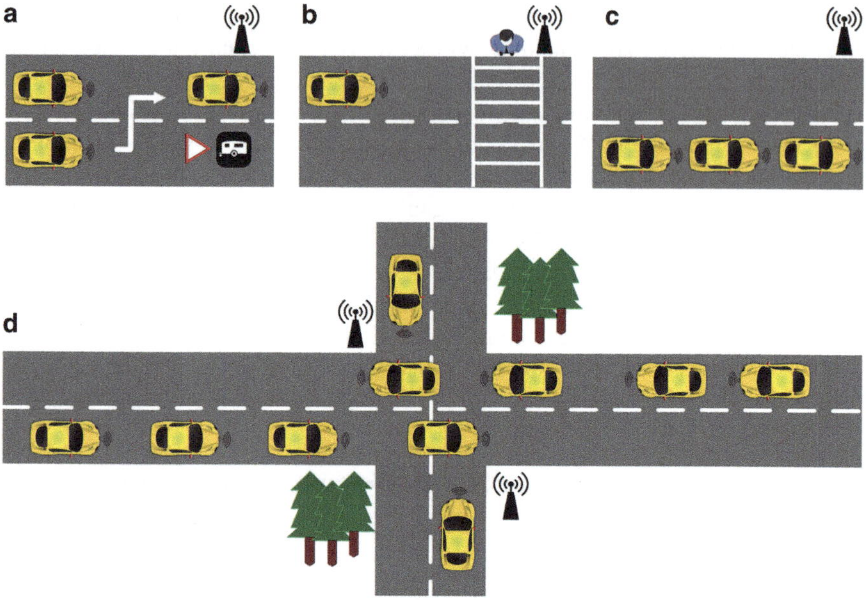

Fig. 1 ITS applications enabled by V2X communication. (**a**) Cooperative lane merge (or change). (**b**) Vulnerable road user (VRU). (**c**) CACC/Platooning. (**d**) Cooperative intersection control

vehicle detects a VRU, it can then cooperate with neighboring vehicles to share this information and prevent hazardous events. Traffic jam warning, emergency vehicle approaching, and forward-collision warning constitute additional use case examples that belong in this group. According to ETSI ITS and 3GPP Rel-14, the minimum frequency of awareness messages is 1 Hz, while a maximum latency of 100 ms is prescribed for this use case group.

Cooperative Sensing In this use case group, the benefits come from sharing information captured by different sources, e.g., radars, laser sensors, and stereo vision from on-board cameras, between vehicles and between vehicles and RSUs in the vicinity, and making decisions in a cooperative manner. The vehicles thus become capable of enhancing their perception of situational awareness beyond the capabilities of short-range on-board sensors, resulting in a more holistic view of their relative position. As use case examples, map sharing, see-through, and bird's-eye view V2X services can be considered in this group. In map sharing, vehicles can exchange raw/processed data of their relative positions to build a collective view of situational awareness. Ideally, the merging of local sensor information with remote information is going to be executed locally and on a remote V2X application server. In the see-through case, a leading vehicle streams the camera-captured view out to a rear vehicle allowing it to see through the forward vehicle to avoid an occlusion area. Similarly, bird's-eye view application facilitates vehicles to exploit streaming information captured by sensors in an intersection and assists in making safe movements. Regarding the QoS requirements, a maximum latency of 100 ms and a downlink data rate of 50 Mbps are required in this use case group.

Cooperative Maneuver In this use case group, vehicles share their local awareness and driving intentions allowing the optimization of their future trajectories, when possible. In the case of cooperative intersection control (Fig. 1d), the planned vehicle trajectories can be coordinated using a centralized intersection controller in a safe and efficient manner. In cooperative lane change use case (Fig. 1a), vehicles collaborate to execute a lane change for either one or a group of cooperative vehicles in a safe and efficient way. In the case of platooning (also referred to as Cooperative Adaptive Cruise Control (CACC)) in Fig. 1c, vehicles are able to dynamically form a group with a leading vehicle that sends data periodically to the rest of the vehicles forming the platoon. In a typical scenario, there is a driver in the leading vehicle directing the platoon followed by trailing vehicles where drivers simply execute the platoon operations based on the received information. Vehicle platooning allows maintaining the smallest possible safe inter-vehicle distance, resulting in an efficient utilization of the road space and thus increase of road capacity.

2.2 *Advanced V2X Use Cases*

As of today, the market push for a widespread commercial availability of high automation vehicles urges the need for advanced V2X solutions in order to achieve the benefits beyond what basic safety applications currently can offer. As shown in Table 1, a set of advanced use cases and necessary key performance indicators (KPIs) have been defined by 3GPP in Rel-15 and Rel-16, aiming at characterizing CCAM services. The scope of these advanced V2X use cases is to further enhance road safety and traffic efficiency and also to offer comfort services to passengers. The following use cases address broad classes of advanced ITS applications:

(a) **Advanced vehicle platooning**: enables vehicles to autonomously form a convoy with a group of vehicles and coordinate their trajectories or maneuvers based on the local perception constructed through on-board sensor data and information received from neighboring vehicles.
(b) **Advanced driving**: the objective is to enable semi- or fully automated driving, letting vehicles to coordinate their trajectories or maneuvers. In principle, each vehicle is expected to exchange its local sensor readings with other vehicles or RSUs in the close vicinity along with respective driving intentions.
(c) **Extended sensors**: the definition and objectives fall into the aforementioned cooperative sensing use case. However, the KPIs are much more stringent in the case of vehicles characterized by a higher degree of automation and with advanced features.
(d) **Remote driving**: enabling a remote driver (via tele-operation) or a V2X application (via V2N) to take over the operation of a remote vehicle in the cases

Table 1 KPI values for advanced ITS applications. (Derived from [6])

Use case type	Max. end-to-end latency[a] (ms)	Data rate (Mbps)	Reliability[b] (%)	Min. communication range (m)
Vehicle platooning	10–500	50–65	90–99.99	80–350
Advance driving	3–100	10–50	90–99.99	360–700
Extended sensors	3–100	10–1000	90–99.99	50–1000
Remote driving	5	UL: 25 DL: 1	>99.999	–

[a] Latency refers to the end-to-end packet delay across all the processing layers involved
[b] Reliability is defined as the percentage of (expected) rate of successful packet deliveries

of assisting those who cannot drive by themselves or a remote vehicle located in a dangerous environment, e.g., icy roads, bad weather conditions, etc.

The KPIs shown in Table 1 demonstrate that such use cases pose much more stringent QoS requirements than basic safety applications. In addition, V2X messages in this set of use cases can be large with varying packet sizes (e.g., payload up to 6.5 KB), thus resulting in the need for higher data rates as well. In the case of high level of automation, the reliability requirements could become even stricter (e.g., 99.99–99.999%) in the first three use cases. As will be described later in the chapter, the existing V2X communication technologies, i.e., 3GPP cellular V2X (C-V2X) and those based on the IEEE 802.11p standard, are not yet able to capture enhanced V2X use cases and satisfy their stringent performance requirements. In this context, emerging 5G systems have been evolving to offer scalable solutions to cater diverse services and devices. In particular, three key service categories have been recognized for 5G: (i) ultra-reliable low-latency communications (URLLC); (ii) enhanced mobile broadband (eMBB); and (iii) massive machine-type communications (mMTC). These categories will drive the development of an unprecedented range of vertical industries, including the automotive sector. Nevertheless, it remains to be seen as to whether 5G systems can become the single enabling wireless technology for V2X solutions.

3 Available Communication Technologies

3.1 Overview and Standardization Landscape

During the last two decades, several radio technologies have been proposed to cover all the different aspects of vehicular communication and support the demanding requirements imposed by different V2X use cases. As the design targets of future V2X services continue to evolve, standardization bodies are working toward novel and innovative approaches to overcome the bottlenecks in terms of performance over the radio interface. In what follows, an overview of the V2X standardization developments to date is presented:

1. **The 3rd Generation Partnership Project (3GPP).** The 3GPP has already raised the need to revisit the design of next-generation cellular networks to efficiently support V2X connectivity. From Rel-14 onward, the 3GPP has been working on the development of C-V2X technologies, often referred to as LTE-V2X as they were initially based on the LTE[4] standard specifications [6–9]. In particular, the 3GPP Technical Specification (TS) 22.185 was a Rel-14 document defining key V2X use cases and service requirements for both safety and non-safety applications [8]. In the context of 3GPP Rel-15, the TS 22.186 was developed with a focus on enhancements of V2X use cases, e.g., vehicle platooning and remote driving, including more rigorous functional requirements for advanced features that could not be achieved by earlier standard specifications [6]. In general, the C-V2X radio enhancements comprise both infrastructure-based solutions in the Uu interface (i.e., between UE[5] and eNodeB[6]) and sidelink-based solutions in the PC5 interface (i.e., between UEs).

   As emerging V2X use cases impose requirements that are hardly met by the existing standard improvements, the V2X development is foreseen to be one of the major topics to be specified in 3GPP Rel-16 and future releases. Ongoing 3GPP efforts aim at enhancing the C-V2X technology in the context of the New Radio (NR) framework which was already standardized in Rel-15 [10]. In particular, study items include the design of the new V2V broadcast, groupcast, and unicast sidelink communication interfaces to support the ever-demanding requirements, e.g., in terms of reliability and latency for remote driving and data rate for cooperative perception. A feasibility study on NR-V2X [11] was recently concluded successfully, and several technical solutions were identified, although NR-V2X is still in its initial stage of development [10]. Spectrum aspects, e.g., which frequency band V2X sidelink should be used, and positioning techniques constitute additional topics under discussion for future standardization activities. The evolution of C-V2X functionalities is summarized in Fig. 2.
2. **Institute of Electrical and Electronics Engineers (IEEE).** Since 2004, the IEEE is working on amendments of their well-established 802.11 family of standards to support wireless access for V2X communication in rapidly changing mobile environments. In this context, the IEEE 802.11p standard defines the data exchange between high-speed vehicles themselves as well as with the roadside infrastructure. The standard operates in the 5.9 GHz frequency band, reserved for ITS services in Europe and the USA,[7] and has a particular focus on safety applications. The IEEE 802.11p incorporates, with some minor

[4]LTE stands for Long-Term Evolution.

[5]UE stands for user equipment.

[6]In LTE terminology, the base station is commonly referred to as Evolved Node B (eNB or eNodeB).

[7]In Japan, a single 9 MHz frequency channel in the 755.5–764.5 MHz band has also been designated for ITS safety-related applications using V2V and V2I communication.

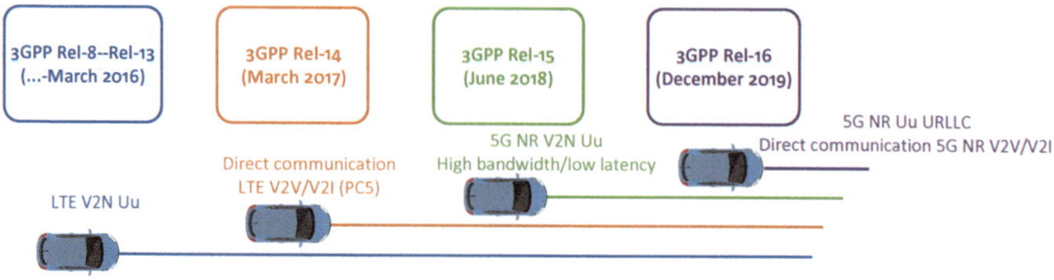

Fig. 2 Evolution of 3GPP C-V2X functionalities and backward compatibility

modifications, the IEEE 802.11a Orthogonal Frequency Division Multiplexing (OFDM) PHY layer and the Medium Access Control (MAC) layer from the IEEE 802.11e standard. The IEEE 802.11p constitutes the underlying radio communication basis for the following mature sets of standards:

a. The ETSI ITS-G5 standard mainly developed in Europe by the ETSI and the European Committee for Standardization (CEN) with their relevant technical committees. ETSI ITS-G5 has undergone a thorough standardization process during the recent years and extensive field trials to test its performance. Recently, ETSI has initiated pre-standardization studies with the purpose of specifying new ITS services to be applicable in the framework of ETSI ITS Rel-2 standard development [12, 13].
b. The Dedicated Short-Range Communication (DSRC) standard suite – also referred to as Wireless Access in Vehicular Environments (WAVE) – mainly developed in the USA by the National Department of Transport and a consortium of automotive manufacturers for interoperability tests. At the protocol stack, DSRC utilizes a slightly modified version of IEEE 802.11p for the PHY and MAC layers, while the suite of IEEE 1609.x standards for WAVE is utilized in higher layers. Above the protocol stack, V2X message sets and related performance requirements are specified by the SAE.
c. In parallel to the standard developments in Europe and the USA, the Japanese research and standardization organization ARIB STD-T109 [14] has developed a standard aiming at driving safety support systems operating in the 700 MHz band. The standard uses a PHY layer very similar to IEEE 802.11p but employs a modified MAC layer.

The IEEE 802.11 community is currently working on enhancing 802.11p technology in the context of the P802.11bd project,[8] which aims at developing the standard amendments in the PHY and MAC layers for the next-generation V2X systems [15]. Recent development efforts by the dedicated IEEE 802.11bd task group focus on the challenging scenarios with high vehicle densities where

[8]The next-generation IEEE V2X standard will be named IEEE 802.11bd, but until publication (expected during 2021), it is generally referred to as P802.11bd.

the transmissions can be delayed beyond acceptable values due to collisions between ITS stations. At the same time, interoperability, coexistence, and backward compatibility with IEEE 802.11p transmissions must be guaranteed. To further improve channel access performance, example items under discussion include, among others, (i) new message formats for channel estimation; (ii) adaptive retransmissions based on the congestion level; (iii) narrower OFDM numerologies, i.e., subcarrier spacing; (iv) support of mmWave[9] frequency bands; etc.

3.2 *Short-Range Communications: DSRC/ETSI C-ITS*

The DSRC [16] and the European C-ITS[10] [17] communication standards are based on enhancements of the IEEE 802.11a standard (Wi-Fi) that adapt the PHY and MAC layers for the requirements of vehicular networks, i.e., high mobility and short-life communication links. These radio technologies facilitate the formation of wireless ad hoc networks whenever vehicles or RSUs are within the range of each other, thus enabling vehicles to directly communicate with other vehicles (V2V) and with the roadside infrastructure (V2I). The protocol stacks of DSRC and ETSI C-ITS are shown in Figs. 3 and 4, respectively.

The PHY and MAC layers of DSRC rely on the IEEE 802.11p standard [19] with an extension of the MAC layer for multi-channel operation specified in IEEE 1609.4 [20]. DSRC operates in the 5.9 GHz band, which ranges from 5.850 to 5.925 GHz. The wireless channels of DSRC are separated into control

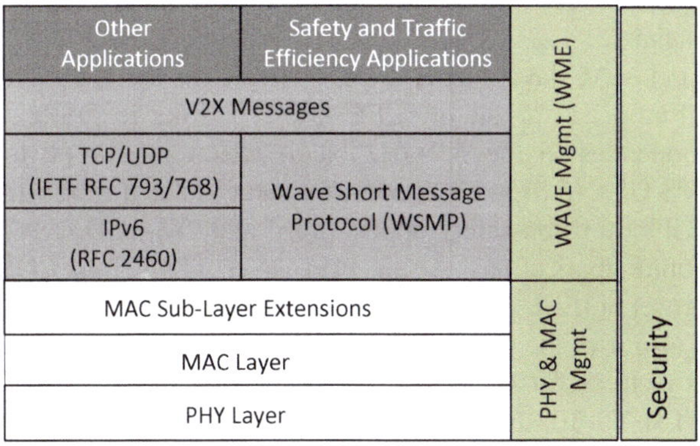

Fig. 3 The DSRC protocol stack ([18])

[9]mmWave stands for millimeter wave.

[10]The European C-ITS is called ETSI C-ITS in the rest of this section.

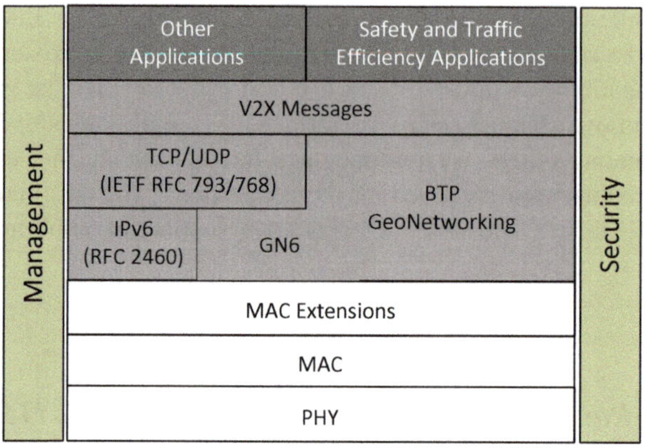

Fig. 4 The ETSI C-ITS protocol stack ([17])

channels (CCH) and service channels (SCH). The switching between channels is coordinated according to the IEEE 1609.4 standard. The PHY layer of IEEE 802.11p uses OFDM, and, compared to Wi-Fi, it reduces the 20 MHz channel bandwidth to 10 MHz and doubles the time parameters of the PHY, thus increasing the performance under the rapidly varying channels of vehicular environments. The MAC layer of IEEE 802.11p is based on the Outside the Context of a Basic Service Set (OCB) mode which reduces latency and facilitates vehicles in close proximity to exchange data immediately, without the prior exchange of control information. The IEEE 802.11p uses the Carrier Sense Multiple Access with Collision Avoidance (CSMA/CA) protocol with the Enhanced Distributed Channel Access (EDCA) scheme, which provides four different access categories for data traffic prioritization using specific parameters for the contention window size and inter-frame spaces per each access category.

The PHY and MAC layers of ETSI C-ITS rely on the ITS-G5 standard [21], which is the European variant of IEEE 802.11p. Like in DSRC, the ETSI C-ITS standard operates in the 5.9 GHz band, which in Europe is divided into three bands: ITS-G5A (5.875–5.905 GHz) for safety-related applications, ITS-G5B (5.855–5.875 GHz) for non-safety applications, and ITS-G5D (5.905–5.925 GHz) reserved for future applications. At the PHY layer, ITS-G5 uses OFDM with the same parameter set of IEEE 802.11p but with adapted spectrum masks. At the MAC layer, ITS-G5 also uses the OCB mode, CSMA/CA, and EDCA. In addition, ITS-G5 introduces features for Decentralized Congestion Control (DCC) methods as specified in ETSI TS 102 687 [22], which aim at maintaining network stability, throughput efficiency, and fair resource allocation to ITS stations.

At the networking and transport layers, DSRC uses the Internet Protocol (IP) in combination with the transport protocols User Datagram Protocol (UDP) and Transmission Control Protocol (TCP) and the WAVE Short Message Protocol (WSMP) specified in IEEE 1609.3. WSMP is a single-hop network protocol optimized with short packet headers. While safety, control, and management messages are transmitted on control channels using WSMP, non-safety messages are transmitted on service channels using IP and TCP/UDP. In ETSI C-ITS, the networking and transport layers also rely on IP and TCP/UDP for the transmission of non-safety messages, while the GeoNetworking protocol and the Basic Transport Protocol (BTP) are used for the transmission of safety-related messages. The GeoNetworking protocol is specified in the ETSI EN 302 636 [23] standard. It is an ad hoc routing protocol for multi-hop communication with geographical addressing. In particular, it uses the geographical coordinates to forward packets based on the vehicle's knowledge of its own position and the neighbors' positions. It further facilitates multi-hop routing with the establishment and maintenance of network routes in a dynamic environment with frequent topology changes. The BTP, specified in [24], is a connectionless and unreliable end-to-end packet transport protocol similar to UDP on top of GeoNetworking.

The standards at the facilities layer of DSRC and ETSI C-ITS specify a set of requirements and functionalities to support applications. These standards include V2X messaging protocols, position management, data fusion in Local Dynamic Map (LDM), etc. In DSRC, the SAE J2735 [25] standard defines the syntax and semantics of V2X messages, e.g., the basic safety message (BSM) is sent periodically at a maximum rate of 10 Hz and conveys state information of the vehicle, including position, dynamics, status, and size. In ETSI C-ITS, the cooperative awareness message (CAM) specified in ETSI EN 302 637-2 [26] is the equivalent to the BSM in the DSRC protocol stack. It is a periodic message that provides status information to neighboring vehicles and RSUs. Its rate can vary between 1 and 10 Hz depending on vehicle dynamics and the congestion status of the wireless channel. In addition, the Distributed Environmental Notification Message (DENM), specified in ETSI EN 302 637-3 [27], is an event-triggered message controlled by the application, e.g., for collision avoidance. When a vehicle detects a road hazard, it sends a DENM that conveys safety information in a geographical region. In addition, the ETSI C-ITS protocol stack specifies the LDM in the ETSI EN 302 895 [28] standard. The LDM is a database of time- and location-referenced moving or stationary objects that influence road traffic, e.g., traffic signs and pedestrian walking.

In DSRC and ETSI C-ITS, the application layer is not fully standardized yet. Instead, ETSI C-ITS identifies a basic set of applications (in ETSI TR 101 638) which are classified into four groups: active road safety, cooperative traffic efficiency, cooperative local services, and global Internet services. In DSRC, the security layer relies on the IEEE 1609.2 standard, which provides authentication and optional encryption of messages based on digital signatures and certificates. In order to protect privacy, certificates do not contain driver's information, and the certificate authority can link the certificate to the driver's identity. In addition, in order to avoid the tracking of vehicles, each vehicle changes its certificate frequently and uses it only for a limited time. The approach in the security layer of ETSI C-ITS [29] is very similar to DSRC. The security and data privacy mechanisms of ETSI C-ITS are based on the security architecture defined in ETSI TS 102 940, as well as ETSI TS 102 941 for confidentiality and ETSI TS 102 942 for data integrity.

3.3 Cellular-Based: LTE-V2X, NR-V2X

Recently, the 3GPP has developed the first set of cellular standards for V2X communication. Today's realization of C-V2X is based on the LTE-V2X [30] standard specified in 3GPP Rel-14, and it will evolve into the future NR-V2X standard to be specified by the end of 2019 in 3GPP Rel-16. C-V2X is gaining support from the leaders of the automotive and telecom industries, which has led to worldwide C-V2X trials (some examples in [31–33]), while it is already stated that C-V2X offers superior performance than IEEE 802.11p-based radio technologies [34]. In this section, we provide an overview of the current C-V2X standard specified in 3GPP Rel-14 and Rel-15, as well as a brief outlook into NR-V2X (Rel-16).

The protocol stack of C-V2X is shown in Fig. 5. As it can be observed, the lower layers are specified by 3GPP for radio access, whereas the upper layers (i.e., applications, facilities, networking and transport, and security) are reused from the core standards used in DSRC and ETSI C-ITS. This allows a one-to-one mapping of the already existing applications that were already developed for DSRC and ETSI C-ITS and ensures interoperability with the emerging C-V2X applications. The C-V2X (3GPP Rel-14 and Rel-15) standard provides two different radio communication interfaces that can be used to support all types of vehicular use cases: the Uu interface and the PC5 interface.

The Uu interface uses the conventional cellular link between the UE and the eNodeB and operates in commercial licensed cellular spectrum. A UE using the

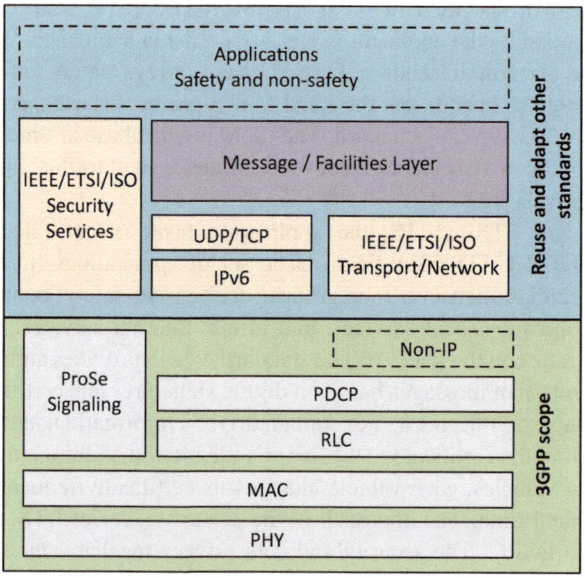

Fig. 5 The cellular V2X protocol stack [35]

Uu interface transmits messages to the eNodeB in the uplink, and messages are sent from the eNodeB to the destination UE in the downlink. The Uu-based communication requires that the UE resides within the coverage area of an eNodeB, and the eNodeB is responsible for the radio resource management. The main advantage of the Uu interface is that it facilitates a large dissemination range of V2X messages by leveraging the cellular core network. However, due to the inherent network delays, the Uu interface is expected to be used for latency-tolerant use cases, such as dynamic high-definition (HD) maps, software updates, infotainment, traffic information, and informational safety.

The PC5 interface allows for direct communication between UEs (e.g., vehicles, RSUs, and other road users) without requiring every message to be routed through the eNodeB. Therefore, the PC5 interface is suitable for time-critical safety use cases that require low latency communication with enhanced range, reliability, and non-line-of-sight performance. The PC5 interface specified in 3GPP Rel-14 and Rel-15 evolves from the device-to-device (D2D) framework standardized in the previous Rel-12 and Rel-13 for proximity services, e.g., emergency communications in case of natural disaster. The UEs can employ the PC5 interface both in the presence and absence of the eNodeB, i.e., with or without cellular coverage. In case of cellular coverage, the eNodeB manages resource allocation and scheduling for transmissions (referred to as sidelink Mode 3). When the UE is outside cellular coverage, it will manage itself the radio resources in an autonomous manner (referred to as sidelink Mode 4). This is achieved through a resource reservation algorithm which requires that the UE senses the channel and processes the results to ensure that other UEs reserve orthogonal resources in time and/or frequency in order to reduce packet collisions.

The sidelink Mode 4 of the PC5 interface can operate without provisioning of a subscriber identity module (SIM), thus not requiring subscription with a mobile network operator. Therefore, PC5-based communication allows the support of critical vehicular safety services when cellular coverage is not available or when the vehicle does not have a cellular subscription. In order to support SIM-less operation, automotive OEMs will have to configure the UE on-board each vehicle with the parameters required to autonomously reserve radio resources, e.g., allocation schemes, sources of synchronization, etc.

The enhancements introduced in C-V2X were basically aimed at handling high relative vehicle speeds and to improve reliability, throughput, and latency. C-V2X is capable of supporting a basic set of vehicular use cases that require the delivery of periodic messages ranging from 1 to 10 Hz periodicity and 50–100 ms end-to-end latency. The main enhancements will be introduced in NR-V2X (3GPP Rel-16) which will provide ultra-reliable, low latency, high-throughput communication to support autonomous and advanced driving use cases. The current design of NR-V2X focuses on the following main areas: design of an enhanced sidelink, enhancements to the NR Uu interface, configuration/allocation of sidelink resources using the NR Uu interface, mechanisms to select the best interface (among LTE sidelink, NR sidelink, LTE Uu, and NR Uu) for V2X message transmission, and coexistence of C-V2X and NR-V2X within a single device.

Table 2 A summary of short-range and C-V2X communication technologies

Technology	Features	Benefits	Use cases
DSRC/ETSI ITS (IEEE 802.11p)	Direct communication	High density/mobility support	Basic safety only
	Self-managed	Direct range up to 250 m	
	Can operate in 5.9 GHz spectrum	Mature for deployment	
	IEEE/ETSI security services	Extensively standardized	
C-V2X		High density/mobility support	
		Direct range over 450 m	Basic safety
	Direct and network communication	Extended coverage via cellular infrastructure	Enhanced safety
	Managed by cellular operators	Address V2X applications in an end-to-end manner	Advanced V2X in autonomous driving
	IEEE/ETSI/ISO security services	Broad and global support ecosystem	
	Can operate in 5.9 GHz spectrum		
	Short-range mmWave support		
		Involvement of key stakeholders	

Table 2 summarizes the key features of short-range and cellular-based V2X technologies, highlighting their benefits and applicable ITS use cases.

4 Challenges for Future V2X Communication

Although both 802.11p-based and C-V2X technologies provide satisfying results for a basic set of vehicular safety applications in low channel load conditions and in favorable propagation environments, they cannot adequately address the stringent QoS requirements imposed by the next-generation automotive use cases, especially when an increasing level of automation is considered [36]. When the vehicular density exceeds a certain limit, i.e., traffic overload in IoV scenarios, the contention-based channel access mechanism of 802.11p results in high collision probability due to the simultaneous transmissions and the hidden terminal problem. Thus, the scalable support of mission-critical services associated with high reliability levels, e.g., vulnerable user protection, is prevented.

In addition, although C-V2X sidelink communication introduces a sensing mechanism to cope with the undesirable effects of channel congestion, it does not completely eliminate the collision risk, especially in the case of high traffic load where all the resources might be sensed occupied and still the transmitter has to choose one of them. Therefore, several challenges remain open for the future C-ITS applications and drive the standardization efforts for both 802.11p-based and C-V2X technologies. In what follows, we provide an overview of the most important challenges that require innovative solutions to enable fully autonomous driving:

1. **Meeting C-ITS communication requirements simultaneously:** The future C-ITS use cases cover a broad range of high-mobility scenarios associated with a multiplicity of – often conflicting – requirements which makes the design of a single network really challenging, both for the radio interface and the system architecture [37]. As the level of driving automation increases, safety-critical services impose stringent requirements in terms of end-to-end latency (less than 3 ms), reliability (higher than 99.999%), and positioning accuracy (down to 5 cm). Achieving ultra-reliable communication with low latency is a major challenge in terms of physical design due to the fundamental trade-off that lies between the two targets, especially given the fast-changing nature of the propagation environment and the level of interference due to the vehicles' high mobility [38]. The support of multiple subcarrier spacing options by the NR offers flexibility in the frame structure, and the selection of the suitable physical layer numerology mainly relies on the resulting latency-reliability trade-off. The utilization of short-range mmWave bands could offer great potential in terms of high data rate, e.g., for HD map exchange or sharing large volumes of contextual data. However, the transmit/receive beam alignment associated to the operation on mmWave frequencies imposes a significant challenge in highly dynamic V2X environments [39]. In addition, sidelink-based V2X communication, which is preferred in many V2X scenarios, may not be sufficient to meet the requirements as a stand-alone solution; therefore, the potential of multi-connectivity combining the sidelink with the uplink/downlink (Uu) link could be explored for enhancing reliability as well as data rate for advanced V2X communication.
2. **Traffic differentiation and QoS management**: The accommodation of the various requirements of emerging V2X applications raises new challenges for efficient traffic differentiation, interference handling, and resource allocation. Although the dynamic sharing of the same resources for different services is certainly beneficial in terms of spectral efficiency, it brings challenges on the system design that has to optimally multiplex different service flows with different QoS requirements [40]. For example, providing URLLC for V2X mission-critical services and maintaining high data rates for eMBB services are contradictory requirements and become even more challenging in resource-constrained scenarios. Similarly, different types of interference arise, especially considering all different V2X communication forms, frequent handovers, and the potential operation over unlicensed bands. The interference due to co-channel operation or from adjacent channels should be efficiently mitigated in order to increase reliability levels.

3. **Precise localization of vehicles and road users**: Ubiquitous, accurate, and real-time knowledge of road users' position and vehicles' trajectory is a crucial requirement and enabler for many V2X use cases, e.g., lane merge and platoon formation. High mobility and ever-changing network topologies in V2X communication make it difficult to achieve high accuracy (i.e., centimeter level) for absolute and relative positioning, trajectory alignment, etc. [41]. The V2X dynamics, including high Doppler and delay spread due to moving transmitters, receivers, and scatterers, create a harsh propagation environment that hinders the ubiquitous and real-time tracking of road users. The high frequency bands considered for V2X communication, e.g., the mmWave band, may further deteriorate the impact caused by the Doppler spread, frequency error, and phase noise. In addition, the delay to process and/or feed location information to the corresponding server is not negligible. To overcome these issues, radio-assisted positioning techniques leverage triangulation algorithms running in multiple base stations with Multiple-Input Multiple-Output (MIMO) beamforming to achieve accurate localization. In addition, the use of smaller antenna array sizes in the frequency range above 6 GHz would exploit the available spatial information to enrich the time measurements (conventional LTE positioning) with angle measurements and achieve accuracy levels of below 1 m.
4. **Accurate channel modeling**: The availability of appropriate channel models for V2X communications constitutes a fundamental prerequisite for the V2X interface design. Due to the short coherence time and the limited coherence bandwidth, both channel estimation and data equalization are challenging in highly mobile environments. In [42] and [43], the state of the art in vehicular channel measurements and related models is presented. However, the presented modeling approaches are scenario-dependent and cannot be generalized to characterize a variety of scenarios (e.g., urban, rural, and highways) encountered in real-life vehicular environments. In addition, the modeling methodologies to date do not yet consider the direct communication between two moving vehicles in a dense urban environment, or the characteristics of V2X-specific network elements like RSUs. Overall, the choice of the appropriate channel model along with the proper combination of parameters for each of the V2X use cases constitutes an open challenge for radio designers.
5. **Distributed computing and network slicing**: Vehicles are progressively being equipped with an increasing number of sensors for object detection, velocity measurement, virtual imaging, or generation of HD maps. In emerging V2X use cases, there is a gradual trend to deploy higher-performance computing and storage devices on board; automated driving will transform vehicles into powerful computing and networking hubs for increased safety. However, it is still unclear how and where (i.e., locally, on RSUs, or centrally in the cloud) to efficiently process the large amount of on-board generated data. Mobile Edge Computing (MEC) has been recently proposed as an innovative computing paradigm to overcome the limited computation and storing capabilities of on-board units [44]. By deploying cloud-like infrastructure at the vicinity of the vehicles, efficient content (e.g., HD maps) caching can be achieved, reducing the data

streams infused to the network while a short response delay can be provided, e.g., for cooperative driving. In addition, Network Function Virtualization (NFV) enables dynamic computing and storing resource management among different MEC servers leading to a scalable architecture. Since software-defined network (SDN) control modules in MEC servers can separate the control plane from the data plane, multiple wireless access networks can interwork to support the increased traffic data in IoV scenarios, and various radio resources can be abstracted and reallocated to base stations.

6. **Security and privacy**: Future V2X communication will support diversified cooperative applications and services where autonomous vehicles will need to exchange various types of sensitive data, such as vehicle ID, position, and speed. A secure and privacy-aware architecture is thus required to guarantee the level of identities and data protection by ensuring the authentication of the message senders in vulnerable V2X scenarios, e.g., tracking of instantaneous vehicle locations, generation of false alerts and accidents, congestion, etc. [45]. At the same time, V2X security and privacy mechanisms should have minimum impact on the ongoing communication, e.g., the introduced latency for certified message signatures should not violate the end-to-end latency threshold.

5 Conclusions

The automotive industry is undoubtedly one of the key drivers of emerging 5G systems and beyond, with its unique features in terms of heterogeneity of end-users, stakeholders, and technologies, its diversified use cases, demanding application requirements, and unprecedented performance challenges. In this chapter, we have highlighted the fundamental role of wireless connectivity in the ongoing ITS transformation toward fully connected IoV systems with increased level of automation. The IoV relies on the cooperation between road users, vehicles, and RSUs to provide not only conventional V2X safety and infotainment applications but also advanced transport-related services, including autonomous and green driving. Examples of V2X use cases have been presented along with a classification of the main application requirements. In addition, the basic characteristics of the two present-day wireless technologies that are capable of supporting V2X communication are discussed. The stringent needs and fundamentally different characteristics of emerging ITS render indispensable the evolution of the existing technologies and require a major mentality shift on the way networks operate nowadays. To that end, we have analyzed the main research challenges related to the V2X radio interface and system design, and useful insights can be drawn for future research directions.

The potentials of emerging ITS have been acknowledged by the industrial and academic communities, and specifications targeting at ubiquitous V2X connectivity are developed by standard development organizations. Ongoing standardization efforts aim at enhancing the features and functionalities of IEEE 802.11- and cellular-based technologies to achieve the needs of advanced V2X use cases.

Besides the technical challenges of the new automotive landscape, the V2X ecosystem brings together a diverse set of stakeholders: the automotive industry, road infrastructure operators, mobile network operators, standards-developing organizations, policy makers, and end-users. A plethora of open research questions exists for which the answers have to be explored in concert with different V2X stakeholders. Close synergies among them are thus required to shape the future mobility concept and ultimately deliver a brand-new experience for drivers, travellers, and other road users. At the same time, a regulatory framework will be necessary to address ethical, legal, environmental, and safety aspects, while significant effort will be needed to foster end-users' acceptance, a prerequisite for the successful market launch of advanced V2X services.

Acknowledgments This work has been partially supported by the H2020 5GCroCo project under Grant agreement No. 825050, by the CHIST-ERA FIREMAN project funded by the Spanish Government (PCI2019-103780), by the Spanish MINECO under Grant SPOT5G (TEC2017-87456-P), and by the Generalitat de Catalunya under Grant 2017 SGR 891.

References

1. World Health Organization. Road traffic injuries. Available [Online]. https://www.who.int/news-room/fact-sheets/detail/road-traffic injuries
2. ETSI. Automotive Intelligent Transport Systems (ITS). Available [Online]. https://www.etsi.org/technologies/automotive-intelligent-transport
3. ETSI EN 302 665 V1.1.1 (2010–09) Intelligent Transport Systems (ITS); communications architecture. European Telecommunication Standards Institute, Sophia Antipolis
4. ETSI TR 102 638 V1.1.1 (2009–06) Intelligent Transport Systems (ITS); vehicular communications; basic set of applications; definitions. European Telecommunication Standards Institute, Sophia Antipolis
5. 3GPP, Study on LTE support for V2X services (Release 14), 3rd generation partnership project, Sophia Antipolis, technical report 3GPP TR 22.885 V1.0.0, Sept 2015
6. 3GPP, Service requirements for enhanced V2X scenarios (Release 15), 3rd generation partnership project, Sophia Antipolis, technical report 3GPP TR 22.186 V15.0.0, Mar 2017
7. 3GPP, 3GPP TR 22.886 v16.2.0: study on enhancement of 3GPP support for 5G V2X services (Release 16), Dec 2018
8. 3GPP TS 22.185 V14.3.0, Service requirements for V2X services – stage 1 (release 14), Mar 2017
9. 3GPP TS 22.261 V16.4.0, Service requirements for the 5G system; stage 1 (release 16), Jun 2018
10. 3GPP TR 38.885 V1.0.1, Study on vehicle-to-everything (release 16), Feb 2019
11. 3GPP Study Item Description RAN#80, Study on NR V2X, Jun 2018
12. ETSI TR 103 298, Intelligent Transport Systems (ITS); Platooning; Pre-standardization study, Apr 2016
13. ETSI TR 103 299, Intelligent Transport Systems (ITS); Cooperative Adaptive Cruise Control (C-ACC); Pre-standardization study, Apr 2016
14. ARIB STD T109-v1.2, 700 MHz band intelligent transport systems, Dec 2013
15. IEEE P802.11, Next generation V2X study group project authorization request, Available [Online]. http://www.ieee802.org/11/Reports/tgbd_update.htm

16. Kenney JB (2011) Dedicated short-range communications (DSRC) standards in the United States. Proc IEEE 99(7):1162–1182
17. Festag A (2014) Cooperative intelligent transport systems standards in Europe. IEEE Commun Mag 52(12):166–172
18. Festag A (2015) Standards for vehicular communication – from IEEE 802.11p to 5G. Elektrotech. Inform. 132(7):409–416
19. Wireless LAN Medium Access Control (MAC) and Physical Layer (PHY) Specifications amendment 6: wireless access in vehicular environments. IEEE, Standard 802.11p-2010 (2010)
20. Intelligent Transportation Systems Committee, IEEE trial-use standard for wireless access in vehicular environments (WAVE) – multichannel operation, IEEE, technical report, Feb 2011
21. ETSI EN 302 663 (2013) Intelligent Transport Systems (ITS); Access layer specification for Intelligent Transport Systems operating in the 5 GHz frequency band. V1.2.1
22. ETSI TS 102 687 (V1.1.1) Intelligent Transport Systems (ITS); decentralized congestion control mechanisms for Intelligent Transport Systems operating in the 5 GHz range; Access layer part
23. ETSI EN 302 636-3, Intelligent Transport Systems (ITS); Vehicular communications; GeoNetworking; part 3: network architecture
24. ETSI EN 302 636-5-1 V2.1.0 (2017–05), Intelligent Transport Systems (ITS); vehicular communications; GeoNetworking; part 5: transport protocols; sub-part 1: basic transport protocol
25. SAE J2735_201603 (2016–03) Dedicated Short Range Communications (DSRC) message set dictionary
26. ETSI (2014) Intelligent Transport Systems (ITS); vehicular communications; basic set of applications; part 2: specification of cooperative awareness basic service, EN 302 637-2, Nov 2014
27. ETSI (2014) Intelligent Transport Systems (ITS); vehicular communications; basic set of applications; part 3: specifications of decentralized environmental notification basic service, EN 302 637-3, Nov 2014
28. ETSI EN 302 895 (V1.1.1) Intelligent Transport Systems (ITS); vehicular communications; basic set of applications; Local Dynamic Map (LDM), Sept 2014
29. Papadimitratos P, Buttyan L, Holczer T, Schoch E, Freudiger J, Raya M, Ma Z, Kargl F, Kung A, Hubaux J (2008) Secure vehicular communications: design and architecture. IEEE Commun Mag 46(11):100–109
30. 3GPP, Initial cellular V2X standard completed, Release 14 initial C-V2X specification, 26 Sept 2016
31. Continental, Cellular V2X: continental successfully conducts field trials in China. Available [Online]. https://www.continental-corporation.com/en/press/pressreleases/2017-12-18-cellular-v2x-116994
32. FierceWireless, NTT Docomo, Ericsson and Qualcomm to carry out C-V2X trails in Japan, Press Release, 12 Jan 2018
33. Qualcomm, Groupe PSA and Qualcomm advance C-V2X testing for communication between vehicles, 21 Feb 2018
34. 5G Automotive Association, The case for cellular V2X for safety and cooperative driving, white paper, Nov 2016
35. Qualcomm, Accelerating C-V2X commercialization, Sept 2017. Available [Online]. https://www.qualcomm.com/media/documents/files/accelerating-c-v2x-commercialization.pdf
36. Naik G, Choudhury B, Park J (2019) IEEE 802.11bd & 5G NR V2X: evolution of radio access technologies for V2X communications. IEEE Access 7:70169–70184
37. Chen S, Hu J, Shi Y, Peng Y, Fang J, Zhao R, Zhao L (2017) Vehicle-to-everything (V2X) services supported by LTE-based systems and 5G. IEEE Commun Standards Mag 1:70–76

38. Kalalas C, Alonso-Zarate J (2020) Massive connectivity in 5G and beyond: technical enablers for the energy and automotive verticals. In: Proceedings of 6G Wireless Summit 2020 (6G Summit), Levi
39. Lien S, Kuo Y, Deng D, Tsai H, Vinel A, Benslimane A (2019) Latency-optimal mmWave radio access for V2X supporting next generation driving use cases. IEEE Access 7:6782–6795
40. Mei J, Wang X, Zheng K (2019) Intelligent network slicing for V2X services toward 5G. IEEE Netw 33(6):196–204
41. Kuutti S, Fallah S, Katsaros K, Dianati M, Mccullough F, Mouzakitis A (2018) A survey of the state-of-the-art localization techniques and their potentials for autonomous vehicle applications. IEEE Internet Things J 5(2):829–846
42. Matolak DW (2013) V2V communication channels: state of knowledge, new results, and what's next. In: Berbineau M, Jonsson M, Bonnin J-M, Cherkaoui S, Aguado M, Rico-Garcia C, Ghannoum H, Mehmood R, Vinel A (eds) Communication technologies for vehicles. Springer, Berlin/Heidelberg, pp 1–21
43. Viriyasitavat W, Boban M, Tsai HM, Vasilakos A (2015) Vehicular communications: survey and challenges of channel and propagation models. IEEE Veh Technol Mag 10(2):55–66
44. Porambage P, Okwuibe J, Liyanage M, Ylianttila M, Taleb T (2018) Survey on multi-access edge computing for Internet of Things realization. IEEE Commun Surv Tutorials 20(4):2961–2991
45. Ahmed KJ, Lee MJ (2018) Secure LTE-based V2X service. IEEE Internet Things J 5(5):3724–3732

UAV-Enabled IoT Networks: Architecture, Opportunities, and Challenges

Shahriar Abdullah Al-Ahmed, Tanveer Ahmed, Yingbo Zhu, Obabiolorunkosi Olaoluwapo Malaolu, and Muhammad Zeeshan Shakir

1 Introduction

Unmanned Aerial Vehicle (UAV), also widely known as drones, have proven very competent to numerous applications due to their low cost, flexibility and mobility. Initially, they have been used for military surveillance which now can be used by the public and other commercial applications and sectors. Some diverse applications of UAVs can be mentioned as providing medical supplies, environmental monitoring (e.g. air/water pollution, weather monitoring, forest fire detection and industrial applications), delivering products, rescue operation and emergency search [67, 82]. While delivering these services, we require a full duplex communication system for controlling and monitoring the environment from a remote distance. UAVs with on-board sensors, antennas and software are able to meet this requirement. Thus, UAVs are considered as part of the Internet of Things (IoT) [43].

IoT is considered as one of the emerging technologies in which independent smart devices can be utilised in any environment to monitor and exchange data amongst themselves. It is expected that there will be 25 billion IoT devices with unique identification by the end of 2020 [43]. These devices require optimal placement along with connectivity that provide a high Quality of Service (QoS). UAVs can be utilised here as a wireless infrastructure to provide connectivity for data transfer between IoT devices and control centre. In addition, optimal placement of the UAVs with integrated IoT platform (i.e. on-board sensors, cameras, etc.) can collect on-board data from anywhere because UAVs are able to reach most places including difficult to reach spots. Multiple UAVs can be connected together in order

S. A. Al-Ahmed (✉) · T. Ahmed, Y. Zhu · O. O. Malaolu · M. Z. Shakir
School of Computing, Engineering and Physical Sciences, University of the West of Scotland, Paisley, Scotland
e-mail: Shahriar.Al-Ahmed@uws.ac.uk; Muhammad.Shakir@uws.ac.uk

N. H. Mahmood et al. (eds.), *Wireless Networks and Industrial IoT*,
https://doi.org/10.1007/978-3-030-51473-0_14

to create UAV-enabled IoT networks to provide scalability. Wireless connectivity with relaying from UAV-to-UAV opens another dimension for IoT platforms where UAV network itself becomes wireless on-board IoT infrastructure. Notwithstanding the opportunities, this area is not free from challenges. Some challenges like 3-D placement of UAV-enabled IoT sensors to provide the best coverage, ad-hoc network topology, resource management, energy efficiency, safety and security will cause many disruptions while developing, organising and maintaining the UAV-enabled IoT networks. Furthermore, the most difficult challenge is air-to-air and air-to-ground channel modelling at various speeds, weather conditions and antenna direction for link establishment. This chapter focuses on many challenges and some suitable solutions for UAVs acting as wireless IoT systems and UAV networks collecting data from ground IoT devices which are considered as UAV-enabled IoT networks.

2 Overview of UAV-Enabled IoT Networks

In this section, we discuss about the UAV system, UAV regulations in different countries and UAV networks which are part of UAV-enabled IoT networks.

2.1 UAV System

As the name suggests, UAV is an aircraft without any pilot on-board but with a ground-based or on-board controlling system to be managed by humans or computers [28]. Most general UAVs consist of unmanned aircraft, payload, the human element, control elements and data link communication for duplex communication [50]. Nowadays, the UAVs are equipped with IoT devices for many applications which are making UAVs more powerful than before [66].

UAV has many distinct features, for example, altitude, length, width (wing span), weight, range, maximum payload, endurance and so forth. Based on the application, federal law and requirement of the QoS, one needs to use the relevant type of UAV. In general, there are three types of UAV based on altitude: low altitude platforms (LAPs), medium altitude platforms (MAPs) and high altitude platforms (HAPs) which can hover from a few hundred meters to 20 km [3, 39, 67].

HAP: These UAVs are used for Line-of-Sight (LoS) connectivity for a large-scale operation area. Airships or planes are good examples of this type of platform. The longevity for this type of platform to stay airborne could be from a few hours to a few years based on the fuel capacity, type, power constraints, etc. [3].

MAP: These platforms are used for relaying between HAP and LAP and stay airborne for a few hours. These platforms are also used by military operations due to their speed and endurance capabilities [3].

LAP: These types of platforms are in huge demand by everyone for their communication payload capabilities and rapid deployment advantages. These UAVs are also capable of carrying small sensors or IoT devices on-board [3].

Again, the UAVs can also be categorised into two types based on their wings: fixed-wing and rotary-wing. The first type of UAV can travel at a faster speed to cover larger distances, while the second type of the UAV can stay in steady positions [82]. For IoT applications, UAVs with rotary-wings or LAPs are preferable because these UAV can be dynamically and efficiently positioned while using very little power for the IoT to UAV communications [61].

2.2 *UAV Regulations*

UAV regulations have to be brought into attention when the UAV-enabled IoT networks need to be used for any kind of applications. There are many concerns related to the use and deployment of the UAVs with on-board IoT devices or UAVs to interface with ground IoT devices such as security, public safety and collision avoidance with UAV or airplane or any other object in the sky [62]. Every country has its own regulation system being continuously developed. In Table 1, we show some UAV regulations in different countries based on some criteria like applicability and operational restrictions mainly for LAP platforms [30].

UAV communication regulations have not been unnoticed by the communication regulatory bodies. The Electronic Communications Committee (ECC) within the European Conference of Postal and Telecommunications Administration (CEPT) has formed a group in 2015 where the group developed a report called ECC 268 in 2018 [30]. This report supported standardisation of dedicated frequencies for UAVs especially for professional use cases and un-licensed spectrum for non-professional UAV applications or short-range communications. In the USA, the Aerospace Industries Association (AIA) sent a petition to the Federal Communications Commission (FCC) for making a law for secure UAV communications. Public remarks need to be taken into account for decision-making by FCC [30].

2.3 *UAV Networks*

UAV communication network or UAV networks can be defined as the communication between multiple UAVs utilising its flexible functionalities and resources. An airborne network or swarm of UAVs when working together via relay and mother or head nodes can accomplish many challenging tasks [73]. For example, farmers can monitor a farm by deploying multiple UAVs with on-board IoT platform. The same deployment method can be used to monitor forest fire scenarios and can be referred to as airborne sensing [70, 85]. UAV networks are also able to be used for collecting different kinds of data from the ground IoT devices or sensors. In all cases, the UAVs will provide reliability, wide coverage and low latency with greater flexibility.

Table 1 UAV deployment regulations in some countries

Country	Limitations
UK	Highest altitude 122 m Lowest distance to people 50 m Lowest distance to airport N/A Operational time: N/A
Europe	Highest altitude 120 m Lowest vertical distance to people's property 20 m Lowest distance to airport N/A Operational time: N/A
USA	Highest altitude 122 m Lowest distance to people N/A Lowest distance to airport 8 km Operational time: N/A
Australia	Highest altitude 120 m Lowest distance to people 30 m Lowest distance to airport 5.5 km Operational time: daytime only
South Africa	Highest altitude 46 m Lowest distance to people 50 m Lowest distance to airport 10 km Operational time: daytime and clear weather
China	Highest altitude 120 m Not allowed in densely populated area Lowest distance to airport N/A Not allowed in no-fly-zones (e.g. Beijing) Operational time: daytime only

2.4 Applications and Use Cases

The fundamental advantage of UAVs which is flexibility brings lots of use cases and applications. Some applications and challenges are briefly discussed here:

- *Crowd surveillance*: Crowd can be monitored with appropriate IoT devices fitted in UAVs for any large events instead of sending large security services. Any incident can be monitored by the security staff from a centralised location. The incident or suspicious person can be identified, and photos can be taken until a security agent is reached to a certain location [56].
- *Real-time road traffic observation and other uses*: UAVs can be used to monitor real-time traffic conditions or road accidents and send traffic or relevant information to a central server where road traffic agencies and commuters can analyse the data and pick their best route [26]. Moreover, UAVs with

on-board IoT or sensors to measure temperature, humidity and other ambient environment conditions can be used to get accurate weather or environment information for a wide range of audiences such as holidaymakers, farmers, mountain hikers, etc.
- *Disaster management*: For natural disasters in any area including remote areas, cost-effective, reliable, quick and efficient infrastructure is required to carry out risk assessment, evacuate people, search for victims and establish communication. UAV-enabled IoT networks can help to identify disabled people and may assist to find out the missing items from its electromagnetic emissions of any sufferer concealed under the destroyed building or heavily dense forest [58].
- *Earthquake cases*: If any earthquake hits any region, UAV-enabled IoT networks can be used to hover around that area and continue to collect data to analyse the damage, pollution, weather and environmental information. UAVs can also be used to deliver food or medicines to the victims by using on-board sensors; for instance, cameras or location sensors can be used for location estimation. During the East Japan earthquake in 2011, UAVs were utilised for measuring radiation levels of Fukushima nuclear power plant out of many other purposes [68].
- *Smart City*: In a rapidly growing population areas, UAV-enabled IoT networks can provide excellent opportunities to create smarter cities. Such networks can monitor traffic, pollution levels and weather. Moreover, it is possible to perform rescue operation and offer support for police, and they can also be utilised for geo-spatial and surveying, civil security control, fire fighting and so forth [54].

3 UAV-Enabled IoT Systems

This section presents the architecture of the UAV network along with the brief discussions on routing design and protocols for UAV networks.

3.1 Architecture

The UAV-IoT network architecture completely relies on the application or use case and requirement. Basically, there are two types of UAV-enabled IoT networks. First one is with on-board sensors or IoT devices which uses UAV-to-UAV communication to connect with the remote control centre. The second one is where the UAVs collect data from ground IoT devices or sensors and use IoT-UAV communication for data collection and UAV-to-UAV communications to establish connection with remote control centre. Here, the ground IoT devices or sensors can be static (e.g. smart home monitoring) or moving (e.g. cars, trains or electric vehicles). In most

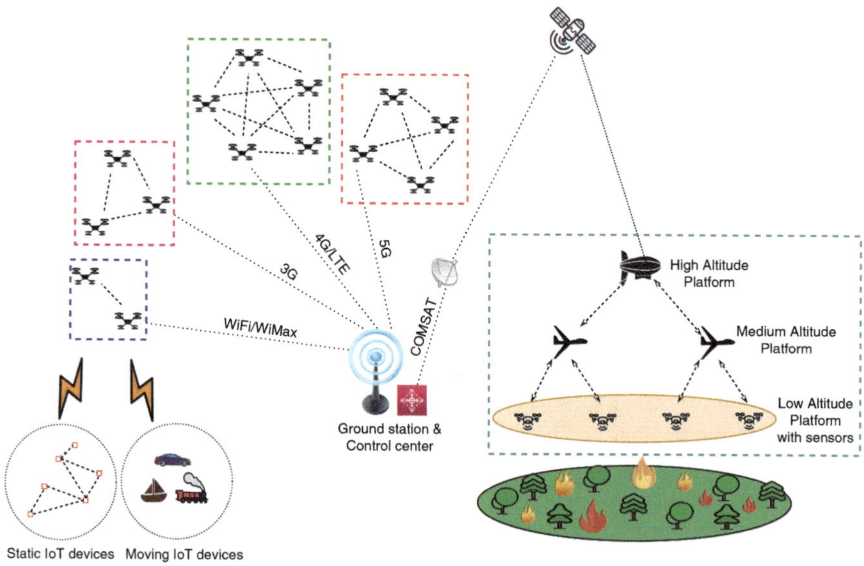

Fig. 1 UAV-enabled IoT networks architecture for various heterogeneous types of the IoT services

cases, the UAV network could be based on a single layer or comprises several layers with one mother or head UAV and ground control systems. One important feature of this network is to reorganise and maintain communication if one of the UAV nodes fails [13]. Other features like collision avoidance for multiple UAVs, route planning and optimisation can be used to get the best result of the UAV deployment. The main topology for the UAV wireless network architecture is considered as star or mesh. Figure 1 shows the illustration of the heterogeneous UAV-enabled IoT network based on [11, 58, 73].

There are many challenges for this type of UAV-enabled IoT networks architecture. Self-organisation, while increasing or decreasing the number of UAVs in the cluster, is a very challenging task while maintaining the security and safety of the public. Full duplex communication to the ground base station is another challenge as many UAVs will try to connect at the same time. Other challenges can be mentioned such as route planning and optimisation, interference management and reliable connection from UAV-to-UAV under diverse weather conditions.

3.2 UAV Communication System

The communication systems assist the UAVs and IoT to manage, control and acquire data in UAV-enabled IoT networks. This system could be between UAV-to-UAV, UAV-Ground or UAV-IoT and Ground-UAV for simplex or full duplex

Table 2 Types of communication mechanism for UAV-based IoT systems

Protocol	Type	Physical layer specs	Data rate	Transmission range
BLE 4.0	WPAN	2.4 GHz, FHSS/FSK	<1 Mbps	<60 m
Z-Wave	WPAN	800–900 MHz, FSK/GFSK	<100 Kbps	<100 m
Zigbee	WPAN	2.4 GHz, OQPSK	<250 Kbps	<100 m
IEEE 802.11a/b/g/n	WLAN	2.4 GHz–5 GHz, DSS/OFDM	<600 Mbps	<250 m
IEEE 802.11p (WAVE)	WLAN	5.9 GHz, OFDM	<27 Mbps	<1 km
LoRa	LPWAN	433–915 MHz, CSS	<50 Kbps	<15 km
SigFox	LPWAN	868–928 MHz, DBPSK/GFSK	<100 bps	<20 km
IEEE 802.16 (WiMAX)	MAN	2–66 GHz, MIMO-OFDMA	<2–75 Mbps	<56 km
NB-IoT	Cellular	Operator defined, OFDMA/SC-FDMA	<200 Kbps	<Cell range
LTE-M	Cellular	Operator defined, OFDMA/16-QAM	<1 Mbps	<Cell range
LTE-A/4G	Cellular	700–2500 MHz	<1 Gbps	<Cell range
5G	Cellular	600 MHz-6 GHz/24–86 GHz	<10 Gbps	<Cell range
Satellite (LEO/MEO/GEO)	WAN	1.53–31 GHz, FDMA/TDMA	<1 Gbps	Worldwide

communication depending on the applications or use cases and requirements. A wide range of communication mechanisms and network protocols can be found in [72]. Some emerging types of suitable communication systems for UAV-enabled IoT networks have been summarised in Table 2 from [36, 43].

3.3 *Routing Protocols*

Routing protocols are necessary in order to maintain efficient transmission of packets between UAV nodes. It has major challenges especially when a large number of UAVs are deployed. The nodes are dynamic as well as may be uneven due to the distribution methods. The range restriction between the base station and nodes can also be taken as a challenge [13]. All of these challenges may cause packet loss and link failure in limited bandwidth scenario which will lead to inefficient network deployment.

UAV networks have their features for routing depending on the requirement and deployment in any scenario. However, it is still very important to avoid packet loss

Table 3 Different routing design for UAV-to-UAV communication

Routing design	Description
Unicast	Direct hop to hop
Broadcast	Flooding messages over the network
Multicast	One hop to many hops
Geocast	Multicast based on geographical location
Cooperative routing	UAV node assist each other by creating a relay
Path discovery	Selecting the best path from all the possible path from source to destination
Single path	Singe path is used calculating predefined routing tables
Multiple paths	Multipath routing table gives choices if there is any problem in the network

and perform relay selection to maintain a secure network. Some routing design techniques have been very briefly discussed in Table 3 from [13].

The other routing design for data communication can be mentioned as grid-based routing, quorum-based routing, store-carry and forward, greedy forwarding and prediction [13]. Routing protocols are also important along with routing design. Some routing protocols that have been studied are topology-based, cluster-based, deterministic and stochastic routing protocols. Many studies have been carried out for routing protocols, and their summary can be found in [13].

4 Propagation Channel Modelling

In this section, we discuss regarding the propagation characteristics of UAV. Modelling of UAV propagation channel can be classified into two broad categories:

1. Empirical channel modelling
2. Analytical channel modelling

4.1 Empirical Channel Modelling

Due to cruising of UAV, channel parameters can change frequently from time to time. Many research have been carried out in order to get the most accurate propagation channel model. We can classify the empirical channel model into the following three types:

1. Air-to-Air path loss model: for UAV-to-UAV communication including with or without on-board IoT devices or sensors
2. Air-to-Ground path loss model: for communication between UAV and IoT devices as well as ground base stations.
3. Ground-to-Air path loss model: for IoT devices to UAV uplink communication.

Air-to-Air Propagation Channel Model

The difference between air-to-air channel modelling and air-to-ground channel modelling is simple and easily manageable. Air-to-air channel modelling has very-low-multi-path fading and is less dependent on the ground surface. However, it has a higher rate of Doppler effect due to significant relative velocity within the UAVs [46]. Several studies have been carried out to characterise air-to-air channel modelling. The air-to-air channel modelling used in between UAVs can be found in aerial wireless sensor networks [4], UAV swarm networks [34], flying ad-hoc networks [16] and wireless backhaul networks [38]. The UAV propagation characteristics depend upon the environmental condition, flight direction, alignment of the LoS, ground reflections and relative velocities. The authors in [4] empirically characterised and used 802.15.4 (ZigBee) wireless sensor networks and showed that the air-to-air communication is better in terms of path loss and the signal strength increases [4] with altitude and decreases with distance.

The authors of [78] have used IEEE 802.11 for multi-hop UAV networks for mainly air-to-air communication, whereas they used the same communication system for single-hop UAV networks for the single UAV and ground station communication. They have also considered log-distance propagation model for higher throughput and longer distance in air-to-ground propagation channel. The authors of [33] carried out their studies both in lab and outdoor environment mainly for altitude dependant multi-path propagation in the air-to-air channel by extending the UAV specific Rice model. The authors of [80] analysed the path loss for horizontal and vertical distances in air-to-air and air-to-ground channels. In addition, the authors in [86] have studied the air-to-air model for UAV-to-UAV using a 3-D Ellipsoid model.

The most recent work of low-altitude UAV air-to-air channel modelling can be found in [46]. In this paper, the authors demonstrated the extension of the log-distance path loss model which is given below:

$$PL(r) = PL(r_O) + 10\varphi \log_{10}\left(\frac{r}{r_O}\right) + X_\delta + X_O \tag{1}$$

where $PL(r)$ denotes the path loss at a certain distance r, φ denotes path loss exponent, X_δ represents the shadow fading which is a zero mean Gaussian distributed random variable and X_O denotes additional fading. The path loss exponent φ changes with the altitude h of the UAV. In expression (1), variable φ can be formulated as:

$$\varphi = a \times h^b + c \tag{2}$$

where $a = 2.598$, $b = -0.5268$ and $c = 1.945$. Furthermore, the authors also used root mean square error ($RMSE$) value of 0.003937 to get low prediction error.

Air-to-Ground Propagation Model

For air-to-ground channel modelling, the signal intensity changes with time depending on the frequency. In terms of signal intensity, the communication model can be divided into two types:

1. Path loss and large-scale fading model
2. Small-scale fading model

Path Loss and Large-Scale Fading Model

Large-scale fading usually occurs when an obstruction comes in between the LoS of UAV and ground base station where it is larger than the wavelength. If there is no obstruction, then the fading effect occurs with the two ray variations from earth surface to multi-path component [41]. In some studies, many researchers discussed the path loss or shadowing (if present) in various scenarios. Air-to-ground channels in LoS system, path loss modelling begins with free-space path loss (FSPL) due to earth surface reflection. In this scenario, the path loss is described by the two ray models. Other measurement results found in the literature use the path loss model where signal is lost proportionally with the distance, mentioned as path loss exponent (PLE) model. The authors in [51] have examined the two ray models whereas the authors in [52, 79] have explored the path loss model and large scale fading in urban environment. It was observed in [79] that PLE for IEEE802.11 was different during hovering and moving of UAV due to different orientations of on-board antennas which produce different antenna patterns and can distort main path loss characteristics which in turn make it difficult or impossible to compensate path loss. Besides, the authors of [32] have proposed the distance and frequency independent path loss model for urban and rural areas, whereas the authors of [71] have suggested the path loss model depends upon distance in the 3-D plane and operating frequency. Reference [12] suggests an altitude-dependent path loss model, and reference [6] introduced angle-dependent PL model.

Based on the LoS probability, one of the simplified path loss model is presented here from [25]. The path loss between a UAV and ground IoT devices relies on the position of the UAV, IoT devices and environmental parameters, for example, rural, suburban, urban and extra-urban. Path loss based on LoS between jth UAV and ith IoT device can be given as:

$$P_{ji_{LoS}} = \frac{1}{1 + \alpha \exp\left[-\beta\left(\frac{180}{\pi}\arctan\frac{r_{ij}}{h_j} - \alpha\right)\right]} \tag{3}$$

In the above expression, α and β denote the constant values dependent on environment, h_j is the height of the UAV and $r_{ij} = \sqrt{(x_i - x_j)^2 + (y_i - y_j)^2 + h_j^2}$ is the horizontal distance between the IoT device and the UAV where $i \in I$, (x_i, y_i)

and $j \in J, (x_j, y_j, h_j)$. In such a situation, the calculation of Non-Line-of-Sight (NLoS) can be written as $P_{ji_{NLoS}} = 1 - P_{ji_{LoS}}$.

Small-Scale Fading Model

This model is applied in narrowband channels or individual multi-path components or tapped delay line wideband models. According to the authors of [49], the depth of small-signal amplitude varies inversely with the bandwidth. Small-scale Stochastic fading models can be developed via analysis and empirical data or through geometrical analysis and simulations. Some commonly used models for small-scale fading distribution are:

Loo Model: For narrowband air-to-ground propagation channel, the authors of [69] studied fading statistics in urban areas using the Loo Model.

Rayleigh Model: It is also known as the Rayleigh scattering environment where the authors of [2] tested relay-based UAV systems. Another study [37] suggests several multiple access ground-to-air channels which can be modelled with Rayleigh fading.

Rician Model: This model is used to approximate the fluctuations in the fading channel with the LoS. The authors of [79] used this model for high-altitude UAVs, and authors of [20] used this model for scattered multi-path environment. For air-to-air channel characterisation, the authors of [33] observed an altitude-dependent Rician K-factor multi path components (MPC) fading due to ground reflected. Their results proved the value of K increases as the altitude increases.

Nakagami-m Model: The authors of [2] found that this model is appropriate for high-altitude applications. The authors in [1] derived an equation to calculate the outage probability of cooperative UAV network using Nakagami-m model. Besides, it has been suggested in [88] that Nakagami-m has the opportunity to estimate the Rician fading.

Ground-to-Air Propagation Model

UAV and IoT devices deployment can co-exist in any environment such as extra urban, urban, suburban and rural. In this type of scenarios, it is difficult to get all kinds of relevant parameters (e.g. altitude, elevation angle) to be known. Therefore, every IoT device has a LoS link probability to a UAV for ground-to-air communication due to randomness of the nature. For this type of uplink communication by utilising Orthogonal frequency-division multiple access (OFMDA) modulation scheme, the suitable expression is similar to expression (3). In that expression, it is obvious that the LoS probability becomes better if the altitude of the UAV is raised. So, the path loss for ith IoT device to jth UAV becomes [19]:

$$L_{ij} = \begin{cases} \mu_1 \left(\frac{4\pi f_c r_{ij}}{c} \right)^{\gamma}, & \text{LoS link,} \\ \mu_2 \left(\frac{4\pi f_c r_{ij}}{c} \right)^{\gamma}, & \text{NLoS link,} \end{cases} \tag{4}$$

In the above expression, f_c denotes the carrier frequency, c represents the speed of light, γ stands for the path loss exponent and μ_1 and μ_2 denote the extreme path loss coefficients where $\mu_2 > \mu_1 > 1$ in LoS and NLoS circumstances. Bearing in mind, the probability of NLoS is $P_{ij_{NLoS}} = 1 - P_{ij_{LoS}}$. Also, there is a fact that it is impossible to figure out which path loss model is encountered by the IoT device-UAV uplink communication. For this reason, the average path loss is taken into consideration. Then, the average path loss using (3) and (4) can be expressed as:

$$\overline{PL}_{ij} = P_{ij_{LoS}} L_{ij_{LoS}} + P_{ij_{NLoS}} L_{ij_{NLoS}} \tag{5}$$

$$= P_{ij_{LoS}} \mu_1 \left(\frac{4\pi f_c r_{ij}}{c} \right)^{\gamma} + P_{ij_{NLoS}} \mu_2 \left(\frac{4\pi f_c r_{ij}}{c} \right)^{\gamma}$$

$$= \left[P_{ij_{LoS}} \mu_1 + P_{ij_{NLoS}} \mu_2 \right] \left(\kappa_o r_{ij} \right)^{\gamma} \tag{6}$$

where $\kappa_o = \frac{4\pi f_c r_{ij}}{c}$. Apparently, the average channel gain between IoT device and UAV is $\overline{G} = \frac{1}{\overline{PL}_{ij}}$. So, average $\overline{G}$ can be used as to model interference and to compute desired link. For all IoT devices-UAV, signal to interference and noise ratio (SINR) will have the advantage of not considering separate LoS and NLoS links [61].

4.2 *Analytical Channel Modelling*

Analytical channel models are useful for channel characterisation with certain assumptions and parameters. It can also predict the performance of the communication systems. They are classified into three main types [41]:

1. Deterministic models
2. Stochastic models
3. Geometric based stochastic models

Deterministic Model

These models are environmental dependent and contain information including obstruction materials, buildings electrical parameters and terrain topography. They can be developed by ray-tracing software which simulates the path loss and shadowing effects. The authors of [22] performed altitude-dependent channel modelling in the air-to-ground propagation for the suburban environment. The authors in [29] designed air-to-ground characterisation for frequencies ranging from 200 MHz to 5 GHz considering the altitude between 100 m to 2 km.

Stochastic Model

They are designed for the Tapped Delay Line (TDL) system with multiple taps. For each tap, the fading statistics of multi-path components can be derived from channel impulse response which can analyse empirically from measurement and numerically by simulation software. The authors in [52] developed wideband stochastic models for different environments, and the authors of [81] developed stochastic models for the narrowband assumptions for air-to-ground channel.

Geometry-Based Stochastic Model

Geometry-based Stochastic modelling technique can provide spatial-temporal channel characteristics with stochastic output in a 3-D geometric simulated environment. Its accuracy depends upon the simulation of a virtual environment with geometrical shapes (i.e. cylindrical, elliptical) where the communication nodes within the scattering region follow a certain probability distribution. The authors of [64] studied geometric-based channel model for the analysis of air-to-ground radio communication.

5 Challenges for UAV-Enabled IoT Networks

Like cellular networks, resource management and energy efficiency are very crucial for any UAV network and pose many challenges. These challenges will occur due to UAVs flight time, path planning, deployment, mobility and air-to-ground and air-to-air interference [43]. Therefore, coordinating and optimising resources like fleet, frequencies, directional antenna and integrated sensors in complex UAV networks can contribute to overcome these challenges. Although there are a number of studies related to UAV-enabled IoT networks, still there are some issues which require further investigation. This section sheds lights on challenges, their possible solutions and research directions related to UAV-enabled IoT networks.

5.1 Optimal Deployment

Optimal 3-D deployment of UAV-enabled IoT networks faces some critical and fundamental challenges due to the mobility and flexibility of UAV in the air. The topology of the network will depend on the 3-D deployment method which will allow the network to perform at its best by providing maximum coverage and capacity. Deploying multiple UAVs opens another door of significant attention. In addition, some factors like the geographical location, environment and the mobility status of the ground IoT devices in urban, suburban and rural areas also need to be taken into consideration [39, 53]. To overcome these challenges, an optimal transport theory method has been presented in [59] for moving UAVs to collect information from ground IoT devices. Machine learning-based multiple UAV networks have been presented in [47]. Reference [21] presented an efficient technique of multiple UAV deployment for IoT communication based on game theory and distributed position optimisation algorithm. Furthermore, UAV deployment strategy for other applications such as cellular communication can also be applied here using centralised, distributed and heuristic optimisation algorithms.

5.2 *Trajectory Design*

The trajectory design of UAVs is another prime challenge. Energy constraints, number of UAVs, environment, type of tasks, collision avoidance and time of flight are the main factors to be calculated efficiently while designing the trajectory path. The other main factors such as the location of the IoT device, type of IoT (e.g. moving or static, transmission system, etc.) and the characteristics of the UAV (i.e. type, weight and capacity) also remain major issues while designing trajectory path. One benefit of moving UAVs is that it can increase the required coverage than the static UAVs [48]. This is also a challenge to determine how much coverage can be provided by moving UAVs. To resolve the issues, the authors in [84] have presented a little complex UAV trajectory optimisation method to collect data from wireless sensor network on the basis of Travelling Salesman Problem (TSP). For static and dynamic scenarios, the authors in [42] have presented UAV trajectory optimisation algorithm founded on quantisation theory. Authors in [15, 23] have studied the UAV path planning inspired by machine learning, centralised algorithms which could be applied here.

5.3 *Resource Management*

The UAVs with integrated or on-board sensors have some challenges of collision. UAVs equipped with IoT devices such as sensors and cameras should be self-organised to avoid this problem. Again, the interference can be minimised by adjusting the beam width of the directional antennas along with the altitude for the coverage requirement [43]. The authors in [40] have introduced a heuristic algorithm for optimal 3-D placement of the UAV which reduces the quantity of the UAV to maintain a certain level of QoS. The same authors also suggested lowering the altitude to mitigate the interference; however, there is a trade-off between altitude and coverage. Concerning path planning, it is very challenging as there are many factors associated with path planning such as battery power, flight duration and obstacle avoidance. Path planning is considered as an optimisation problem with one or many goals as stated above as per requirement.

5.4 Energy Efficiency

To get the maximum output from any battery-operated infrastructure, energy efficiency has to be taken on-board seriously. Simple IoT devices like health monitoring or soil monitoring sensors have limited battery power on the ground. These devices can transmit the signal up to a limited distance. On the other hand, UAVs are also battery operated and have their energy consumption for controlling, hovering itself and maintaining the network which is also dependent on the environment and operational parameters such as airspeed during adverse weather condition. Furthermore, additional energy consumption due to the on-board IoT sensors create another issue. Several studies have been carried out to minimise the energy consumption of these UAV networks. The authors in [59] have demonstrated in their results that 56% transmit power reduction is possible by exploiting optimal transport theory compared to fixed Voronoi deployment method for IoT to UAV uplink communication. The authors in [55] have proposed appropriate UAV selection for specific IoT requirements depending on the conditions such as geographical location, energy budget and so forth. They have used energy aware and delay aware UAV selection method to ensure energy efficiency. The authors in [76] proposed an interesting approach that is to charge the battery-operated ground IoT devices via radio frequency. The authors demonstrated that uplink communication performances are enhanced by 20% in comparison with other existing methods. Furthermore, the authors in [44] have proposed UAV-IoT association for energy efficiency using regret matching algorithm. For UAV-based IoT network, trajectory optimisation using optimal transport theory is a key concept to minimise energy consumption [59]. The authors in [57] proposed energy-aware, delay aware and fair trade-off for UAV selection which is a similar approach like [55] but for UAV with integrated IoT devices. Energy-aware UAV selection aimed to reduce the total energy consumption of UAV, while delay aware UAV selection mainly reduces UAV operation time. Furthermore, the authors in [34] have discussed many aspects of reducing power consumption, for example, battery weight, payload weight, the transmission power of nodes, energy-efficient routing, making some nodes to go sleep and most importantly energy efficiency in different layers of communication.

5.5 Safety Operation

Every flight that is manned or unmanned has some risks. Like manned flights, the mid-air collision of UAVs may cause severe injuries to the general outdoor or indoor public. This collision may occur due to many external and internal factors. External factors can be interference, environmental conditions, navigational environment, communication and air traffic environment. Moreover, internal factors include mechanical, thermal, electronic, algorithmic, technical, hardware and software [7]. Besides, physical obstacles (birds, trees) may increase the chances of collision in the UAV network. Theft and vandalism are other major concerns especially for low-altitude UAVs as they are a very tempting target that can be grasped using an anti-drone rifle or light dart gun when they are in noticeable flying distance [9, 35]. Acquiring the data and hardware inside them will top up the temptation to steal or destroy any UAVs. Introducing the hostile UAV is another recent idea to ground the UAV where the hostile UAVs attaching with fishing net can catch the other UAVs physically [9].

5.6 Airborne Security

The security threats could pose in both physical and operational part depending on the nature of the network: UAV-to-UAV with integrated IoT sensors or multi-UAVs and IoT sensors to UAV networks. Some threats are given below:

Networking Security: The collaboration amongst multi-UAVs with integrated sensors may suffer from the jamming attack, man-in-the-middle attack, fabrication attack and so on [87].

Memory and Computation: The limitation of the computational facility and memory does not allow UAVs to take a high level of cryptology on-board [9]. On the other hand, the IoT devices have very limited or no computational facilities which becomes another major security threat.

Limitation of Power: Attacking the power system will deactivate the node in the network as UAVs have very limited power.

User Knowledge: Insufficient user knowledge may be another huge threat for using UAV integrated IoT devices or IoT devices only. Attackers will be able to intercept, control and destroy the whole system very easily because of inappropriate deployment.

External Attacks: There are several kinds of attacks that vary from physical layers to application layers by unwanted users or hackers. In the physical layer, jamming attacks and sniffing attacks can be performed wherein the link layer selfishness usage could bring similar aftermath like Denial of Service (DoS) attack. Wormhole attacks, blackhole attacks and impersonation and repudiations are common in the link layer [17]. At last, some useful tools such as firewalls are not present in the application layer which will lead to insecure and unfiltered messages[16]. There are countermeasures for these types of concerns. However, these issues are still open to research and need to be developed further in the future.

5.7 Privacy Issues

UAVs are causing numerous privacy issues and making it more serious due to the IoT extension. Some privacy issues are given below:

Illegal surveillance: UAVs can take photos or images secretly while flying over a certain area. In the current policy, the UAV must have a registered name [87]. Again, a webcam installed indoor/outdoor can be accessible by unauthorised users while uploading to the UAV network.

Harmful software: Unsafe and harmful software or plugin may leak the information from the UAVs to the eavesdroppers.

Privacy disclosure: The metadata and location of the shooting of any taken photos can be acquired by unwanted users [87]. It can happen with the IoT sensors installed on-board.

6 UAV-IoT Communication System: Mathematical Models to Address the Challenges

Looking at the propagation channel modelling of UAV-IoT communication system in the network, this section emphasises on mathematical tools to design, analyse and optimise the various challenges discussed in previous section for UAV-enabled wireless networking. Some challenges, for example, trajectory design and energy efficiency , have the opportunity to be optimised using mathematical tools and algorithms like machine learning, optimisation theory and optimal transport theory for the design of UAV-IoT communication systems.

6.1 Machine Learning

Machine learning empowers the framework to improve and maximise the output by enabling them to do their job automatically without any human in the loop model. The basic three types of machine learning technologies are supervised learning, unsupervised learning and reinforcement learning. The future challenges and future opportunities for these types of machine learning are given below:

(a) *Supervised Learning*: In simple words, supervised learning is used to classify and predict the output by comparing it with model data and input. This type of machine learning can be used for UAV network communication link classification for ultralow latency communication [73, 74], image detection, rescue operation, weather forecast, agricultural sector forecasting and management (crop management, soil management, water and irrigation management, species detection and breeding, etc.), IoT to UAV data classification and many more.
(b) *Unsupervised Learning*: This type of machine learning is used to find matched patterns without labelled output from the given input data. This type of machine learning can be used to detect the presence of unknown UAV, fault diagnosis, anomaly data detection, enhancing the security at physical level and so on [43].
(c) *Reinforcement Learning*: This type of learning helps UAVs to learn from its past experience. Some of the applications of reinforcement learning and deep neural network include UAV optimal placement, autonomous UAV for avoiding the collision, remote sensing, image sensing, energy efficiency

6.2 Optimisation Theory

Optimisation theory can contribute with the deployment and movement of the UAVs to connect with the maximum number of IoT devices on ground. Path planning is straightforwardly related to trajectory optimisation. In general, finding the ideal flight path for a UAV is viewed as a difficult objective since it is influenced by different components, for example, energy limits, flight time and obstacle avoidance. Path planning is generally considered as an optimisation issue with many objectives depending on the criteria of interest. In light of their limited battery constraint, UAVs are not regularly capable of accommodating long persistent wireless coverage in scenarios. Their energy autonomy is exceptionally influenced by the UAV role, flight path and climate conditions [62].

6.3 Optimal Transport Theory

Optimal transport theory is a field in arithmetic that reviews situations where goods are moved between different areas. This theory can solve many complex situations of the UAV-IoT wireless framework. Examples can be given such as minimising the total required power to transmit data to the UAVs and maintaining IoT device requirements, IoT device association, resource allocation and the optimal trajectory to ensure the guaranteed uplink transmission in UAV networks [60]. Thus, we will acquire the maximum performance of any framework in terms of latency, throughput and energy efficiency. By exploiting the new concepts from probability and statistics theories, this theory empowers capturing generic distributions of wireless UAVs, which in turn permits a more profound central examination of network performance optimisation than the heuristics algorithms.

6.4 Stochastic Geometry

Stochastic geometry is one of the powerful tools to analyse the performance of any system like ad hoc or terrestrial [62]. For UAV-IoT communication, the Matern cluster point process is one of the suitable tools to evaluate the performance. The other tools are the Matern hardcore process, Poisson Boolean model and Binomial and Poisson cluster which also assist to identify the performance of these UAV networks.

6.5 Game Theory

The combination of machine learning and game theory can contribute to the core of distributed decision-making in UAV networks. Game theory is a standard model or tool to investigate distributed resource management (i.e. UAVs, ground station, IoT devices) and trajectory/path planning optimisation. Besides, other issues such as energy efficiency, hovering time, spectrum allocations and optimal 3-D placement may trigger the use of a multi-game approach. Other models such as stochastic differential games can be applied to control and manage communication for autonomous UAV systems, coalitional game theory to form swarms of UAV for cooperative operations and matching theory to settle the network planning issues. The other relevant game theories can be considered are contract theory and matching theory for network planning [62].

6.6 Other Algorithms

Heuristic algorithms such as A* algorithm, Dijkstra's algorithm and Floyd's algorithm can be exploited in order to design the trajectory path of the UAV [31]. Some intelligent optimisation algorithms such as ant colony optimisation, particle swarm optimisation, genetic algorithms and artificial neural networks can also be used in order to deploy multi-UAVs to work cooperatively in a UAV network.

7 Conclusions

In this chapter, we have presented an overview of the UAV system along with UAV networks, a possible architecture, the communication system and communication modelling for the architecture such as air-to-air, air-to-ground and ground-to-air out of many elements for UAV-enabled IoT networks. For each case, we have provided a short discussion, some issues and possible solutions with an illustration. We have also discussed some important concerns of this system deployment, for example, networking, resource management, energy efficiency as well as safety security and privacy aspects. Table 4 provides a summary of the key challenges, possible solutions, important references and mathematical tools that are discussed. Future research implementation opportunities include, but not limited to, UAV antenna placement and antenna orientation, artificial intelligence integration for their operations, ultra-reliable and low latency machine to machine communication, and UAV integration with Long Range (LoRa) protocols. UAV-enabled IoT networks also have a huge opportunity to be applied in the marine sector to facilitate the wireless communication over seawater and air-to-ground communication. Last but not the least, some other emerging technologies such as mmWave communication, 3-D beamforming, mobile edge computing, wireless power transfer (e.g., lasers) and caching, will need further research to enhance the overall system efficiency.

Table 4 Objectives, challenges and tools in the future direction

Objective	Challenges	References	Tools and future directions
Deployment	Energy-efficient deployment		Machine learning
	Optimal 3-D deployment		Game theory
	Multiple UAV deployment	[10, 19, 21, 31, 39, 47, 59–62]	Optimisation theory
	Maximise the coverage		Centralised algorithm
	Maintain the topology		Distributed motion control algorithm
	Track IoT devices on ground		Facility location theory
Trajectory design	Energy-efficient trajectory design		
	Time of flight		
	Collision avoidance		Optimisation theory
	Location of the IoT devices		Optimal transport theory
	Static or moving IoT	[15, 18, 23, 31, 42, 48, 65, 77, 84]	Heuristic algorithms
	Coverage provided by static or moving IoT		Machine learning
	Transmission systems		
	Connecting multiple UAVs		

(continued)

Table 4 (continued)

Objective	Challenges	References	Tools and future directions
Channel modelling	Air-to-air channel model Air-to-ground channel model Real-world measurement predicated on environment factors Doppler effect	[14, 43, 45, 61, 75, 83]	Extensive measurement Ray tracing techniques Machine learning
Resource manage-ment	Fleet management Low bandwidth Frequency selection Antenna directions Dealing with interference Performance analysis	[24, 27, 40, 43, 62, 73, 74]	Game theory Machine learning Centralised algorithms Optimal transport theory
Energy efficiency	Limited battery power of the UAV Energy consumption by on-board IoT Energy consumption for communication Energy consumption due to environmental and operational factors	[34, 44, 55, 57, 59, 76]	Machine learning Optimisation theory Optimal transport theory Game theory
Safety operation	Injuries to the public due to collision Theft and vandalism Breaking into UAV for data and hardware	[5, 7–9, 43, 63]	Machine learning System integrity Accountability of action Training and education
Airborne security	Networking security Memory and computation Limitation of power User knowledge External attacks	[5, 8, 9, 16, 17, 43, 63, 87]	Machine learning Air policing Authorised access Information confidentiality Securing unmanned systems
Privacy issues	Illegal surveillance Harmful software installed Privacy disclosure	[5, 8, 43, 63, 87],	Location access control UAV tracking system

References

1. Abualhaol IY, Matalgah MM (2006) Outage probability analysis in a cooperative UAVs network over Nakagami-m fading channels. In: IEEE Vehicular Technology Conference. IEEE, pp 1–4
2. Abualhaol IY, Matalgah MM (2010) Performance analysis of multi-carrier relay-based UAV network over fading channels. In: 2010 IEEE Globecom Workshops. IEEE, pp 1811–1815
3. Ahmadi H, Katzis K, Shakir MZ (2017) A novel airborne self-organising architecture for 5G+ networks. In: 2017 IEEE 86th Vehicular Technology Conference (VTC-Fall). IEEE, pp 1–5
4. Ahmed N, Kanhere SS, Jha S (2016) On the importance of link characterization for aerial wireless sensor networks. IEEE Commun Mag 54(5):52–57
5. Akram RN, Markantonakis K, Mayes K, Habachi O, Sauveron D, Steyven A, Chaumette S (2017) Security, privacy and safety evaluation of dynamic and static fleets of drones. In: 2017 IEEE/AIAA 36th Digital Avionics Systems Conference (DASC). IEEE, pp 1–12
6. Al-Hourani A, Gomez K (2017) Modeling cellular-to-UAV path-loss for suburban environments. IEEE Wirel Commun Lett 7(1):82–85
7. Allouch A, Koubaa A, Khalgui M, Abbes T (2019) Qualitative and quantitative risk analysis and safety assessment of unmanned aerial vehicles missions over the internet. IEEE Access 7:53392–53410
8. Altawy R, Youssef AM (2016) Security, privacy, and safety aspects of civilian drones: a survey. ACM Trans Cyber Phys Syst 1(2):1–25
9. Altawy R, Youssef AM (2017) Security, privacy, and safety aspects of civilian drones: a survey. ACM Trans Cyber Phys Syst 1(2):7
10. Alzenad M, El-Keyi A, Lagum F, Yanikomeroglu H (2017) 3-D placement of an unmanned aerial vehicle base station (UAV-BS) for energy-efficient maximal coverage. IEEE Wirel Commun Lett 6(4):434–437
11. Alzenad M, Shakir MZ, Yanikomeroglu H, Alouini M-S (2018) FSO-based vertical backhaul/fronthaul framework for 5G+ wireless networks. IEEE Commun Mag 56(1):218–224
12. Amorim R, Nguyen H, Mogensen P, Kovács IZ, Wigard J, Sørensen TB (2017) Radio channel modeling for UAV communication over cellular networks. IEEE Wirel Commun Lett 6(4):514–517
13. Arafat MY, Moh S (2019) Routing protocols for unmanned aerial vehicle networks: a survey. IEEE Access 7:99694–99720
14. Bacco M, Berton A, Gotta A, Caviglione L (2018) IEEE 802.15. 4 air-ground UAV communications in smart farming scenarios. IEEE Commun Lett 22(9):1910–1913
15. Bayerlein H, De Kerret P, Gesbert D (2018) Trajectory optimization for autonomous flying base station via reinforcement learning. In: 2018 IEEE 19th International Workshop on Signal Processing Advances in Wireless Communications (SPAWC). IEEE, pp 1–5
16. Bekmezci I, Sahingoz OK, Temel Ş (2013) Flying ad-hoc networks (FANETs): a survey. Ad Hoc Netw 11(3):1254–1270
17. Bekmezci İ, Şentürk E, Türker T (2016) Security issues in flying ad-hoc networks (FANETs). J Aeronaut Space Technol 9(2):13–21
18. Bellingham JS, Tillerson M, Alighanbari M, How JP (2002) Cooperative path planning for multiple UAVs in dynamic and uncertain environments. In: Proceedings of the 41st IEEE Conference on Decision and Control, 2002, vol 3. IEEE, pp 2816–2822
19. Bor-Yaliniz RI, El-Keyi A, Yanikomeroglu H (2016) Efficient 3-D placement of an aerial base station in next generation cellular networks. In: 2016 IEEE International Conference on Communications (ICC). IEEE, pp 1–5
20. Cai X, Gonzalez-Plaza A, Alonso D, Zhang L, Rodríguez CB, Yuste AP, Yin X (2017) Low altitude UAV propagation channel modelling. In: 2017 11th European Conference on Antennas and Propagation (EUCAP). IEEE, pp 1443–1447
21. Dai H, Zhang H, Li C, Wang B (2020) Efficient deployment of multiple UAVs for IoT communication in dynamic environment. China Commun 17(1):89–103

22. Daniel K, Putzke M, Dusza B, Wietfeld C (2010) Three dimensional channel characterization for low altitude aerial vehicles. In: 2010 7th International Symposium on Wireless Communication Systems. IEEE, pp 756–760
23. Dogancay K (2012) UAV path planning for passive emitter localization. IEEE Trans Aerosp Electron Syst 48(2):1150–1166
24. Ejaz W, Azam MA, Saadat S, Iqbal F, Hanan A (2019) Unmanned aerial vehicles enabled IoT platform for disaster management. Energies 12(14):2706
25. El Hammouti H, Ghogho M (2018) Air-to-ground channel modeling for UAV communications using 3D building footprints. In: International Symposium on Ubiquitous Networking. Springer, pp 372–383
26. Elloumi M, Dhaou R, Escrig B, Idoudi H, Saidane LA (2018) Monitoring road traffic with a UAV-based system. In: 2018 IEEE Wireless Communications and Networking Conference (WCNC). IEEE, pp 1–6
27. Erdelj M, Król M, Natalizio E (2017) Wireless sensor networks and multi-UAV systems for natural disaster management. Comput Netw 124:72–86
28. Fahlstrom P, Gleason T (2012) Introduction to UAV systems. John Wiley & Sons, Chichester
29. Feng Q, Joe McGeehan, Tameh EK, Nix AR (2006) Path loss models for air-to-ground radio channels in urban environments. In: 2006 IEEE 63rd Vehicular Technology Conference, vol 6, pp 2901–2905. IEEE
30. Fotouhi A, Qiang H, Ding M, Hassan M, Giordano LG, Garcia-Rodriguez A, Yuan J (2019) Survey on UAV cellular communications: practical aspects, standardization advancements, regulation, and security challenges. IEEE Commun Surv Tutorials 21:3417–3442
31. Fu Z, Yu J, Xie G, Chen Y, Mao Y (2018) A heuristic evolutionary algorithm of UAV path planning. Wirel Commun Mob Comput 2018
32. Goddemeier N, Daniel K, Wietfeld C (2010) Coverage evaluation of wireless networks for unmanned aerial systems. In: 2010 IEEE Globecom Workshops. IEEE, pp 1760–1765
33. Goddemeier N, Wietfeld C (2015) Investigation of air-to-air channel characteristics and a UAV specific extension to the rice model. In: 2015 IEEE Globecom Workshops (GC Wkshps). IEEE, pp 1–5
34. Gupta L, Jain R, Vaszkun G (2015) Survey of important issues in UAV communication networks. IEEE Commun Surv Tutor 18(2):1123–1152
35. Hodgkins K (2015) Anti-drone shoulder rifle lets police take control of UAVs with radio pulses
36. Jawhar I, Mohamed N, Al-Jaroodi J, Agrawal DP, Zhang S (2017) Communication and networking of UAV-based systems: classification and associated architectures. J Netw Comput Appl 84:93–108
37. Jiang F, Swindlehurst AL (2012) Optimization of UAV heading for the ground-to-air uplink. IEEE J Sel Areas Commun 30(5):993–1005
38. Kaadan A, Refai HH, LoPresti PG (2014) Multielement FSO transceivers alignment for inter-UAV communications. J Lightwave Technol 32(24):4785–4795
39. Kalantari E, Shakir MZ, Yanikomeroglu H, Yongacoglu A (2017) Backhaul-aware robust 3D drone placement in 5G+ wireless networks. In: 2017 IEEE International Conference on Communications Workshops (ICC Workshops). IEEE, pp 109–114
40. Kalantari E, Yanikomeroglu H, Yongacoglu A (2016) On the number and 3D placement of drone base stations in wireless cellular networks. In: 2016 IEEE 84th Vehicular Technology Conference (VTC-Fall). IEEE, pp 1–6
41. Khawaja W, Guvenc I, Matolak DW, Fiebig U-C, Schneckenberger N (2019) A survey of air-to-ground propagation channel modeling for unmanned aerial vehicles. IEEE Commun Surv Tutorials 21:2361–2391
42. Koyuncu E, Shabanighazikelayeh M, Seferoglu H (2018) Deployment and trajectory optimization of UAVs: a quantization theory approach. IEEE Trans Wirel Commun 17(12):8531–8546
43. Lagkas T, Argyriou V, Bibi S, Sarigiannidis P (2018) UAV IoT framework views and challenges: towards protecting drones as "Things". Sensors 18(11):4015

44. Lhazmir S, Kobbane A, Chougdali K, Ben-Othman J (2019) Energy-efficient associations for IoT networks with UAV: a regret matching based approach. In: Proceedings of the 9th International Conference on Information Communication and Management. ACM, pp 132–136
45. Liu M, Yang J, Gui G (2019) DSF-NOMA: UAV-assisted emergency communication technology in a heterogeneous internet of things. IEEE Internet Things J 6(3):5508–5519
46. Liu T, Zhang Z, Jiang H, Qian Y, Liu K, Dang J, Wu L (2019) Measurement-based characterization and modeling for low-altitude UAV air-to-air channels. IEEE Access 7:98832–98840
47. Liu X, Liu Y, Chen Y (2019) Reinforcement learning in multiple-UAV networks: deployment and movement design. IEEE Trans Veh Technol 68(8):8036–8049
48. Liu X, Liu Y, Chen Y, Hanzo L (2019) Trajectory design and power control for multi-UAV assisted wireless networks: a machine learning approach. IEEE Trans Veh Technol 68(8):7957–7969
49. Malik WQ, Allen B, Edwards DJ (2007) Impact of bandwidth on small-scale fade depth. In: IEEE GLOBECOM 2007-IEEE Global Telecommunications Conference. IEEE, pp 3837–3841
50. Marshall DM, Barnhart RK, Hottman SB, Shappee E, Most MT (2016) Introduction to unmanned aircraft systems. CRC Press, Boca Raton
51. Matolak DW, Sun R (2014) Antenna and frequency diversity in the unmanned aircraft systems bands for the over-sea setting. In: 2014 IEEE/AIAA 33rd Digital Avionics Systems Conference (DASC). IEEE, pp 6A4–1
52. Matolak DW, Sun R (2017) Air–ground channel characterization for unmanned aircraft systems—Part III: the suburban and near-urban environments. IEEE Trans Veh Technol 66(8):6607–6618
53. Mayor V, Estepa R, Estepa A, Madinabeitia G (2019) Deploying a reliable UAV-aided communication service in disaster areas. Wirel Commun Mob Comput 2019
54. Mohammed F, Idries A, Mohamed N, Jameela Al-Jaroodi, Jawhar I (2014) UAVs for smart cities: Opportunities and challenges. In: 2014 International Conference on Unmanned Aircraft Systems (ICUAS), pp 267–273. IEEE
55. Motlagh NH, Bagaa M, Taleb T (2016) UAV selection for a UAV-based integrative IoT platform. In: 2016 IEEE Global Communications Conference (GLOBECOM). IEEE, pp 1–6
56. Motlagh NH, Bagaa M, Taleb T (2017) UAV-based IoT platform: a crowd surveillance use case. IEEE Commun Mag 55(2):128–134
57. Motlagh NH, Bagaa M, Taleb T (2019) Energy and delay aware task assignment mechanism for UAV-based IoT platform. IEEE Internet Things J 6:6523–6536
58. Motlagh NH, Taleb T, Arouk O (2016) Low-altitude unmanned aerial vehicles-based internet of things services: Comprehensive survey and future perspectives. IEEE Internet Things J 3(6):899–922
59. Mozaffari M, Saad W, Bennis M, Debbah M (2016) Mobile internet of things: can UAVs provide an energy-efficient mobile architecture? In: 2016 IEEE Global Communications Conference (GLOBECOM). IEEE, pp 1–6
60. Mozaffari M, Saad W, Bennis M, Debbah M (2016) Optimal transport theory for power-efficient deployment of unmanned aerial vehicles. In: 2016 IEEE International Conference on Communications (ICC). IEEE, pp 1–6
61. Mozaffari M, Saad W, Bennis M, Debbah M (2017) Mobile unmanned aerial vehicles (UAVs) for energy-efficient internet of things communications. IEEE Trans Wirel Commun 16(11):7574–7589
62. Mozaffari M, Saad W, Bennis M, Nam Y-H, Debbah M (2019) A tutorial on UAVs for wireless networks: applications, challenges, and open problems. IEEE Commun Surv Tutorials 21:2334–2360
63. Nassi B, Shabtai A, Masuoka R, Elovici Y (2019) SoK-security and privacy in the age of drones: threats, challenges, solution mechanisms, and scientific gaps. arXiv preprint arXiv:1903.05155

64. Newhall WG, Reed JH (2002) A geometric air-to-ground radio channel model. In: MILCOM 2002. Proceedings, vol 1. IEEE, pp 632–636
65. Rucco A, Aguiar AP, Hauser J (2015) Trajectory optimization for constrained UAVs: a virtual target vehicle approach. In: 2015 International Conference on Unmanned Aircraft Systems (ICUAS). IEEE, pp 236–245
66. Seiber C, Nowlin D, Landowski B, Tolentino ME (2018) Tracking hazardous aerial plumes using IoT-enabled drone swarms. In: 2018 IEEE 4th World Forum on Internet of Things (WF-IoT). IEEE, pp 377–382
67. Shah SAW, Khattab T, Shakir MZ, Hasna MO (2017) A distributed approach for networked flying platform association with small cells in 5G+ networks. In: GLOBECOM 2017-2017 IEEE Global Communications Conference. IEEE, pp 1–7
68. Siminski J (2014) Fukushima plant's radiation levels monitored with an UAV. The Aviationist
69. Simunek M, Fontán FP, Pechac P (2013) The UAV low elevation propagation channel in urban areas: Statistical analysis and time-series generator. IEEE Trans Antennas Propag 61(7):3850–3858
70. Singh PJ, de Silva R (2018) Design and implementation of an experimental UAV network. In: 2018 International Conference on Information and Communications Technology (ICOIACT). IEEE, pp 168–173
71. Tavares T, Sebastiao P, Souto N, Velez FJ, Cercas F, Ribeiro M, Correia A (2015) Generalized LUI propagation model for UAVs communications using terrestrial cellular networks. In: 2015 IEEE 82nd Vehicular Technology Conference (VTC2015-Fall). IEEE, pp 1–6
72. Triantafyllou A, Sarigiannidis P, Lagkas TD (2018) Network protocols, schemes, and mechanisms for internet of things (IoT): features, open challenges, and trends. Wirel Commun Mob Comput 2018:1–24
73. Wang D, Al-Ahmed SA, Shakir MZ (2019) Optimized link distribution schemes for ultra reliable and low latent communications in multi-layer airborne networks. IEEE Trans Ind Inform 16
74. Wang D, Al-Ahmed SA, Shakir MZ (2019) The optimization of uRLLC in Multiple Layers B5G Airborne Networks by Polychromatic Sets and SVM. unpublished
75. Wang J, Jiang C, Wei Z, Pan C, Zhang H, Ren Y (2018) Joint UAV hovering altitude and power control for space-air-ground IoT networks. IEEE Internet Things J 6(2):1741–1753
76. Wei Y, Bai Z, Zhu Y (2019) An energy efficient cooperation design for multi-UAVs enabled wireless powered communication networks. In: 2019 IEEE 90th Vehicular Technology Conference (VTC2019-Fall). IEEE, pp 1–5
77. Wu Q, Zeng Y, Zhang R (2018) Joint trajectory and communication design for multi-UAV enabled wireless networks. IEEE Trans Wirel Commun 17(3):2109–2121
78. Yanmaz E, Hayat S, Scherer J, Bettstetter C (2014) Experimental performance analysis of two-hop aerial 802.11 networks. In: 2014 IEEE Wireless Communications and Networking Conference (WCNC). IEEE, pp 3118–3123
79. Yanmaz E, Kuschnig R, Bettstetter C (2011) Channel measurements over 802.11 a-based UAV-to-ground links. In: 2011 IEEE GLOBECOM Workshops (GC Wkshps). IEEE, pp 1280–1284
80. Yanmaz E, Kuschnig R, Bettstetter C (2013) Achieving air-ground communications in 802.11 networks with three-dimensional aerial mobility. In: 2013 Proceedings IEEE INFOCOM. IEEE, pp 120–124
81. Zaman MA, Mamun SA, Gaffar M, Alam MM, Momtaz MI (2010) Modeling VHF air-to-ground multipath propagation channel and analyzing channel characteristics and BER performance. In: 2010 IEEE Region 8 International Conference on Computational Technologies in Electrical and Electronics Engineering (SIBIRCON). IEEE, pp 335–338
82. Zeng Y, Zhang R, Lim TJ (2016) Wireless communications with unmanned aerial vehicles: opportunities and challenges. IEEE Commun Mag 54(5):36–42
83. Zhan C, Zeng Y, Zhang R (2017) Energy-efficient data collection in UAV enabled wireless sensor network. IEEE Wirel Commun Lett 7(3):328–331
84. Zhan C, Zeng Y, Zhang R (2018) Trajectory design for distributed estimation in UAV-enabled wireless sensor network. IEEE Trans Veh Technol 67(10):10155–10159

85. Zhang Q, Jiang M, Feng Z, Li W, Zhang W, Pan M (2019) IoT enabled UAV: network architecture and routing algorithm. IEEE Internet Things J 6(2):3727–3742
86. Zhang X, Liu J, Gu F, Ma D, Wei J (2019) An extended 3-D ellipsoid model for characterization of UAV air-to-air channel. In: ICC 2019-2019 IEEE International Conference on Communications (ICC), pp 1–6. IEEE
87. Zhi Y, Fu Z, Sun X, Yu J (2019) Security and privacy issues of UAV: a survey. Mob Netw Appl 25:1–7. https://doi.org/10.1007/s11036-018-1193-x
88. Zhou L, Yang Z, Zhou S, Zhang W (2018) Coverage probability analysis of UAV cellular networks in urban environments. In: 2018 IEEE International Conference on Communications Workshops (ICC Workshops). IEEE, pp 1–6

Index

N. H. Mahmood et al. (eds.), *Wireless Networks and Industrial IoT*,
https://doi.org/10.1007/978-3-030-51473-0

L

GPSR Compliance

The European Union's (EU) General Product Safety Regulation (GPSR) is a set of rules that requires consumer products to be safe and our obligations to ensure this.

If you have any concerns about our products, you can contact us on ProductSafety@springernature.com

In case Publisher is established outside the EU, the EU authorized representative is:

Springer Nature Customer Service Center GmbH
Europaplatz 3
69115 Heidelberg, Germany

Batch number: 10732218

Printed by Printforce, the Netherlands